Suresh Chan
Ashok Gulati
Edit

Economic Reforms
and Food Security
The Impact of Trade
and Technology in South Asia

Pre-publication
REVIEWS,
COMMENTARIES,
EVALUATIONS . . .

"This well-crafted set of readings tackles the full set of economic reforms as they relate to agricultural, growth, and food security. South Asia encompasses close to half of the world's poor and this book is must-reading for anyone interested in poverty reduction and agricultural growth. The book correctly posits that the two most important causes of high agricultural growth rates are trade and technology. Trade is vital. If agriculture is to play a major role in poverty reduction it must grow faster than domestic demand. Technology is vital because land is tightly constrained in most low-income countries and technology must raise yields to break the land constraint. While the book gives special attention to these two vital components, it gives at least cursory attention to all aspects of agricultural growth and its relation to poverty reduction and food security. Not only are there specific chapters on each substantive element but several chapters provide an integrated overview of the specific problems and potentials of each South Asian country."

John W. Mellor, PhD
Former Director, International
Food Policy Research Institute;
Chief Economist,
United States Agency
for International Development;
Professor, Cornell University

More pre-publication
REVIEWS, COMMENTARIES, EVALUATIONS . . .

"**I** read *Economic Reforms and Food Security: The Impact of Trade and Technology in South Asia* with great interest. It is a fresh approach to better understanding the role of agriculture in the development process of South Asia. The book updates information on the promising reform agenda occurring in the region, and identifies challenging hurdles still facing this reform and the implications on the competitiveness of the agricultural sector. There are several chapters analyzing recent developments in agricultural productivity, public investment in the sector, what is good and what is bad about this investment and the distortions it may cause, and the complementarity between the market and the state in sustaining agricultural growth.

The book examines the role of agriculture and rural development in reducing rural poverty in South Asia and presents a useful framework for policymakers and practitioners to better target poverty reduction through well-designed investment guided by appropriate policy analysis. It attempts to address the appropriate role government can play in creating a conducive environment for investing in the sector without distorting the incentives for sustainable growth.

Poverty is defined beyond the narrow sense of income and consumption to include food security and better environmental management, and the institutional and policy reforms needed to achieve the millennium development goals in poverty reduction in the next decade are also covered. The useful discussion of the state of natural resources is an important contribution to further understanding the serious challenges facing the region in the years ahead.

The chapters are based on empirical work and extensive field studies, and provide excellent reading of both the research methodologies and findings of the country-based studies. This book makes a valuable contribution toward enriching understanding of international and regional agriculture in the age of globalization. The book presents extensive discussions of the role of public policies in agriculture and how to create the conducive environment needed to integrate agriculture in developing countries with global trade."

Shawki Barghouti, PhD
Agricultural Research and Portfolio Advisor, Agriculture and Rural Development Department, The World Bank

Food Products Press®
International Business Press®
Imprints of The Haworth Press, Inc.
New York • London • Oxford

Economic Reforms and Food Security

The Impact of Trade and Technology in South Asia

FOOD PRODUCTS PRESS®
Crop Science
Amarjit S. Basra, PhD
Senior Editor

Economic Reforms and Food Security
The Impact of Trade and Technology in South Asia

Suresh Chandra Babu
Ashok Gulati
Editors

Food Products Press®
International Business Press®
Imprints of The Haworth Press, Inc.
New York • London • Oxford

For more information on this book or to order, visit
http://www.haworthpress.com/store/product.asp?sku=5302

or call 1-800-HAWORTH (800-429-6784) in the United States
and Canada or (607) 722-5857 outside the United States and Canada

or contact orders@HaworthPress.com

Published by

Food Products Press® and International Business Press®, imprints of The Haworth Press, Inc.,
10 Alice Street, Binghamton, NY 13904-1580.

Cover design by Kerry E. Mack.
Cover design concept by Anjan Dutta.

Library of Congress Cataloging-in-Publication Data

Economic reforms and food security : the impact of trade and technology in South Asia / Suresh
Chandra Babu, Ashok Gulati, editors.
     p. cm.
"A selection of the presentations made at the South Asia Regional Conference held in New Delhi in
April 2002"—Foreword.
    Includes bibliographical references and index.
    ISBN 1-56022-256-5 (hard : alk. paper)—ISBN 1-56022-257-3 (soft : alk. paper)
    1. Food supply—South Asia—Congresses. 2. Agriculture and state—South Asia—Congresses.
3. Agriculture—Economic aspects—South Asia—Congresses. 4. Agricultural innovations—South
Asia—Congresses. 5. Free trade—South Asia—Congresses. I. Babu, Suresh Chandra. II. Gulati,
Ashok, 1954-
HD9016.S642E36 2005
338.1'954—dc22
                                          2004012056

CONTENTS

PART VI: FOOD SECURITY INTERVENTION IN SOUTH ASIA

PART VII: EMERGING ISSUES

ABOUT THE EDITORS

Suresh Chandra Babu, PhD, is Senior Research Fellow and Senior Training Advisor at the International Food Policy Research Institute, Washington DC, where he conducts research, outreach, and capacity strengthening activities in the areas of global food and nutrition security. Prior to joining IFPRI in 1992 as a research fellow, Dr. Babu was a research economist at Cornell University, Ithaca, New York. He spent many years in Southern Africa on various capacities. He was Senior Food Policy Advisor to the Malawi Ministry of Agriculture on developing a national level food and nutrition information system; an evaluation economist for the UNICEF-Malawi working on designing food and nutrition intervention programs; Coordinator of UNICEF/IFPRI food security program; and a senior lecturer at the Bunda College of Agriculture, Malawi.

Dr. Babu currently heads IFPRI's Training and Capacity Strengthening Program and jointly coordinates its South Asia Initiative. He serves as a member of Capacity Strengthening Working Group of the United Nations System's Forum for Nutrition, Scientific Committees of Southern Africa Food Policy Network, African Capacity Building Foundation and Millennium Ecosystem Assessment, and IUCN's Commission on Ecosystem management.

Dr. Babu has conducted development research for bilateral and international organizations including the Food and Agricultural Organization (FAO) of the United Nations, the World Bank, UNCTAD, UNICEF, GTZ, and USAID. Dr. Babu has authored or co-authored more than fifty refereed journal papers and book chapters. He is an associate editor of United Nations University's *Food and Nutrition Bulletin* and the *African Journal of Food and Nutritional Sciences,* and is co-editor of *Food Systems for Improved Human Nutrition: Linking Agriculture, Nutrition, and Productivity* (Haworth).

Ashok Gulati, PhD, joined IFPRI as Director of the Markets, Trade, and Institutions Division in January, 2001. Prior to joining IFPRI, he was a NABARD Chair Professor at the Institute of Economic Growth in Delhi, India. Before that, he was Chief Economist at the National Council of Applied Economic Research in Delhi. At the time he joined IFPRI, he was a member of the Economic Advisory Council of the prime minister of India, a member of the Advisory Council of the chief minister of Andhra Pradesh, and Chairman of the Committee on Agriculture of the Federation of Indian Chambers of Commerce and Industry.

Dr. Gulati's publications include *Trade Liberalization and Indian Agriculture, Subsidy Syndrome in Indian Agriculture,* and *Financial and Institutional Reforms in Indian Canal Irrigation.* Throughout his research career, he has been an applied economist with strong involvement in analysis and policy advice. His research specializes in issues such as agricultural markets, trade liberation, WTO and trade negotiations, globalization, pricing policies, veterical linkages between farms and firms, and the role of infrastructure and institutions in making markets function efficiently.

CONTRIBUTORS

Akhter U. Ahmed, senior research fellow, International Food Policy Research Institute, Washington, DC.

Pratap S. Birthal, senior scientist, National Centre for Agricultural Economics and Policy Research, New Delhi, India.

Carlo del Ninno, senior economist, The World Bank, Washington, DC.

Choni Dendup, chief marketing officer, Planning and Policy Division, Ministry of Agriculture, Royal Government of Bhutan, Thimpu, Bhutan.

S. Mahendra Dev, professor and director, Center for Economic and Social Studies, Hyderabad, India.

Eugenio Díaz-Bonilla, executive director for Argentina and Haiti, Inter-American Development Bank, Washington, DC.

Paul Dorosh, senior rural development economist, The World Bank, Washington, DC.

Peter Hazell, director, Development Strategy and Governance Division, International Food Policy Research Institute, Washington, DC.

Anwarul Hoda, member, Planning Commission, Government of India, and former deputy director-general of the World Trade Organization.

P. K. Joshi, research fellow/South Asia Coordinator, International Food Policy Research Institute, Washington, DC.

Saman Kelegama, executive director, Institute of Policy Studies of Sri Lanka, Colombo, Sri Lanka.

Ruth S. Meinzen-Dick, senior research fellow, International Food Policy Research Institute, Washington, DC.

Deki Pema, planning officer, Ministry of Agriculture, Royal Government of Bhutan, Thimpu, Bhutan.

Per Pinstrup-Andersen, former director general, International Food Policy Research Institute, and Winner of the 2001 World Food Prize.

Sarfraz Khan Qureshi, senior policy analyst, Centre for Poverty Reduction and Income Distribution, Government of Pakistan, and former director, Pakistan Institute of Development Economics, Islamabad, Pakistan.

K. V. Raju, associate professor, Agricultural Development and Rural Transformation Unit, Institute for Social and Economic Change, Bangalore, India.

Anitha Ramanna, visiting research associate, Indira Gandhi Institute of Development Research (IGIDR), Mumbai, India.

Sherman Robinson, professor, Department of Economics, School of Social Sciences, University of Sussex, Brighton, United Kingdom.

Mark W. Rosegrant, director, Environment and Production Technology Division, International Food Policy Research Institute, Washington, DC.

Quazi Shahabuddin, director general, Bangladesh Institute of Development Studies, Dhaka, Bangladesh.

Abusaleh Shariff, chief economist and head, Human Development Division, National Council of Applied Economic Research (NCAER), New Delhi, India.

Suman Sharma, lecturer, Department of Economics, Tribhuvan University, Kathmandu, Nepal.

Manmohan Singh, The Honorable Prime Minister of India, Government of India, New Delhi, India.

Panjab Singh, director, Centre for Extension Education; professor, School of Agriculture, Indira Gandhi National Open University; former secretary, Department of Agricultural Research and Education (DARE), Government of India; and director-general, Indian Council of Agricultural Research (ICAR), Ministry of Agriculture, New Delhi, India.

Laxmi Tewari, researcher, National Center for Agricultural Economics and Policy Research (NCAP), New Delhi, India.

Marcelle Thomas, research analyst, International Food Policy Research Institute, Washington, DC.

Sangeeta Udgaonkar, advocate and Infosys scholar, National Law School of India University, Bangalore, India.

Klaus von Grebmer, director, Communications Division, International Food Policy Research Institute, Washington, DC.

Vijay S. Vyas, is a member of the Economic Advisory Council to the Prime Minister; professor emeritus and chairman, Governing Board, Institute of Development Studies, Jaipur, India; and former senior advisor, Agriculture and Rural Development Department, The World Bank, Washington, DC.

Regional Share (Percentage) of the World's Poor Living on < $1 per Day

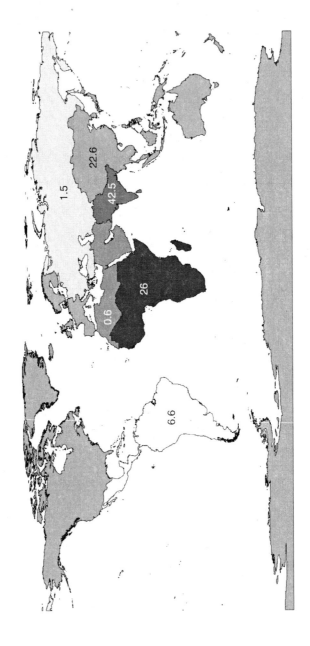

1.5

22.6

42.5

0.6

26

6.6

Source: Global Economic Prospects 2003 The World Bank, Washington, DC.
Note: Figures are for 1999, and based on the world's population living on less than $1.08 a day at 1993 international prices (equivalent to $1 in 1985 prices adjusted for purchasing power parity).

Foreword

Even after dramatic improvements in food availability at the national level in South Asian countries, due to increased agricultural production over the past three decades, the region remains plagued by food insecurity and poverty. Recent economic reforms in agricultural and other sectors in the region, implemented over the past fifteen years and to different degrees by each country, have made a major impact on the living standards of the population there and have further integrated South Asia into the global economy. Yet the challenge of eliminating poverty, hunger, and malnutrition remains.

Continuing with these reforms for liberalization and developing new and effective policies for food security and agricultural development will benefit from policymakers receiving a continuous flow of research-based information regarding the effect current reforms have had on the agricultural sector and the welfare of the poor, and what the consequences of alternative policies might be. Policy research and analysis in South Asia have played important roles in shaping the policymaking process there. To further help forge the links between research and policymaking, *Economic Reforms and Food Security: The Impact of Trade and Technology in South Asia* brings together researchers, from both inside and outside the region, and South Asian policymakers and advisors to share findings, concerns, and perspectives.

This book examines the approaches that have been taken in South Asia for agricultural growth and food security, discusses the critical questions each of the countries face, and suggests future courses of action. Trade and technology are subjects of some focus because they have recently emerged as two of the main factors that will determine the prospects of agriculture-led development in the region. The book also takes an inventory of the major research that has been conducted on the diverse aspects of food security and identifies the information gaps that must be filled if policies are to be more effective in the future.

The chapters of this book represent a selection of the presentations made at the South Asia Regional Conference held in New Delhi in April 2002. The conference was cosponsored by the Indian Council for Agricultural Research, the Indian Council for International Economic Relations, and the International Food Policy Research Institute (IFPRI). For the past twenty-five years, the International Food Policy Research Institute has been working with several partners in South Asia on a variety of research projects concerning the region. These studies, on agricultural production, irrigation strategies, fertilizer use, market liberalization, and food subsidies, have been a source from which the governments in South Asia have drawn important policy lessons. The conference and this collection together mark the beginning of a renewed commitment of IFPRI, under its South Asia Initiative, launched in 2001, to address the food policy issues of the region. It is the hope of IFPRI that this work will stimulate productive discussion and sound decision making in the current policy environment of South Asia.

Joachim von Braun, Director General
International Food Policy Research Institute

Isher Judge Alhuwalia
Chairperson, IFPRI, Board of Trustees
and Former Director and Chief Executive, ICRIER

Mangala Rai
Secretary, Department of Agricultural Research and Education,
Ministry of Agriculture, Government of India,
and Director-General, Indian Council of Agricultural Research

Acknowledgments

We would like to thank the many institutions and individuals who helped to make the conference on which this book is based a success, as well as those who contributed to the production of this volume. The conference was made possible thanks to the cosponsorship of the Indian Council for Research on International Economic Relations (ICRIER), the Indian Council for Agricultural Research (ICAR), and the International Food Policy Research Institute (IFPRI). Additional financial support received from USAID-New Delhi, USAID-Washington, and the Ford Foundation, New Delhi, is gratefully acknowledged.

We are indebted to Isher Judge Ahluwalia, chairperson of the IFPRI Board of Trustees, for the leadership role she took as executive director of ICRIER in organizing the conference, and we greatly appreciate the commitment she has demonstrated to the South Asia region and her support of this work. We are also pleased to have had the support of Joachim von Braun, Director-General of IFPRI, and Mangala Rai, Secretary of the Department of Agricultural Research and Education of the government of India, and Director General of ICAR. Our thanks also go to Per Pinstrup-Andersen, former Director-General of IFPRI, for opening the conference, and Manmohan Singh, former Finance Minister and Parliament Member, India, and now the Prime Minister of India, for presenting the keynote address.

A large thanks goes to Ashwin Bhouraskar of IFPRI for the excellent job he did in editing the chapters and providing research assistance for some of them. Shirley Raymundo and Valerie Rhoe of IFPRI provided essential assistance in organizing the conference, and the staffs of ICAR and ICRIER also deserve our gratitude for the support they gave in the conference's arrangement.

The editors are obliged in a special way to the contributors of this volume. They have shared with us their thoughts and insights, which today are opening up new areas of inquiry regarding South Asia's food security. We also appreciate deeply the commitment to addressing this issue that more than 100 individuals showed by traveling from near and far to attend the conference. The participation of these researchers, policymakers, and civil society representatives produced a vibrant and stimulating discussion of the many issues affecting food security in South Asia.

Finally, we thank our families for the patience and understanding they showed while we were preparing for the conference and editing this volume.

Introduction

Economic Reforms and Food Security in South Asia: An Overview of the Issues and Challenges

Suresh Chandra Babu
Ashok Gulati

BACKGROUND

Since the beginning of the post–World War II period and through the initial era of development, in which a concerted international effort was undertaken to eliminate poverty in the world, South Asia has made steady and significant progress in terms of economic growth and food security. Yet due in large part to population growth, the region continues to hold the largest share of the world's poor, with a substantial number of its inhabitants living in poverty and hunger. Whereas poverty is deeper in many sub-Saharan African countries compared to South Asia, the sheer number of poor in the latter has posed a challenge for the governments of the region and international donors. This challenge is particularly great now as the international community is striving to achieve the Millennium Development Goals for poverty reduction. Indeed, the situation in South Asia implies that for the Millennium Development Goals to be reached investments will need to be targeted to a large extent to the region. At the same time, trends in the region warrant optimism about its future. Moderate investments alone could make a major impact on poverty in the region in a relatively short period of time, lifting a substantial number of people out of poverty. This book, in its examination of the experiences of Bangladesh, Bhutan, India, Pakistan, and Sri Lanka, and the issues and challenges the countries have faced with the ongoing economic reforms, seeks to demonstrate, with a special focus on

Ashwin Bhouraskar, Senior Research Assistant at IFPRI, provided research assistance for this chapter.

1

trade and technology, that by helping the South Asian countries achieve their economic, social, and technological goals the international development community would also make a significant advance toward reaching the Millennium Development Goals.

Over the past decade, each of the countries of South Asia has implemented an array of economic reforms, within the framework of comprehensive packages and with varying degrees of intensity, to spur economic growth and alleviate poverty. A major policy question that has been discussed in policy circles therefore, at both the regional and international levels is whether the economic reforms have in fact reduced poverty and food insecurity in South Asia (Ahluwalia, 2002). Over the past several decades, the countries of the region have no doubt been serious about alleviating poverty and achieving food security, and have spent considerable resources for achieving these goals. Moreover, economic growth in the region has been rather good, with an annual average growth rate for the region of more than 5 percent over the past twenty years (see Table I.1 and Figure I.1).

However, the impact of the economic reforms on poverty and food security seems to be below the expectation of policymakers. For most of the region's countries, for which data are available, more than a third of the population lives below the poverty line, and the figure is higher for rural areas (Table I.2). Moreover, for all of the countries listed, except Sri Lanka, a third or more live on less than a dollar a day. Given the high population of the region, which constitutes about 22 percent of the world figure (Table IA.1), a large number of the world's poor are concentrated in South Asia. Of the total number of people in the world living on less than a dollar a day, 42.5 percent live in South Asia (Figure I.2).

Improving food security in the region has not occurred without challenges. Although gross national income (GNI) per capita grew in terms of current U.S. dollars and purchasing power parity, especially the latter, for

TABLE I.1. Average annual growth of GDP and GDP per capita, 1980-2000

Country	1980-1990	1990-2000	1999	2000
Bangladesh	4.3 (1.7)	4.8 (3.0)	4.9 (3.1)	5.9 (4.1)
Bhutan	7.6 (5.4)	6.5 (3.4)	7.0 (3.9)	7.0 (3.9)
India	5.8 (3.5)	6.0 (4.1)	7.1 (5.2)	3.9 (2.0)
Nepal	4.6 (2.3)	4.9 (2.4)	4.4 (1.9)	6.5 (3.9)
Pakistan	6.3 (3.5)	3.7 (1.2)	3.7 (1.2)	4.4 (1.9)
Sri Lanka	4.0 (2.5)	5.3 (3.9)	4.3 (2.8)	6.0 (4.3)

Source: World Bank, 2002.

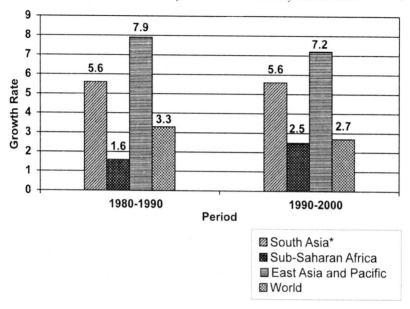

FIGURE I.1. Average annual GDP growth of selected regions and world, 1980-2000. (*Source:* World Bank, 2002.) *Includes Afghanistan and the Maldives. *Note:* Figures for regions and world are weighted averages.

the region from 1995 to 2001, it was uneven and in some years only modest. For Pakistan, there were actually declines in GNI per capita (see Table IA.2 at the end of this chapter). Reducing whatever benefit the general increase in GNI per capita provided to the region's poor was the overall rise in food prices at an equal or higher rate. Table I.3 shows the increase in the food price index from 1995 to 2000 and selected intervals before this period for countries for which the data are available.

In addition, food-grain availability remained low, fluctuated, and, for some countries, even declined over the period (Table IA.3). This trend, which affects the poor, is important to note against the more favorable pattern of the change in calorie supply from all foods, which has been due to the diversification of production and consumption to higher-value goods that higher-income groups are able to enjoy. It can be argued that the instability in the supply of calories from basic cereals may be due partly to the fluctuations in cereal production that have occurred in each of the countries (Table IA.4). However, based on the data, South Asia did not perform poorly when compared to some of the other regions in the developing world. Certainly, food security and development have been at lower levels

TABLE I.2. Poverty in South Asia and selected regions, 1991-2000*

	National poverty line		International poverty line
	Population below the poverty line (%)		Population living on < $1/day (%)**
	National	Rural	
Bangladesh	35.6	39.8	29.1
India	35.0	36.7	44.2
Nepal	42.0	44.0	37.7
Pakistan	34	36.9	31.0
Sri Lanka	25	—	6.6
South Asia			40.0
Sub-Saharan Africa			48.1
East Asia and Pacific			14.7
Latin America and Caribbean			12.1

Sources: For national figures, World Bank, 2003; for regional figures on population percentage living on < $1/day, the Millennium Development Goals Web site: <www.developmentgoals.org/Data.htm> "Country tables."
*Latest single year. The figures for each country were collected in different years. The survey years in which national and international poverty figures were collected for a country were not the same in all cases.
**Population on < $1/day in purchasing power parity terms: $1.08/day at 1993 international dollars (equivalent to $1 a day in 1985 prices adjusted for PPP).

in sub-Saharan Africa. Yet South Asia has still to achieve the results of East Asia and the Pacific.

The South Asian economies are and will, for the foreseeable future, remain dependent to a large degree upon agriculture for growth and poverty alleviation, given agriculture's prominent share in the GDP of the region's countries (Table I.4) and the large percentage of poor in the rural areas. The region, on the whole, performed relatively well over the past two decades, compared to others in the developing world (Figure I.3). Nevertheless, Bangladesh and Sri Lanka's achievements were rather low, and growth has been unsteady, with significant variation from year to year. Nepal and Bhutan experienced declines from the first decade to the second (Table IA.5).

Clearly, policies that promote sustained agricultural growth and benefit the poor by involving small farms are needed. These policies, moreover, will need to achieve growth in an environmentally sustainable manner. Past practices to boost agriculture have produced adverse effects on the environment, and new lands for farming are severely limited. Indeed, new agricul-

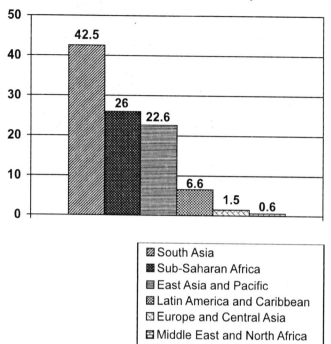

FIGURE I.2. Percentage of word's poor living on < $1/day in six selected regions in 1999. (*Source:* World Bank, 2003.) *Note:* Population living on > $1/day in purchasing power parity terms: $1.08/day at 1993 international dollars (equivalent to $1 a day in 1985 prices adjusted for PPP).

TABLE I.3. Food price indices for the South Asian countries (1995 = 100), 1980-2000

	1980	1985	1990	1995	1996	1997	1998	1999	2000
Bangla-desh	30.7	52.2	80.5	100	101.3	105.9	116.9	—	—
India	24.6	37.9	57.6	100	108.5	114.8	132.0	134.2	136.5
Nepal	21.8	33.1	58.4	100	110.7	116.2	134.1	144.4	—
Pakistan	—	39.3	56.6	100	109.3	122.4	129.4	134.5	138.8
Sri Lanka	19.2	33.8	61.7	100	119.2	132.1	146.6	152.4	159.2

Source: World Bank, 2002.

TABLE I.4. Share (percentage) of agriculture in national GDP in South Asian countries

	1980	1990	1999	2000
Bangladesh	49.6	29.4	25.3	24.6
Bhutan	56.7	43.2	34.6	33.2
India	38.6	31.3	26.2	24.9
Nepal	61.8	51.6	41.3	40.3
Pakistan	29.5	26.0	27.0	26.3
Sri Lanka	27.6	26.3	20.7	19.5

Source: World Bank, 2002.

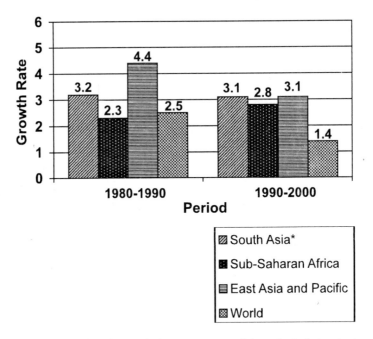

FIGURE I.3. Agricultural growth (average annual) in selected developing regions, 1980-2000. (*Source:* World Bank, 2002.) *Includes Afghanistan and Maldives.

tural technologies and the sustainable use of natural resources, including the institutional structures through which certain resources, such as water for irrigation, are used will play a pivotal role in South Asia's potential to achieve food security. For this reason technology for agriculture is one of the key issues this book focuses on. The other key component is trade. There is general agreement among the authors in this book that liberalized trade, along with the other economic reforms that have accompanied it, have the potential to improve food security and benefit small farmers. One of the primary ways it can do this is if policies support the diversification of agricultural production into higher-value goods. The South Asian nations have in fact been moving in this direction, and a number of the contributors discuss what the implications are for further trade liberalization. Yet given the level of poverty in the region and the acute and chronic food insecurity among the poor at the household level, domestic market reforms and targeted interventions will be necessary to support the new policies for technology development and trade. This introduction presents an overview of the issues and challenges, identified by the chapters of this volume, that the South Asian countries face as they move along the path of economic reforms.

OUTLINE OF THE BOOK

The first set of chapters, presented in Part I, provide an overview of the issues involved in the economic reform process: trade, technology, and food security. In Chapter 1, Manmohan Singh addresses the fundamental question presented previously. Have economic reforms enhanced food security? Singh observes that the economic reforms initiated in the early 1990s in India have contributed to reducing the two major development constraints, the shortage of food and foreign exchange. He calls for developing a policy framework and policies that can be effectively implemented considering the social, economic, institutional, and political factors.

Chapter 2, by Per Pinstrup-Andersen, provides a long-term view on how food security can be achieved in South Asia. It discusses several general policies that are imperative for achieving food security in South Asia. Some of the arguments that have been made, and which Pinstrup-Andersen reiterates, are that public investment for improving rural infrastructure and agricultural productivity is fundamental for rural growth, investment in human resources in terms of education and health will form a solid foundation for long-term economic growth, and access to productive resources and employment will reduce poverty levels more quickly. For marginal areas where poverty is concentrated, pro-poor technologies are needed to alleviate hunger. Programs and policies that reduce the vulnerability of the poor while the country is on the path to development will also be required. While infrastructure will help in the marketing of crops, it will also foster rural industrialization and the development of agroprocessing, and thus convert increased agricultural productivity into higher rural incomes. The efficient management of natural resources and the improvement of environmental conditions cannot be neglected, especially in less-favored areas. But perhaps what is most important is good governance for carrying out these policies.

Although all of the countries of the region have either achieved or increased self-sufficiency in food production, the presence of high levels of food insecurity at the subregional and household level remains a development challenge. The chapter by Vijay Vyas (Chapter 3) puts forth what is perhaps the second most central question for policymakers in the region: if the governments of South Asia committed themselves long ago to eliminating hunger and poverty, have expended vast resources to achieve it, and the countries have been experiencing satisfactory growth, then why has no major reduction in food insecurity occurred? Vyas points out that while advances have been made toward ensuring food security at the national level, food insecurity still remains at the household level. Moreover, though extreme hunger exists in only a few areas, chronic malnutrition has persisted and is widespread. He illustrates the remaining gap in part by tracing the changes in the definition of food security from one that was relatively narrow to a definition that has expanded to include issues of food and nutrition access at the household level, and by suggesting what the implications of the expanding definition of food security are for policy. After providing a survey of the development approaches, policies, and programs that have been implemented in the region at various levels and have influenced food security, Vyas concludes by suggesting that because the countries of the region all share the goal of food security and possess several characteristics in common, regional collaboration and knowledge exchange for this goal could provide them with mutual benefits.

Trade Liberalization

Part II of the volume consists of chapters that address food trade liberalization and food security in the region. Trade liberalization has been a major issue as the countries of South Asia have pursued economic reforms. Bangladesh, India, Nepal, Pakistan, and Sri Lanka are already members of the WTO and Bangladesh has least-developed-country status. Bhutan is in the process of joining the WTO. The extent to which agricultural trade has been liberalized in the region varies among the countries. Also, liberalization in the countries has occurred at different times. In Sri Lanka, a more open agricultural trade regime was developed in the 1970s, when the regimes in the other countries remained largely protectionist and discriminatory against agricultural goods. Not until 1997 did some of the anti-agricultural bias of these regimes begin to decline. In the pre-1997 period, the markets for most agricultural goods were largely insulated from the world market, and internal supply and demand, support prices, and input prices determined domestic prices. Moreover, exchange rate overvaluation hurt established and potential primary export industries. With the trade reforms in 1997, which were designed mainly for manufactured goods, the impact on agriculture came mainly from the currency devaluation that preceded or accompanied the reforms, rather than directly. Currency devaluation helped the export commodities but did not affect the domestic prices of products. The measures directly affecting agriculture were uneven and limited in scope, and the 1997 reforms brought only a modest reduction in the bias against agriculture. In fact, even the signing of the Agreement on Agriculture (AOA) did not have much immediate impact on agricultural trade policies. Most countries bound tariffs high and passed the AMS test with their low domestic support prices for food grains. Besides Sri Lanka, the region remained essentially closed to imports (Pursell, 2002).

Since 1997, however, Pakistan has radically liberalized trade in its agricultural goods, except oilseeds, and Bangladesh became more open in 2002. India, however, despite the challenge in the WTO to its import licensing, which amounted in effect to an import ban, has tried to remain protectionist through a variety of means (Pursell, 2002). Table IA.6 provides the tariffs and nontariff barriers on selected commodities in South Asia. With the increasing globalization of the economy and the requirement of AOA member countries to abide by the agreements, the major question for the South Asian countries is how they can take advantage of the new trade opportunities and minimize the negative impacts trade liberalization may bring. Any involvement in global trade will heavily affect the agricultural sectors of the region since the countries are still largely dependent on agriculture. In general, since 1980, all of the countries for which data are available have greatly increased their food imports and exports, yet only India and Sri Lanka have had positive food trade balances (Table IA.7).

But what effect might trade liberalization have on food security? And, is the WTO-AOA equipped to address food security concerns? Chapter 4, by Díaz-Bonilla, Thomas, and Robinson, looks at the questions the WTO-AOA must consider for it to properly address the concerns of developing countries. The first question is: What is the relevance of the current WTO classification of countries with respect to food security? The second is whether the current agreements, which use this classification to define WTO commitments, are satisfactorily handling food security concerns through special and differential treatment. The authors point out that if the categories do not adequately represent food security concerns, the differential treatment under the WTO will most likely not handle those concerns in a meaningful way. Trade imbalances have continued to exist between industrialized and developing countries. Industrialized countries have enough legal room under the WTO to subsidize their own agriculture sectors and the financial resources to provide subsidies, while South Asian countries still lack the financial, human, and institutional resources to invest in food security. This chapter concludes by saying that this imbalance needs to be corrected by ensuring that agriculture trade negotiations proceed parallel to increased funding for agriculture and rural development, food security, and rural poverty alleviation by international and bilateral organizations.

Anwarul Hoda and Ashok Gulati, in "Indian Agriculture, Food Security and the WTO-AOA" (Chapter 5), examine the case of India's participation in the agreement's negotiations and the issues it faces. The key question that has been raised in India is whether agricultural trade liberalization would destroy India's food security. Groups arguing that food security is a fundamental human right oppose more open trade because they believe it will cause greater food insecurity. Food security certainly ought to be considered a fundamental right. What Hoda and Gulati explain is that under the WTO-AOA, India can still take measures at the domestic level to ensure food security. Yet they also point out that the country's various food security programs will have to undergo reforms so that they achieve greater efficiency. Domestic institutional reforms in both food-grain procurement and distribution and proper targeting of food subsidies, which could involve a food stamp system, need to be undertaken to ensure better household food security. Of course, securing a level playing field in international negotiations by reducing high levels of support in developed countries must be a priority. This reduction would lead to a worldwide production shift to relatively more efficient economies, thus benefiting more countries in South Asia. Therefore, it is important for South Asian countries to engage themselves more effectively in the negotiations within the WTO-AOA. Moreover, protecting farmers from international price fluctuations and providing alternative sources of income, while also guaranteeing that consumers benefit through increased access to food, should be high on the policymakers' agenda. Further assistance to affected groups of the population during the transition period of trade liberalization is a key challenge for governments in South Asia.

In Chapter 6, Paul Dorosh and Quazi Shahabuddin show how the liberalization of international trade in food grains contributed to national food security in Bangladesh and stabilizing markets after production shortfalls, and in this sense support the general argument of Chapter 5. The experience of Bangladesh illustrates the role that market reforms and private trade in grain markets can play in achieving food security. Following a poor harvest in 1997 and a massive flood in 1998, private traders in Bangladesh were able to import several million metric tons of rice from India due to the loosening of government controls. To be sure, the success was partially dependent on the trade liberalization that had occurred in India in the early 1990s, but the Bangladesh experience shows what the potential benefits of trade liberalization for national food security are because liberalization enabled rapid increases in food supplies when domestic production fell short of demand. However, the contribution that trade liberalization made to short-term food security does not minimize the importance of increasing agricultural productivity and rural economic growth for providing rural poor households with sufficient incomes to acquire food. Furthermore, a flexible trade policy may be needed to protect producer interests and long-term food security, particularly in the face of export subsidies or a steep decline in world food prices.

Technology for Food Security

New technologies are nearly as important for agricultural growth and food security today as the Green Revolution technologies were for the region in the 1960s and 1970s. Owing to the high population density and the need to conserve natural areas, land for agricultural expansion is severely limited. Furthermore, a large share of the rural population, particularly the poor in marginal areas, is unable to produce enough to meet their daily needs. Of all of the regions in the developing world, only South Asia has no land in reserve for agriculture (Table I.5). In addition, while the region already has a considerable amount of land that has been lost to either degradation or urban sprawl—150 million hectares—this quantity is likely to grow

TABLE I.5. Regional variation in land scarcity, 2001 (in million hectares)

	Land in use for cropping	Unsustainably cropped land	Total land lost	Land reserve
South Asia	150	80	150	0
Sub-Saharan Africa	160	0	175	680
East Asia and Pacific	240	0	150	75
Latin America and the Caribbean	165	0	150	700
Middle East and North Africa	95	0	65	20

Source: World Bank, 2003.
Note: Figures are approximate.

as the region has about 80 million hectares currently being cropped in an unsustainable manner. If this land too is lost, South Asia will have the largest quantity of land lost of any region in the developing world.

At the same time, the new technologies must be environmentally sustainable. Green Revolution technologies have brought about an increased use of chemical fertilizers and pesticides, and these have had serious negative impacts on the soil and water resources in the breadbasket areas of the region. The new technologies would be most beneficial if they address the problems small farmers continue to face, such as pests and drought. For these challenges, safe forms of biotechnology that are implemented in a responsible and transparent manner could play a significant role. Most important of all is that these new technologies are available to and benefit the small farmer, and that the institutions involved in technology research and delivery allow for participation.

Issues of technology development, application, and transfer are what concern Panjab Singh in Chapter 7. He presents an overview of how technology and the applied sciences have contributed to agricultural production and food security in the region. He cites the most well-known case, the Green Revolution in India, and like others reminds us that the increase in production would not have been possible without the high-yielding crop varieties that were developed and disseminated to small farmers. Singh goes further to point out that the development and application of new technologies now has a role to play in remedying the natural resource degradation that production under the Green Revolution caused and in enabling environmentally sustainable production in the future. Given the region's increasing population and food security being an important goal of the countries, further yield increases will be absolutely critical, and biotechnology offers some promise, Singh states. But he warns us that technology transfer today is not as free as it was under the Green Revolution. Various constraints to technology transfer exist under the WTO-AOA that could prevent the increases in crop production that are so necessary for food security.

Peter Hazell, in "Future Challenges for the Rural South Asian Economy" (Chapter 8), supports the notion that technological advances can generate higher agricultural yields and address the environmentally harmful effects of past production patterns. But he argues that to sustain the favorable trends in food production and availability and poverty reduction that have been in existence since the mid-1960s and are now stalling, and to generate future growth and employment creation, a new and broader approach must be adopted. Hazell recommends that the South Asian countries take five steps:

1. increase the level and efficiency of public spending in agriculture and rural development,
2. make use of new opportunities for market-driven growth in the agricultural sector that trade liberalization and dietary diversification offer,
3. invest in less-favored regions where the largest share of the poor live,
4. redress the environmentally negative impact of past agricultural growth and make future production environmentally sustainable, and
5. devise policies to enhance the rural nonfarm economy to create employment.

Although he agrees that new technologies are needed for increasing foodgrain production, he states that the future growth of the agricultural sector will be driven less by production of food grains than by high-value commodities, including those from agroindustrial processing and for export.

Several issues confront the development of biotechnology for agricultural development in South Asia: the role of agricultural biotechnology in improving food security, the importance of social choice in considering such technology, and concerns and regulations in terms of consumer preference, food safety, and management of intellectual property. Designing and implementing a policy framework that guides the development of biotechnology in South Asia is important. Furthermore, the countries need to create policies for investing in biotechnology research and set priorities. And, while biosafety systems are in place in some countries, they are absent in others. Where they do exist they need to be more transparent and guided by the regulations established for them. To encourage greater private investment in biotechnology research, intellectual property rights ought to be developed. Yet in the process of conducting and implementing biotechnology research it is important to protect the interests of small farmers and consumers in the region.

A framework is needed for increasing opportunities from technology transfer and reducing constraints on it, and addressing the political economy and legal tensions arising from the new intellectual property regime in South Asia. The new intellectual property regime structure aims to grant protection to firms and breeders for plant varieties that they develop and revise patent laws while at the same time safeguarding the country's genetic resources and farmers' rights and privileges. Research on the effects of various intellectual property regimes on technology generation and transfer, including biotechnology, would be highly beneficial at this point in the region's development.

In Chapter 9, Anitha Ramanna and Sangeeta Udgaonkar address the issues surrounding intellectual property rights (IPRs), which have been receiving increased attention in discussions on agricultural production, trade, and conservation in South Asia. They correctly point out that the extension of IPRs to agriculture will be one of the crucial measures determining the future of food security in the developing countries. The chapter discusses the opportunities and constraints for technology transfer under the new IPR model that is currently emerging in South Asia, under which protection is given to corporate breeders, farmers, and traditional communities. The goal of the new model is to enable the country to acquire technology and promote innovation, and at the same time protect genetic resources, the rights of farmers and traditional communities, and prevent developed nations from creating products from the developing country's resources without compensation. Ramanna and Udgaonkar trace the development of the new model and discuss the particular case of India. They discover that a number of factors and stakeholders influenced the creation of the regime. Based on their findings, they reasonably conclude that food security and agricultural growth in South Asia will depend on the ability of these countries to establish IPR regimes that balance the rights of various actors and same time provide incentives for the exchange of resources.

Water Issues

In addition to high-yielding technologies, irrigation played an important role in boosting agricultural output and reducing hunger in South Asia during the Green Revolution. Future agricultural development will depend equally on water availability. In fact, with the exception of Pakistan, the greater part of arable land in the region's countries remains nonirrigated (Table I.6 and Figure I.4). But irrigation today will have to be delivered through more efficient and participatory institutions, and accompanied by policies that control the demand for water on the part of South Asia's growing urban and industrial sectors. At the microlevel, in areas where rural infrastructure, including irrigation, is severely lacking, such as in the eastern and northeastern regions of India, water availability for farming could lead to improved household food security.

The new irrigation systems will also have to sustain the soil resources on which agricultural production depends. As Panjab Singh and Peter Hazell each explain in their respective chapters, irrigation under the Green Revolution and since then has produced an array of environmental problems for sustained future production, other human uses, and ecological integrity, such as groundwater overdraft, water quality reduction, waterlogging, and

TABLE I.6. Irrigated land in South Asia, 1999

Country	Cropland irrigated (%)
Bangladesh	47.2
Bhutan	25.0
India	34.8
Nepal	38.2
Pakistan	82.0
Sri Lanka	34.8

Source: World Bank, 2002.

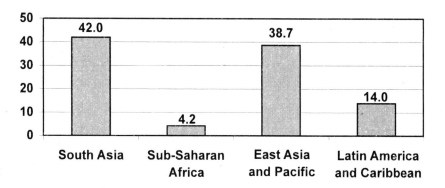

FIGURE I.4. Percentage of cropland irrigated by region, 1999. (*Source:* World Bank, 2002.)

salinization. These problems are acute in the fertile area of Punjab. In addition to these, other water-related problems that threaten future agricultural production today, according to Ruth Meinzen-Dick and Mark Rosegrant (Chapter 10), are the growing demand for water from urban and industrial centers, accompanied by a slowdown of growth in water supply investments and only modest increases in water use efficiency. In their chapter, the authors examine the water use trends in South Asia and, using an integrated global water and food model, assess the impact of water supply on future food production growth and livelihoods. Meinzen-Dick and Rosegrant find that competition for water between different uses will result in a decline in water availability for agriculture, and this decline will reduce the growth of food crop yields. There are solutions, however. Policy measures for demand management could address the potentially negative effects, and targeted investments in agricultural research could lead to new technologies that enable higher production with a reduced water supply.

While Meinzen-Dick and Rosegrant examine the external threats to irrigation for agriculture at the regional level, K. V. Raju and Ashok Gulati in Chapter 11 investigate the weaknesses within India's irrigation sector that might prevent it from being financially and physically sustainable. Not addressing these issues could mean agricultural yields in the future that are below their potential. These problems, specifically, stem from the low price of canal water and increasing irrigation subsidies. Because of the absence of rational pricing, the financial returns from canal irrigation are insufficient to meet normal operation and maintenance costs and funding for the construction of new and ongoing canal networks has declined, leading to increasing costs and lower benefits. Raju and Gulati argue that innovative institutional reforms could reverse these trends and that unless they are adopted the Indian irrigation sector could collapse. One reason why the problem of irrigation subsidies has not received sufficient attention is that the magnitude of the subsidy has been underestimated. Raju and Gulati show, using a different conceptualization of the subsidy and alternative quantitative methods, that the magnitude has in fact been quite large and unsustainable. After looking at the impact of the subsidy in different spheres, the authors propose an agenda for reform. Institutional reforms consisting of involving farmers in management and decision making, establishing an independent regulatory commission, creating transparency, and granting greater autonomy to the irrigation authority would make the irrigation sector efficient and financially sustainable. They state that farmers most likely have the ability to pay a higher price for water but that the quality of service must be improved before such a price can be asked. Improving the sector through reforms and taking advantage of the fact that farmers are able to pay more could create a situation favorable to both the irrigation management authority and farmers.

One method of reducing national water scarcity would be to look beyond the national borders to a regional approach for sharing water. There is the potential for a water-sharing arrangement in South Asia, especially with regard to the regulation of water flow over the year according to the hydrology of the region. Such an approach will help address the problem of food security in more holistic terms. Further research is undoubtedly required to identify appropriate policies that will increase the efficiency of water use in the region.

Market Reforms and Agricultural Diversification

With the move in South Asia toward trade liberalization over the past few decades, agricultural diversification offers new opportunities for the region to increase agricultural exports in the form of high-value goods, such as fruits, vegetables, and livestock and fish products. Indeed, diversification is important for the next generation of growth in South Asian agriculture. Trade liberalization and the rising demand in the developed as well as developing countries that increasing incomes and urbanization have generated provide an opportunity for the South Asian countries to move toward high-value agriculture at a faster pace. Table IA.8 shows the change in food consumption among urban and rural populations in selected South Asian countries from the 1970s to the 1980s. Interestingly, the populations of both areas in most cases increased their consumption of higher-value goods, such as fruits and vegetables, dairy products, and meat.

The key benefits of increasing agricultural diversification are increasing farmers' income, generating employment opportunities, alleviating poverty and undernutrition, and conserving natural resources. But for this opportunity to be exploited, the appropriate infrastructure, institutions, public support and technologies must be in place. In the smallholder-dominated agriculture of South Asia, infrastructure and institutions are necessary to allow procurement or the integration of production and marketing, and agroprocessing at low transaction costs, and to assure crop quality and food safety. Institutional innovation is also needed to obtain the participation of small farmers in the development of high-value goods so that they benefit from trade liberalization. Their involvement is necessary because their marketable surplus is small, adequate capital is unavailable, and their capacity to partake in global markets is weak. New technologies will make it possible to improve food quality and safety standards and can contribute to the value addition of crops. Public support should be in the form of investments for agricultural research and development to encourage high-value goods. Thus far support for diversification has been driven by demand rather than an approach to increase supply. Studies on the effects of diversification would help to more effectively target these public investments. For example, policymakers ought to know what the likely impact of diversification will be on the food security of small and marginal farmers and their households. Research could also look at the possible effects on the countries from various regional trading arrangements.

In "Agriculture Diversification in South Asia: Patterns, Determinants, and Policy Implications" (Chapter 12), P. K. Joshi and colleagues look at the extent, nature, and speed of agricultural diversification, identify its determinants, and assess the implications of diversification on food security, employment, and natural resource use. They examine the trends in diversification in all seven countries of the region from a macrolevel and conduct an analysis across regions in India at the meso level. According to the authors, the countries have been moving toward diversification and taking advantage of the new opportunities it offers. Diversification, as the authors point out, may be especially important since South Asian agriculture is experiencing shrinking landholdings, decelerating technological advances in staple crops, declining investments in agriculture, and increasing natural resource degradation. If carried out appropriately, diversification can overcome many of these challenges. Moreover, it may be the future path in agriculture for raising farm incomes, generating employment, and alleviating poverty. The results presented in Chapter 12 on the patterns of and constraints on diversification would help to craft the right policies for institutional arrangements and infrastructure so that diversification can benefit the smallholder. Indeed, infrastructure has thus far been the key factor in the trend toward producing higher-value goods. Some of the other reforms besides infrastructure development and institutional innovation that P. K. Joshi and colleagues suggest are investments in R&D for high-value goods and domestic market reforms. Though still a "silent revolution," diversification, according to the study, holds promise for reducing food insecurity, generating employment, and alleviating poverty.

A different perspective on market reform is taken in Chapter 13, by S. Mahendra Dev. Here Dev, focusing on India, concentrates on the market for procuring staple grains for the country's public food-grain distribution program. Originating out of the desire to enhance the food security of the poor, the Indian government has been intervening in the food-grain market through price policies to accumulate food stocks for distribution to the poor at below-market cost. But while the intentions have been good, the price policies and the management of the buffer stocks and the government's Public Distribution System (PDS) have led to inefficiency and insufficient availability of food grain to the poor. The development community rightly wonders why in a country in which roughly 40 percent of the population is food insecure, more than 60 million tons of staple grains lies in public stocks. Dev explains how high procurement prices gradually led to the overaccumulation of stocks and what the adverse effects of the government's policies have been for the private trade in food grains and the export sector. He then offers a series of reforms to make the procurement more efficient and improve the PDS. However, although reforms on the "supply-side" would be helpful, Dev also argues that the food security of the poor could be improved by creating employment and thus increasing the purchasing power of the poor.

The author thus raises a major issue in Indian development policy. The difficulty in increasing food security at the household level has been due not only to the slowdown in food-grain production and the inability of food supply to keep up with demand, but also to the low purchasing ability of the poor. Rural development policies have not sufficiently stimulated the demand for food. Increased employment for those living below the poverty line, which will enable increases in agricultural production translate into poverty reduction, are therefore vital. This measure would, no doubt, spur market development as well.

While India, Bangladesh, and Pakistan have been struggling to improve food security at the national level and liberalize their trade, Sri Lanka, examined in Chapter 14 by Saman Kelegama and Suresh Chandra Babu, has long had a high level of food security, ahead of the other South Asian countries as well as of many of the other developing ones, and the most liberalized regime for agricultural trade. This has not meant, however, that Sri Lanka has not had any challenges. Indeed, Sri Lanka's experience demonstrates just how difficult and complicated the path can be to achieving food security, poverty alleviation, and agricultural growth. Food security at the household level still remains a problem for the poor, children, and those displaced by the war. Sri Lanka has fared well in social-sector development, but it can no longer claim to be an exception among developing countries, as the physical quality of life index may not fully reveal the food intake deficiencies that Sri Lankan households face. Other challenges are the stagnation in agricultural production in recent years, which has led to reduced incomes for farmers, the need to further liberalize trade, and the various constraints on the diversification toward higher-value commodities. In addition to these problems, although increased production has reduced the need for imports and Sri Lanka has achieved 90 and 80 percent self-sufficiency in rice and fish, respectively, food imports remain relatively high (although they have declined as a percentage of total imports), amounting to roughly one million tons of wheat annually. The goal of this chapter is to describe this experience and the issues and challenges the country's policymakers face in trying to achieve these objectives. Some of the solutions the authors propose are the dissemination of agricultural technologies to increase production, the design of policies to generate rural employment, and for the short-term, the improvement of targeted poverty alleviation programs. With food insecurity still prevailing, Sri Lanka is in the process of changing the roles of institutions and policies for food security.

Chapter 15 provides a portrait of yet another country in South Asia; Bhutan. The case of Bhutan is an especially interesting one, for although the country is heavily dependent on agriculture for GDP and employment, like all the others in the region, and is pursuing food self-sufficiency as a means to achieve food security, it also places high importance on environmental conservation. In fact, much of the land is kept under forest cover, and this in addition to topographical factors enables only about 8 percent of the land to be used for agriculture. Yet the level of food self-sufficiency the country has achieved—70 percent—is impressive. Reaching this level was possible in part of due to the emphasis the ministry of agriculture placed on input provision, the development of extension systems, an area-based development approach for food production, and administrative decentralization. But the authors, Suresh Chandra Babu, Choni Dendup, and Deki Pema, point out that increasing production to feed a growing population, without bringing more land under cultivation, and raising the incomes of the rural poor will depend on the adoption of new agricultural technologies and the promotion of diversification in agriculture. Reforms will also be needed for marketing and distribution to increase food availability within the country. To ensure that they are on the most efficient course for food security, it would be of benefit to policymakers if they support research on a number of issues. One of these is whether the best option would be to focus on producing more rice, the main staple crop, over export crops. Another important area for research is food availability at the household level. Other priorities for the country include enhanced food security at the household level and the development of agriculture and natural-resource-based industries.

Food Security Interventions and Country Experiences

As the chapters in the previously described sections discuss, broader-level policies for market reform, trade liberalization, diversification promotion, and investing in agriculture research are necessary to achieve and sustain food security in the long-term. Yet given the levels of poverty and hunger in South Asia, the time it will take for the policies to have an effect, and the ethical obligation of the development community to alleviate the suffering that the hungry experience each day, programs designed as interventions to provide food security in the shorter term and for the poorest of the poor deserve equal attention. In addition to the data available regarding poverty, information on life expectancy, child malnutrition, and mortality gives us some understanding of the degree of food insecurity among the poor and at the household level (Table I.7).

Almost 50 percent of children in South Asia are malnourished, with the figures for the individual countries being, in most cases, from 30 to 50 percent. In Bangladesh, nearly two-thirds of the children under five are malnourished. The gap between South Asia and East Asia and the Pacific is striking. However, considerable progress has been made in the region in re-

TABLE I.7. Child malnutrition (weight for age) and mortality, and life expectancy in South Asia and selected regions, 1990 and 2000

	Malnourished children under five (%)*	Under five mortality rate (per 1,000)		Life expectancy at birth (years)	
		1990	2000	1990	2000
Bangladesh	61	136	83	55	61
Bhutan	19	—	—	—	62
India	47	112	88	59	63
Nepal	47	138	105	54	59
Pakistan	38	138	110	59	63
Sri Lanka	33	23	18	70	73
South Asia	49	121	96	58	62
Sub-Saharan Africa	—	159	162	50	47
East Asia and Pacific	10	55	45	67	69

Sources: For child malnutrition, World Bank, 2002; for infant mortality and life expectancy, World Bank, 2003.
*Latest year available, 1994-2000.
Note: Regional and world figures are weighted averages.

ducing child mortality and increasing life expectancy. For the region, child mortality declined by more than 20 percent from 1990 to 2000, and life expectancy rose by nearly 7 percent during this period. In contrast, the situation in sub-Saharan Africa worsened with respect to these two indicators. Yet the figures for South Asia remain high; child mortality in the region is more than twice as high as it is in East Asia and the Pacific, and life expectancy is more than a decade behind this region at the current rate of progress. Clearly, targeted food, nutrition, and health intervention programs are needed.

In fact, South Asia has much to offer the rest of the developing world in terms of model programs for crisis mitigation, poverty alleviation, and food distribution. India's Employment Guarantee Scheme is one example. In nearly every country of the region an array of safety-net programs exist. Part VI of this book deals in part with the various forms of intervention for food security developed in the region. At the same time, it looks at the different experiences of some of the countries in the region in their efforts to achieve food security. Examining the various policies and programs instituted within the country context might be helpful in understanding the reasons why certain policies were adopted and be instructive on whether the same policies could be adopted elsewhere.

Bangladesh's Food for Education scheme, which Akhter Ahmed and Carlo del Ninno discuss in Chapter 16, is a recent and innovative food security intervention program from the region. Its several successes, which the authors document, are likely to add it to the list of successful antipoverty interventions for the developing world. By providing food to poor families conditional upon their enrollment of their children in primary school, the program seeks to develop long-term human capital through education. In the short run, it supplies critical food supplements to help sustain these families. Poverty has kept families in not only Bangladesh, but in all of South Asia, from sending their children to school. But it is well known that without education these children will only inherit the poverty of their parents. The program, Ahmed and del Ninno show, has been successful in increasing school enrollment, promoting attendance, and improving household food security. Although there is room for improvement in the scheme, it is a creative and effective instrument for tackling both the short-term needs of the poor and bringing people out of poverty in the long term that other developing countries could replicate with likely success.

Chapter 17, "Agricultural Trade Policy Issues for Pakistan in the Context of the AOA," by Sarfraz Khan Qureshi, is the first chapter of this part on a specific country experience. It discusses Pakistan's experience in implementing the AOA and addresses such areas as food trade, domestic support, and export subsidies. Pakistan's difficulty, Qureshi illustrates, has been that although agricultural production has increased and contributed significantly to overall growth, and agricultural exports have increased considerably, food imports have been rising and are projected to increase further in the future. Because many of the imports are basic foods vital for the poor and the poverty level remains high, the government has been unwilling to curb them. As in all of the other South Asian countries, producing higher yields to alleviate poverty in the long term will be necessary. However, the author informs us that unfortunately agricultural investments in Pakistan have been on the decline. Nevertheless, on the whole, a potential for the country remains. With little domestic support, Pakistan has been able to develop high-value goods and the country has taken bold steps toward liberalization to encourage agricultural trade and poverty alleviation. Currently, there is a policy to phase out most of the commodity price supports and subsidies and to promote private-sector investment in agriculture. It might also be added that apart from increasing agricultural R&D investments a strong case can be made for increasing public investments in rural infrastructure and improving irrigation systems. The lessons of its experience will be instructive for all the countries in the region.

But if market reform, trade liberalization, and support for increased agricultural production and diversification are the measures required at a fundamental level, how important are program interventions designed to increase food and nutritional security at the household or regional level? And, what role do they play in a country's strategy to achieve food security? In Chapter 18, Suman Sharma illustrates that in Nepal both deeper-level policies, such as for increased agricultural production, and improvements in intervention programs, specifically for food distribution, are needed, and that each plays a role that is critical and complementary to the other. Due to a decline in agricultural productivity over the past three decades, Nepal went from having the highest crop yields among South Asian countries and being a net exporter of food grain to having the lowest yields in the region and being a net food grain importer. The principal cause of the poor agricultural performance, aside from the small size of landholdings, is the unavailability of production technologies. The average holding of a poor farmer cannot produce food even for the household. As a result of low productivity, coupled with inadequate infrastructure in the country and a weak marketing network, food insecurity is at a high level. Severe food shortages exist in some regions, particularly the remote ones, half of all districts are food deficient, and 50 percent of all children suffer from nutritional deficiencies. What schemes does Nepal have to reduce food insecurity? Sharma states that although the country does not have a food aid policy, it does have an array of targeted programs for food distribution and subsidizing the transport of farm inputs. However, the food distribution program, under the Nepal Food Corporation (NFC), has suffered from poor targeting, leakage, and inefficiency. Its reduction of the food deficit in remote areas has been only marginal. To address the country's low agricultural growth rate, the government has adopted a new policy, according to the Agricultural Prospective Plan, to provide a variety of technologies and support services to farmers. It will also provide a conducive environment through publicly supporting infrastructure and agricultural technology development. Yet until the benefits of this policy begin to appear, the public food distribution program will have to be reformed to ensure that those districts experiencing food deficits will have their food needs met. Other reforms should include the creation of a monitoring system for the NFC's food distribution program and infrastructure improvement Also important for food intervention programs in both the short and long term is that they take on the task of reducing food insecurity at the household level.

Abusaleh Shariff's piece, "Household Food and Nutrition Security in India" (Chapter 19), appears to support the general argument of Sharma's chapter. Despite the fact that India has, as Shariff points out, the world's largest officially supported food and nutrition security program, including schemes for public food distribution, food for work, integrated child development services, and maternal nutrition, among others, roughly 70 percent of the population still suffers from calorie and micronutrient deficiencies. Among certain groups, the deficits in some micronutrients, such as iron, and other food elements are dangerously high. Unlike Nepal, India achieved a major increase in per capita availability of food grain, yet household food insecurity remains. In his chapter, Shariff clarifies and discusses in some detail the types of food insecurity that exist in the country: chronic, nutritional, food absorption-related, and transitory. Women, infants, tribal groups, the landless, and those in remote areas have tended to suffer more from these types of food insecurity. But, much like Sharma, Shariff concludes by arguing that although food security and nutrition intervention programs are important and can have a greater impact if they are improved, broader policies to increase food production, lower food prices, enhance availability, and generate employment must be implemented if food insecurity and poverty are to be alleviated. Chapters 18 and 19 in a sense reveal the failure of attempts to reduce hunger through specific targeted schemes when the essential and structural policies to promote agricultural growth and poverty alleviation are missing. At the same time, together with the chapter by Ahmed and del Ninno on Bangladesh's Food for Education program, they illustrate the vital role that targeted interventions have in minimizing suffering and addressing the needs of particular groups in both the short and long run.

Emerging Issues

Part VII consists of chapters that either provide overviews of the topics discussed in the book or address strategies outside the food and agriculture sector that could help in the collective effort to further food security. Chapter 20, by Per Pinstrup-Andersen, offers a solid summation of the arguments on trade and technology made in the previous chapters. In the first section, the author presents a general picture of the issues and opportunities in trade that the countries of South Asia face. Globalization, Pinstrup-Andersen reminds us, can be made to work for the poor and hungry, but the governments of the region will have to develop the appropriate and supportive policies and institutions for the poor to derive the benefits. Certainly some obstacles exist to deriving gains from trade liberalization. One of the greatest is made up of the developed countries' relatively closed markets to the goods of South Asia and the developing world in general, and their heavy subsidization of their respective agricultural sectors and, in some cases, exports. The chapter suggests various measures that the countries of the region can take to either strengthen their position in the WTO or develop trade opportunities elsewhere. These include building coalitions in trade agreements and creating a regional free-trade area. In the second half of the chapter, Pinstrup-Andersen, in discussing the role of technology in South Asian agriculture, convincingly argues that today and in South Asia's future a variety of different technologies—agroecological, conventional breeding, and biotechnologies, developed and implemented through institutions that allow farmer participation—will be needed to further reduce food insecurity and poverty. These production technologies will have to generate additional crop increases while being environmentally sustainable and will need to tackle problems that farmers continue to encounter, such as pest damage and drought. A key step the countries will therefore have to take is increase public investments in agricultural research, as only the governments can provide public goods that meet the needs of poor farmers.

Chapter 21 takes an approach that is different from that of the other chapters here. Most of the chapters, written by researchers or policymakers, discuss the results of policy research on a specific issue or the challenges and opportunities facing a country. The author in this chapter, Klause von Grebmer, instead looks at how improved communication strategies can help those engaged in research have a greater impact on policy so that it helps the poor. Von Grebmer reminds us that policy research is not an end in itself, but rather an activity that is meant to produce some benefit to humankind. It can achieve this goal only if the information it generates is communicated effectively to policymakers and to those institutions that influence decision making; the media. Von Grebmer looks at what makes policy communication efforts successful and what researchers need to keep in mind about the policymaking process. Essentially, researchers need to send the right information in the right form at the right time. This involves presenting results in a manner that is easy to understand and takes into account the political arena, and at the appropriate stage in the policymaking process. This formula also basically applies in efforts to use the media to focus attention on certain issues. The importance of the media lies in the fact, according to the author, that once the media takes up an issue, the likelihood that policymakers will address it increases. Unfortunately, much of the policy research communication occurring in South Asia and the rest of the developing world is weak because it is based on the idea that research results need merely to be provided to the policymaking apparatus after which they are immediately understandable to government officials. Improving research communication strategies in South Asia will make policymaking less ad hoc and more research based and would likely lead to the development of more effective policies for food security, agricultural growth, and poverty alleviation. Policy communication can rightly be considered a major missing element in our efforts.

CONCLUSION

The South Asian countries differ from one another with respect to experience, the development path followed, size, and even primary objectives. Yet for this reason, each country offers important lessons for the others. At the same time, despite their differences, the countries face many of the same interlinked challenges. These are to increase agricultural production in an environmentally sustainable manner, enhance food availability for the poor through domestic marketing reforms and employment creation, which will raise incomes, actively diversify the production of smallholders, which will generate higher returns for them, and liberalize trade to assure food security and channel the benefits of high-value goods to small farmers. Along with these, improved intervention programs are needed to ensure food security at the household level and among the vulnerable groups. A broad consensus exists among the contributors to this work that if agricultural development is to play a key role in further reducing poverty and food insecurity in South Asia, the region's policymakers must continue with the current trend of reforms and embark on new policies in the food, agriculture, and natural resource sectors in South Asia.

To meet these policy challenges, the researchers and policymakers of South Asia could build more effective dialogue and collaboration. Creating a better understanding between the two, through the formation of national and regional networks, will help to identify the critical issues and enable policymakers to understand the impacts of their policies, in short, to make the two groups more responsive to each other. What this will require is researchers and their institutions adopting more effective strategies to communicate their policy recommendations to decision makers and sensitizing policymakers to various issues. Finally, more effective policy analysis, design, implementation, and monitoring and evaluation will not be possible without capacity development for these activities.

Reducing the high levels of poverty, food insecurity, and malnutrition in South Asia is necessary from an ethical standpoint, but it also makes good economic sense, given the high costs in terms of lost production and benefits. If the governments of the region and donors seriously wish to achieve these goals, there will have to be a break with business as usual. The needed investments and changes in policies and institutions will require substantial resources, but a considerable payoff in terms of economic growth, equity, and sustainable management of natural resources will benefit all, especially the poor, in South Asia.

Investing in poverty reduction in the region will also help substantially to reach the Millennium Development Goal of reducing by 2015 the percentage of people living on less than $1 a day in the world in 1990 by a half, to 890 million. With the number of people in this category in South Asia being more than half of the targeted reduction, strong developing-community support for the countries of the region would lead to a significant step being made toward that goal. In the past decade, modest growth in the region brought a sizeable number of people out of poverty. With investments in small farmer-oriented technologies and new agricultural strategies and participatory nutritional and food intervention programs, international donors and the governments of the region could, by increasing growth and alleviating poverty and food insecurity in South Asia, make a serious dent in hunger and poverty in our world.

REFERENCES

Ahluwalia, Montek. 2002. Economic Reforms in India Since 1991: Has Gradualism Worked? *Journal of Economic Perspectives,* 16(3): 67-88.

Food and Agriculture Organization (FAO). FAOStat Online, FAO Statistical Databases. Available at <faostate.fao.org>.

Food and Agriculture Organization (FAO). 1993. *Compendium of Food Consumption Statistics from Household Surveys in Developing Countries, Volume 1: Asia.* Rome: FAO Statistics Division.

Pursell, Gary. 2002. Trade Policy and Agriculture in South Asia. Draft Version.

World Bank. 2002. World Development Indicators 2002 [CD-ROM]. Washington, DC: Author.

World Bank. 2003. *Global Economic Prospects.* Washington, DC: Author.

APPENDIX

TABLE IA.1. Population and density in South Asia and selected regions, 2001

	Population, midyear (millions)	Density (No. per sq. km.)
Bangladesh	133.4	1025
Bhutan	0.828	18
India	1033.4	348
Nepal	23.6	165
Pakistan	141.5	183
Sri Lanka	19.6	304
South Asia	1379.8	289
Sub-Saharan Africa	673.9	29
East Asia and Pacific*	1825.2	115
World	6132.8	47

Source: World Bank, 2003.
*Includes China.

TABLE IA.2. GNI per capita, in current US$* and purchasing power parity**, 1995-2001 (current US$ in bold)

	1995	1996	1997	1998	1999	2000	2001
Bangladesh	**320**	**330**	**340**	**340**	**350**	**370**	**370**
	1290	1340	1390	1420	1500	1590	1680
Bhutan	**430**	**460**	**500**	**530**	**570**	**590**	**640**
	1150	1190	1240	1280	1360	1440	1530
India	**380**	**410**	**420**	**420**	**440**	**450**	**460**
	1850	1970	2020	2080	2240	2340	2450
Nepal	**220**	**230**	**230**	**220**	**230**	**240**	**250**
	1170	1210	1240	1240	1290	1370	1450
Pakistan	**480**	**490**	**480**	**460**	**450**	**440**	**420**
	1730	1770	1740	1730	1790	1860	1920

	1995	1996	1997	1998	1999	2000	2001
Sri Lanka	700	740	790	810	820	850	830
	2770	2870	3010	3090	3250	3460	3560
South Asia	380	410	420	420	430	440	450
	1780	1880	1920	1970	2110	2240	2300
Sub-Saharan Africa	520	530	540	510	490	470	470
	1470	1510	1510	1500	1540	1600	1620
East Asia and Pacific	940	1,080	1,140	1,000	1,010	1,060	900
	3060	3320	3500	3500	3770	4130	4040
World	5,030	5,250	5,290	5,050	5,040	5,170	5,140
	6260	6480	6630	6650	6950	7410	7570

Source: World Bank, 2002.
*Atlas Method
**Current International Dollars
Note: South Asia includes Afghanistan and Maldives.

TABLE IA.3. Cereal supply (1,000 metric tons) in South Asia and selected regions in developing world, 1990-2001

	1992	1993	1994	1995	1996	1997	1998	1999	2000	2001
Bangladesh	20,604	20,042	21,216	20,919	22,149	22,535	23,631	23,693	24,645	25,391
India	150,581	136,806	145,479	156,427	159,657	163,100	152,293	158,692	160,427	166,345
Nepal	3,603	3,996	3,797	3,664	3,906	3,958	4,122	4,264	4,463	4,572
Pakistan	16,947	17,953	19,534	18,532	19,365	20,556	20,645	20,843	21,035	22,128
Sri Lanka	2,437	2,452	2,615	2,688	2,597	2,567	2,526	2,797	2,598	2,650
South Asia	194,205	181,278	192,672	202,265	207,710	212,750	203,251	210,326	213,205	221,125
Sub-Saharan Africa	56,355	58,127	60,570	62,651	64,793	66,646	68,594	70,356	72,540	74,878
East Asia and Pacific	87,098	89,456	92,942	95,993	96,949	97,782	100,856	101,062	103,382	104,623

Source: FAOStat.
Note: South Asia includes Maldives.

TABLE IA.4. Per capita supply of calories per day from cereals and from all foods[a] in South Asia, 1995-2001

	1995	1996	1997	1998	1999	2000	2001
Bangladesh	1,655 (1,986)	1,716 (2,046)	1,714 (2,073)	1,758 (2,117)	1,725 (2,122)	1,757 (2,158)	1,768 (2,187)
Bhutan	—	—	—	—	—	—	—
India	1,550 (2,471)	1,547 (2,469)	1,556 (2,542)	1,430 (2,380)	1,465 (2,499)	1,458 (2,489)	1,487 (2,487)
Nepal	1,663 (2,239)	1,727 (2,285)	1,710 (2,298)	1,766 (2,348)	1,780 (2,442)	1,795 (2,446)	1,796 (2,459)
Pakistan	1,227 (2,385)	1,246 (2,493)	1,289 (2,474)	1,264 (2,440)	1,240 (2,461)	1,219 (2,456)	1,248 (2,457)
Sri Lanka	1,285 (2,247)	1,241 (2,253)	1,220 (2,276)	1,204 (2,306)	1,309 (2,355)	1,215 (2,345)	1,226 (2,274)
South Asia	1,526	1,532	1,542	1,449	1,471	1,466	1,492
Sub-Saharan Africa	1,009.9	1,017.6	1,020.6	1,022.5	1,022.2	1,026.3	1,035.2
East Asia and Pacific	1,630.8	1,622.8	1,606.9	1,633.3	1,618.1	1,625.8	1,621

Source: FAOStat.
Note: South Asia includes Maldives.
[a]Per capita calories from all foods are shown in parentheses.

TABLE IA.5. Average annual growth of the agricultural sector in South Asia and selected regions, 1980-2000

	1980-1990	1990-2000	1999	2000
Bangladesh	2.7	2.9	4.8	7.4
Bhutan	5.1	3.2	3.0	2.5
India	3.1	3.0	1.3	-0.2
Nepal	4.0	2.5	2.7	5.0
Pakistan	4.3	4.4	1.9	6.1
Sri Lanka	2.2	1.9	4.5	1.7
South Asia*	3.2	3.1	—	—

Source: World Bank, 2002.
*Includes Afghanistan and Maldives.
Note: Figures for regions and world are weighted averages.

TABLE IA.6. Tariffs and nontariff barriers on selected agricultural, livestock, and processed foods in South Asia, August 2002

	India	Pakistan	Bangladesh	Sri Lanka	Nepal	Bhutan
Rice (common)	87.2+STE	10	26	60	10	20
Durum (hard) wheat	50+STE	25	11	0+STE	10	20
Maize	19.6 or 56 (TRQ)	10	3.5	0	10	20
Poultry	35.2	25	36	30	10	10
Fresh milk and cream	35.2	25	36/55.9	30	10	30
Potatoes	35.2	10	36+S	60	10	20
Onions	5	10	26	60	10	20
Apples	56	25	75.8	30	10	20
Coconuts	76.8	20	26	30	10	20
Roasted coffee in bulk	108	20	36	30	25	20
Cardamon	76.8+QR	20	102.3	30	10	20
Copra	76.8+STE	10	26	30	10	20
Processed palm oil	85+TV	S	26/36+NT	30	15	30

Source: Pursell, 2002.

Notes: The products included in this table have been chosen because they are derived from rural livestock or agricultural activities that are important in all or at least some of the South Asian countries, and are also important in consumption. In India, Bangladesh, and Sri Lanka the tariff rate is the total estimated protection rate of other import taxes as well as the customs duty. All the tariffs are the MFN rates, i.e., they are not preferential rates. Bhutan has a free trade agreement with India so there is no tariff on imports from India.

A slash between two or more tariffs means that the rates indicated apply to different products or specifications in the heading.

+S means that the tariff is the higher of the ad valorem rate or a specific duty.

S means that there is a specific tariff only.

TV means that the tariff is based on a specified "tariff value" rather than cif prices, or the higher of cif prices and the tariff value.

TRQ means there is a tariff rate quota, under which the lower tariff applies to an import quota, and the upper rate to any imports in excess of the quota.

STE means that a state trading enterprise (usually a public sector enterprise such as the Food Corporation of India) controls imports.

QR means that that there is some form of quantitative restriction (e.g., an import ban, import licensing, or an import quota), the principal or major purpose of which is to protect domestic production. The import of some products is banned or restricted for religious reasons: these cases have not been noted as QRs. QR* means that only some of the products within the general heading are subject to QRs. For example, in Bangladesh the import of shrimps and fish livers is banned, but the import of all other fish, live or processed, is free of QRs.

TABLE IA.6 *(continued)*

In all the countries the products in the table require some form of health, safety, sanitary or phytosanitary clearance to be imported. This has not been noted except in a few cases where NT indicates that there is information that suggests protection is probably a major purpose and effect of the controls (see text discussion). NT* means that this applies to some but not all products within the general heading.

In Bhutan there is import licensing of all imports except imports from India, including livestock and agricultural and processed food products. Information on the actual restrictiveness of this system has not been obtained.

Tariffs for a general product heading are given when these rates apply to most products under the heading. Important products in that product group and products within the group that have different tariffs or other import conditions (e.g., QRs) than the general rate are indicated below the heading.

TABLE IA.7. Food trade in South Asia: Value of food exports* and imports** (in US$ millions) for 1980-1999

	1980		1990		1999	
	Exports	**Imports**	**Exports**	**Imports**	**Exports**	**Imports**
Bangladesh	95	614	239	686	—	—
Bhutan	16	9	—	—	18	36
India	2,418	1,336	2,800	766	5,263	3,245
Nepal	17	15	28	102	126	243
Pakistan	616	697	520	1,309	1,108	1,742
Sri Lanka	501	415	681	512	968	901

Source: World Bank, 2002.
*freight on board (fob); ** cost, insurance, and freight (cif).

TABLE IA.8. Changes in food consumption expenditure (%) patterns in South Asia*

	Bangladesh		India		Pakistan		Sri Lanka		
	Urban	Rural	Urban	Rural	Urban	Rural	Urban	Rural	Estate
Early 1970s									
Cereals	51	61	49	56	25	30	27	35	40
Fruits/vegetables	6	6	8	7	10	9	8	9	8
Meat	4	2	4	3	9	6	3	2	1
Fish/seafood	8	7	—	—	1	1	10	7	3
Milk and dairy	4	2	10	10	18	19	4	2	3
Oils and fats	7	6	7	5	14	15	8	9	5

| | Bangladesh | | India | | Pakistan | | Sri Lanka | | |
	Urban	Rural	Urban	Rural	Urban	Rural	Urban	Rural	Estate
Late 1980s**									
Cereals	42	55	26	41	22	28	35	41	50
Fruits/vege-tables	11	10	13	11	13	11	8	8	8
Meat	5	3	7	6	11	7	4	1	1
Fish/seafood	12	8	—	—	1	1	11	8	4
Milk and dairy	4	3	18	15	20	20	6	3	3
Oils and fats	4	4	9	7	9	13	9	9	8

Source: FAO (1993).

* Note that percentages do not add to 100 percent as other food expenditures such as stimulants and miscellaneous account for the balance. The years of the surveys are as follows: Bangladesh (1973/1974 and 1988/1989); India (1972/1973 and 1986/1987); Pakistan (1979 and 1987/1988); and Sri Lanka (1973 and 1981/1982).

**Except for Sri Lanka, where the latest survey data relate to 1981/1982.

PART I:
ECONOMIC REFORMS, TRADE, TECHNOLOGY, AND FOOD SECURITY

Chapter 1

Have Economic Reforms
Enhanced Food Security?

Manmohan Singh

With regard to economic reforms and food security in India, until 1990 it was fashionable to say that there were two major constraints on India's development: food and foreign exchange. Today, however, more than ten years later, these constraints do not appear to be factors restraining India's development. Presently, the country's foreign exchange reserves are more than the value of eight to nine months of imports, and the balance of payments on current accounts is in reasonably good shape—at less than 1 percent of the GDP. Furthermore, as an indication of the country's ability to manage fluctuations in food output, the public-sector system today holds close to 60 million tons of food grains. Therefore, whatever may bedevil Indian growth prospects, shortages of foreign exchange and of food, in the classical sense of the term, do not appear to be such limiting factors on development as they were in the past.

Nevertheless, several elements of the present situation are cause for concern. Although in the past ten years the national growth rate has been about 6 percent per year on average, India has not been able to increase it, particularly the industrial growth rate. The current abundance of foreign exchange reserves is partly a consequence of this inability. Had there been higher growth expectations, there would have been pressure on these reserves.

Similarly, with regard to food, although available data from all accounts show that poverty in the absolute sense of the term, as austerely defined by our planning commission, has decreased—though probably not at the same speed as was visible in the 1980s—the proportion of people living below the poverty line in India is still between 26 and 27 percent of the total population. Moreover, this is in spite of the fact that an abundance of food stocks exists. In terms of market demand for food there is no shortage, but a large number of people still suffer from malnutrition and their food intake is not what it ought to be. The persistence of poverty and hunger is also an indicator of the gaps in the nation's economic performance. Because the Western-type institutions of social security are not well developed in India, the best social security measure is the promotion of employment. Available data show that although the Indian economy has grown at roughly 6 percent per year, the employment trends are not encouraging. Achieving food security in India, in other words, is not only a matter of ensuring that food supply keeps pace with demand, but also that policy instruments which would generate adequate levels of employment, especially for those below the poverty line, are in place so that within a reasonable period of time chronic malnutrition, which plagues a considerable proportion of the population, can be eliminated.

Another aspect of concern is that while the public sector system has a large surplus of food grain, data for the last decade indicate that the growth of food grain production has slowed down significantly. What do these food surpluses represent? Has the production problem in agriculture, particularly with regard to food grains, been resolved? The available data show that that is not the case. Has the purchasing power of the population not grown fast enough? There is mixed evidence on this; although progress has been made, it has not been as great as desired. Understanding the Indian economy, particularly the agricultural sector, requires in-depth analysis. What appears to be the case is that although total production has not gone up, the proportion of stocks held by the public sector has risen steeply, whereas the proportion of stocks traditionally held in the private sector has gone down. Therefore, the current situation should not lull us into complacency with regard to the production problem in agriculture. Finally, this problem also reiterates the need to increase the purchasing power of the poor through the development of job opportunities.

India is a country of great diversity. It has approximately fifteen agroclimatic zones, and it should not be assumed that all of them face the same problems. One challenge before India's policymakers is to work out viable policy packages, taking into account the different social, economic, and physical constraints that each of these zones face.

It is often the case that an economist will put forth a policy as beneficial and then add the qualification that whether the policy is actually implemented is a matter of political will. The role of political will is referred to frequently in discussions of both international and domestic policy concerns. However, a sound policy framework cannot divorce the design of a sensible economic policy from the ability to implement it. It cannot be said that a policy is good but that it is badly implemented. It is necessary for Indian policymakers to look at their policy framework and to design policies based on what can be effectively implemented, given the social, economic, institutional, and political factors that exist. This is an area where much scope exists for further analytical work.

When policies do not work, it is the fault of the implementing agencies in the government sector, the neglect of institutional reforms, or the failure of original policy formulation to take adequate account of the underlying economic constraints. Whatever the reason, what is perhaps most important, if a policy paradigm requiring large-scale investments is being considered, is that the policies formulated have credibility. People will not invest significant resources in new directions unless they are confident that the government has the will, the ability, and the incentive to implement those policies and steer the course. Therefore, the credibility of the policy paradigm as understood by the actors who are affected by these policies is very important. It is equally essential that attention be paid to the sort of expectations the principal economic actors possess. It is well known, for example, that the impact of a policy based on imperfect knowledge is not the same as it would be if the knowledge were perfect. The difference between rational and nonrational expectations and how each can influence the outcome of a policy is also understood. All these factors need to be looked at when economic policies to meet the varied needs of Indian agriculture are being formulated. Very often economists talk about one equilibrium solution. In practice, the world is much more complicated. It is quite possible that a particular policy design may not lead to one equilibrium but that several possibilities of equilibrium exist at different levels of production and distribution. These possibilities must be kept in view when designing policies in India in the new millennium.

However, policies are influenced not only by the domestic environment but also by the globalized world. Now that agriculture is being increasingly integrated into the evolving global economy, there are both opportunities and risks. India's objective should be to work out policies that take full advantage of the opportunities that the globalized economic environment of trade and investment offer while minimizing the associated risks. These risks are considerable, particularly for those sections of the farming population that live on the edge of subsistence. Competition, as it has been said, is a double-edged sword. It helps those who are strong and able to take advantage of the opportunities that a more competitive environment creates, but it also tempers those who have neither the will nor the resources to take advantage of the new opportunities. Therefore, at the international level India has to take note of the opportunities and risks that the new trade regime creates. Because agriculture is a subject matter of further discussion in the WTO, it is also essential that India knows what sort of international regimes are likely to emerge when it comes to dealing with agricultural problems with the developed countries. Since the Great Depression, market forces have not been allowed to operate to a very considerable extent, even in the developed countries. Price, trade distortions, and domestic subsidies have proliferated. In this context, Indian policymakers might want to know what sort of regime the country will face if structural changes are made to align the agricultural sector with the new world trading system. Clearly, these subsidies in the developed countries continue to proliferate. If international trade regimes are not able to discipline the powerful countries in the world trading system, then India's farmers, particularly small and marginal ones, who in general are risk averse, will be even more reluctant to make the farm investments necessary to take advantage of the opportunities that do exist.

Therefore, institutions such as the International Food Policy Research Institute have great opportunities to be of assistance to developing countries and also an obligation to ensure that the new global arrangements are positive-sum games in which the have-nots of the world will have equal opportunity to be partners in progress and will be able to participate in the fruits of enhanced global growth, and to determine what needs to be done to make that possible. They can also help to reduce the risks that Indian farmers face as they contemplate changing their production. The concerns discussed here are surely worthy of consideration.

Chapter 2

Food Security in South Asia: A 2020 Perspective

Per Pinstrup-Andersen

INTRODUCTION

Government policy, institutions, and appropriate technology are of critical importance in efforts to assure food security for all in a manner compatible with sustainable management of natural resources. To be successful, action by governments must be based on a comprehensive understanding of the interaction between technology, policy, and institutions. Even the best of technology may fail to achieve its social objectives if the policies are inappropriate or if appropriate institutions are missing. Lack of investments in rural infrastructure, low-price policies for agricultural products, and high prices for fertilizers and other inputs, poorly functioning markets, inappropriate allocation mechanisms for water, and unclear property rights are examples of policies and institutions that may render new technology useless for farmers and society as a whole.

FOOD SECURITY SITUATION AND OUTLOOK TO 2020

According to the latest report from the Food and Agriculture Organization of the United Nations (FAO), 303 million people in South Asia are food insecure, i.e., they lack sustainable access to sufficient food to lead healthy and productive lives. This figure represents nearly one of every four people in this region. Alarmingly, it is an increase from 1991, when 288 million people (26 percent of the region's population) lived in food insecurity, and from the level of 1970, when 267 million people were food insecure. On the positive side, the proportion of South Asians who are food insecure has

Marc J. Cohen, Special Assistant to the Director General of IFPRI, assisted in preparing this chapter.

fallen considerably from thirty years ago, when it was 37 percent (FAO, 2000, 2001). Presently, about 40 percent of the food-insecure people in the developing world live in South Asia.

Malnutrition among preschool children is of particular concern. The nations of South Asia have made steady, albeit slow, progress in reducing the incidence of child malnutrition since the mid-1980s, and the absolute number of malnourished preschoolers has fallen as well. But nearly half of the children under the age of five in the region remain chronically malnourished, and the 85 million malnourished South Asian preschoolers account for more than half of all malnourished preschoolers in the developing world (Rosegrant et al., 2001b). Malnutrition is associated with about half of the 99 deaths per 1,000 live births that occur each year among South Asian children under five years of age (World Bank, 2001). For those who survive, it usually means irreversible damage to their physical and mental development. Adults whose growth has been stunted by childhood malnutrition are 2 to 9 percent less productive than nonstunted adults (Gillespie and Haddad, 2000). Countless people in this region will not grow up to be scientists, software engineers, creative artists, political leaders, entrepreneurs, or productive farmers and workers because of this scourge.

A critical factor behind South Asia's high rates of child malnutrition are birth weights of less than 2.5 kilograms for the affected children: 21 percent of the children are born with low birth weights, accounting for 64 percent of the world's low-birth-weight newborns. This is usually the result of poor maternal nutrition both before conception and during pregnancy. In effect, malnutrition is directly transmitted from one generation to the next (Allen and Gillespie, 2001). Cultural practices contribute to this situation: because families of daughters must pay bridegrooms a dowry, girls tend to receive less care and food than boys. Girls therefore have higher mortality rates than boys, and those who survive grow up malnourished and likely to have low-birth-weight babies (Meinzen-Dick et al., 1997).

Generally speaking, if a person consumes an adequate level of calories, he or she will also take in enough protein. However, this does not guarantee adequate consumption of vitamins and minerals. Insufficient intake of these micronutrients—often called "hidden hunger"—affects vast numbers of people, with serious public health consequences. In South and Southeast Asia, 76 percent of pregnant women and 63 percent of preschool children are anemic, and around 50 percent of the world's anemic women live in South Asia. Deficient iron in the diet is the leading cause of anemia. The risk of maternal mortality among anemic women is 23 percent higher than that of nonanemic mothers. Their babies are more likely to be premature, have low birth weights, and die as newborns. The incidence of anemia is also high among South Asian school-aged children. Anemia can impair child health and development, limit learning capacity, impair immune systems, and reduce work performance. Iron-deficiency anemia is estimated to reduce productivity by up to 17 percent for heavy manual labor. Even when iron deficiency does not progress to anemia, it can reduce work performance. These effects of iron deficiency result in annual economic losses estimated at $5 billion for the South Asia region (ACC/SCN and IFPRI, 2000; Gillespie and Haddad, 2000).

Insufficient intake of vitamin A among children in developing countries is the leading cause of preventable severe visual impairment and blindness and contributes to infections and death. Pregnant women who are vitamin A deficient face increased risk of mortality and mother-to-child HIV transmission. The limited data available indicate that Bangladesh, India, and Nepal have much higher prevalences of clinical vitamin A deficiency among preschool children than other Asian countries, while the prevalence in Sri Lanka is much lower. About half of Pakistani preschoolers suffer from subclinical vitamin A deficiency (Allen and Gillespie, 2001).

Currently, enough food is available in South Asia to provide everyone with 2,416 calories per day, or more than enough to meet minimum requirements, if the food were distributed according to need (Rosegrant et al., 2001b). Food insecurity persists not because of lack of food but because the people who are food insecure are too poor to afford the food that is available and lack access to the resources to produce adequate food for themselves. According to the World Bank, 522 million South Asians live on the equivalent of less than US$1 per day and account for 40 percent of the people in the region and 44 percent of all poor people in developing countries. Nearly 1.1 billion South Asians (84 percent of the people in region) live on the equivalent of less than $2 per day. During the 1990s, the percentage of the region's people living in poverty declined a bit, but due to population growth, the number of poor people stayed about the same. In India and Bangladesh, poverty has declined more quickly in urban areas than in the countryside, due to slow agricultural growth, landlessness, and high rural unemployment. Pakistan and Sri Lanka have made very slow progress in reducing poverty (World Bank, 2002).

Without changes in national and global policies, IFPRI projects that substantial food insecurity will persist in South Asia in the year 2020. Child malnutrition will fall by 31 percent, but over 63 million preschoolers will remain malnourished, accounting for nearly half of the developing world's children under five suffering from malnutrition (Rosengrant et al., 2001a). India alone will be home to 44 million malnourished preschoolers, accounting for 34 percent of the developing world total (Rosegrant et al., 2001b). Overall food insecurity is likely to remain substantial as well. If economic growth is sluggish between now and 2015, the World Bank projects that one of every four South Asians will remain poor. However, more rapid, broad-based growth could reduce the regional poverty rate to 18 percent (World Bank, 2002).

There is nothing inevitable about these rather bleak prospects. With concerted action by governments, international agencies, nongovernmental organizations (NGOs), business and industry, and individuals, backed by adequate resources and appropriate changes in institutions and policies, it is possible to accelerate progress toward achieving sustainable food security for all by 2020.

DRIVING FORCES

To be effective, the design and implementation of policies and institutions must take into account a set of driving forces, some of which are new or emerging while others have been with us for a very long time. Nine driving forces believed to be particularly important in the context of policies for food security are described here. These nine are believed to be of global importance, but the relative importance will vary among countries. The nine driving forces are as follows.

Accelerating Globalization, Including Further Trade Liberalization

Globalization offers developing countries significant new opportunities for broad-based economic growth and poverty alleviation, but it also carries significant risks. Continued protection of domestic agriculture and increasing food safety concerns in industrialized countries may limit access to their markets by developing countries. The most critical issue is how globalization can be guided to benefit low-income people, particularly their food and nutrition situation as well as the impact on natural resources. Without appropriate accompanying policies and institutions at both the national and international levels, globalization may either bypass or harm many poor people in developing countries.

Sweeping Technological Changes

New technological advances in molecular biology and information and communications offer potential benefits for poor people that may advance food security and improve the sustainability of natural resources management. However, there are serious concerns over whether poor and food-insecure people will have access to these technologies, many of which currently are developed by the private sector and focused primarily on nonpoor people in industrialized countries. Although past public agricultural research tailored to solving problems of small-scale farmers and low-income consumers in developing countries has been effective in expanding productivity, protecting the environment, and increasing food security, rapid changes in the financing, management, and organization of agricultural research may require new policy interventions to further enhance the benefits obtainable by low-income people. Without such changes in policies and institutions, the current and expected future technological revolutions may leave the poor and food insecure further behind.

Degradation of Natural Resources and Increasing Water Scarcity

Degradation of natural resources is rampant in many resource-poor areas of developing countries, particularly those areas with fragile soils, irregular rainfall, relatively high population concentration, and stagnant productivity in agriculture. Natural resource degradation is also occurring in agricultural areas exposed to misuse of modern farming inputs. Natural resource degradation often is a consequence of poverty, but it also contributes to poverty. Such a downward spiral is found in many locations where low-income people reside. Water scarcity is emerging as the most constraining factor for food security in many regions in the future. Failure to effectively deal with the natural resource issue in the quest to achieve food security for all will not result in sustainable solutions.

Emerging and Reemerging Health and Nutrition Crises

The tragic pandemic of HIV/AIDS, the persisting threats from malaria, the reemergence of tuberculosis, the widespread prevalence of micronutrient deficiencies, and a variety of chronic diseases caused in part by obesity compromise food and nutrition security in many developing countries. This global health crisis is impoverishing those affected and contributing to rising health care costs, labor shortages, and declining asset bases. It is causing loss not only of human lives but also of opportunities. Achieving a food-secure world for all calls for a healthy population.

Rapid Urbanization

Most of the population increase in coming years will occur in cities and towns of developing countries; by 2020, a majority of the developing world's population will live in urban areas. This will present new challenges to provide employment, education, health care, and food. Although current actions must continue to focus on the rural areas where the majority of the poor and food insecure reside, future policy actions must pay increasing attention to the growing poverty and food insecurity in urban areas.

The Rapidly Changing Structure of Farming

A number of factors such as the aging of the farm population, the feminization of agriculture, labor shortages, and depleting asset bases resulting from the HIV/AIDS crisis, and the decreasing cost of capital relative to labor are conspiring to result in rapid changes in the structure of farming in many developing countries. These rapidly emerging factors call for new and innovative approaches to agricultural policy and rural institutions. Small-scale family farms, which traditionally have been considered the backbone of much of developing-country agriculture, are under threat as labor scarcity caused by out-migration and disease becomes more pronounced, while globalization and domestic investment in infrastructure improves markets and makes capital available for larger production units. The future of small-scale farming is increasingly uncertain.

Continued Conflict

Violent conflicts continue to cause severe human misery in a large number of developing countries and, unfortunately, have affected several nations in South Asia. The impact on food security, nutrition, and natural resource management are severe. Although humanitarian assistance may be effective in providing food and shelter for the many millions of refugees and displaced persons, policy action is needed to deal with the underlying causes and the resulting impact on the people in war-torn and neighboring areas. Achieving sustainable food security for all is unlikely to be possible in the midst of conflict.

Climate Change

Future policy action to achieve sustainable food security for all must incorporate the likely consequences of the ongoing climate change and associated fluctuations in weather patterns. Policies and institutions will be needed to counter or compensate for negative effects. While agriculture may contribute to or reduce the concentration of carbon dioxide (CO_2) in the air, the primary responsibility of future agricultural policies will be to find ways to accommodate food, agriculture, and natural resources as the climatic change continues.

The Changing Roles and Responsibilities of Key Actors

The diminished and changing role of national governments in many developing countries, which has been under way over the past couple of decades, is likely to continue in the future. Given the importance of public goods, what is the most appropriate role of national governments in efforts to achieve the sustainable food security for all? Local governments along with the private sector and civil society, including NGOs, are taking on an increasing number of responsibilities for activities previously undertaken by national governments. Local communities, frequently with the help of community-based NGOs, are demanding an increasing say in policies and programs that affect them. At the global level, transnational corporations and broad coalitions of civil society organizations are taking on more prominent roles in policy debates and in actual national and international policy formulation. New emphasis on exposing corruption where it occurs is likely to contribute to the ongoing changes in the roles and responsibilities of the various actors.

POLICY PRIORITIES

Achieving sustainable food security for all will depend on policy action and institutions that address the causes of food insecurity, malnutrition, and unsustainable management of natural resources within the context of the driving forces, including those mentioned previously. The specific policies that will be most appropriate will vary according to local and national circumstances, but rapid pro-poor economic growth, empowerment of poor people, and effective provision of public goods are essential. IFPRI research suggests a number of steps that governments in South Asia could take to achieve pro-poor economic growth. Development assistance donor agencies should support efforts to take these steps.

Public Investment Priorities

IFPRI forecasts that South Asia is likely to make public investments of $148 billion over 1997-2020 in irrigation, rural roads, education, clean water, and agricultural research. Boosting this figure by an additional $50 billion would mean that the number of malnourished children in the region will fall by an additional 13 million (Rosegrant et al., 2001b). Although that is a substantial sum, on an annual basis it represents less than 1 percent of current government spending in South Asia (World Bank, 2002). IFPRI research shows that from 1970 to 1995, improvements in female education and increased food availability accounted for nearly 70 percent of the reductions in child malnutrition in the developing world, with improvements in the health environment and in women's social status relative to that of men accounting for the rest (Smith and Haddad, 2000). At present, however, in South Asia the net enrollment rate for girls is 20 percent lower than for boys (Watkins, 2001). IFPRI has also found that in rural India, public investment in rural roads and agricultural research and extension have a substantial impact on both agricultural growth and poverty reduction. Government spending on education has a substantial impact on rural poverty reduction as well, largely as a result of the increases in nonfarm employment and rural wages (Fan, Hazell, and Thorat, 1999). Experience in Bangladesh has shown that food-for-education programs can be effective in achieving the dual goal of better education and improved food security (IFPRI, 2001).

Investment in Human Resources

Healthy, well-nourished, literate citizens are an essential precondition for successful pro-poor economic growth. Universal access to primary and preventive health care is essential to achieving a healthy population. Several strategies, taken together, hold potential for success in fighting micronutrient malnutrition: food fortification and supplementation with needed micronutrients, nutrition education to promote healthy diets, and development of staple crops rich in iron and vitamin A.

Access to clean water and safe sanitation are also critical for good health and good nutrition. Contaminated food and water are sources of much illness and death in developing countries. Policies and institutions are needed for improving sanitary conditions, storage, transport, processing, conservation of food, and other cost-effective ways to prevent food- and water-borne illnesses.

Improving Access to Productive Resources and Employment

Pro-poor economic growth can take place only if poor people have access to productive resources and employment. As urbanization proceeds, the need for jobs in urban areas is growing. With a majority of poor people living in rural areas, however, their access to productive resources and employment is critical. More productive agriculture is vital, for productivity gains in agriculture can boost the incomes of rural people both on and off the farm. To the extent that gains in agricultural productivity lead to lower food prices, they will benefit poor consumers. Increasing productivity in agriculture can also slow the pace of rural-to-urban migration. In addition, a healthy agricultural economy offers farmers incentives to conserve the natural resource base.

Productive resources must go not only to men but also to women, for gender equality is an important contributor to food security. Women play important roles as producers of food, managers of natural resources, income earners, and caretakers of household food and nutrition security. Research has shown that giving women the same access to physical and human resources as men increases agricultural productivity dramatically. In Bangladesh, we have found that assets in the hands of women increase the share that households spend on children's clothing and education, as well as reduce the rate of illness among girls (IFPRI, 2000).

Policies for Pro-Poor Technological Change

Rapid technological developments in molecular biology, information, communication, and energy are placing new demands on government policy to exploit the new opportunities for the benefit of poor people and their food security and for sustainable management of natural resources, while managing new risks and uncertainties. The impact of the new technology on low-income people and their food security will to a very large extent depend on accompanying policies. The choice, design, and implementation of appropriate policies to guide technology development and use for the benefit of poor people are critical. Currently, action by governments, the for-profit private sector, and civil society tends to be excessively influenced by ideology and unsubstantiated claims about risks and opportunities. Lack of appropriate facilitating and regulatory policies and related low levels of public investment in public-goods creating research are major reasons why potential benefits from the new technology are not reaching low-income people in developing countries to the extent they should. It is encouraging, therefore, that in March 2002, India gave approval to commercial planting of genetically modified cotton, following appropriate biosafety reviews. This has the potential to boost the incomes of small cotton growers in resource-poor areas, as it has in China and South Africa, and therefore to improve their families' food security.

Policy is urgently needed on intellectual property rights questions, particularly as they relate to institutional requirements in developing countries; biosafety and food safety regulations; facilitation of markets for improved seed; solar panels, cell phones, and other new technology; allocation of public and private funds to research, including traditional agricultural research; and a variety of facilitation and regulatory policy issues. Policy is needed to guide high-priority biological research and development to solve critical problems facing small farmers and poor consumers and to help integrate new technology into agroecological approaches (Pinstrup-Andersen and Schiler, 2001). Policy is also needed to strengthen the ability of new communications technology, satellite-based cell phones, and solar-panel-based generation of energy, to improve rural infrastructure in low-income countries and remote regions.

One of the challenging policy questions is how food security for the poor can benefit from property rights that are moving toward more exclusive patents for biological technology. A related set of policy questions is how such new patents will affect diet diversity and biodiversity, how traditional knowledge, plant materials, and experience can be protected against exclusive patenting, with for example, farmers rights legislation, and how this would affect food security. The recent passage of the Protection of Plant Varieties and Farmers' Rights Legislation by the Indian parliament represents an attempt to balance the need to offer incentives for innovation to private plant breeders with provisions on benefit sharing for individual and community holders of traditional knowledge and on the rights of farmers to save and exchange seeds.

Molecular-biology-based innovations in agriculture and health as well as new technology in information, communication, and energy continue to be focused on markets in high-income countries. Policy is needed to exploit opportunities for both public- and private-sector investments with high social returns in low-income countries. Much of the recently developed new technology may be important elements of a strategy to reduce poverty, food security, and unsustainable use of natural resources in developing countries. However, it must be adapted to the conditions within which small farmers and poor consumers operate and supported by appropriate policies and institutions.

Increasing Globalization

Current trends toward increasing globalization, including international trade liberalization, the opening up of economies in both developed and developing countries, more integrated international capital markets, and a freer flow of labor, information, and technology are likely to continue. The implications for food security, agriculture, and natural resources in developing countries are important. Unfortunately, they are not well understood.

Globalization may take different forms. The nature of future globalization, rather than globalization itself, is the key question. The challenge to food policy for developing countries is twofold: first, to design and guide globalization to reduce poverty, improve food security of low-income people, and promote sustainable productivity increases in developing country agriculture; and second, to implement domestic policy changes in developing and developed countries required to avoid negative and maximize positive effects of globalization on developing countries in general and poor people in particular. Alternative approaches to compensate potential losers such as food aid need to be considered, as should changes in competitiveness of small-scale agriculture.

The interaction between developing countries' domestic policies and globalization, including but not limited to trade liberalization and regional integration, is of critical importance. A low level of price transmission is but one indication that poor countries may fail to capture potential benefits, yet this interaction is of critical importance if poor countries are to benefit from globalization. A combination of lack of information and a rapidly polarizing debate in various parts of the world may result in policies that prohibit large potential benefits from materializing for low-income countries and people. Failure of industrialized nations to open up their markets for goods and services from developing countries, use of a variety of nontariff barriers such as requirements regarding social conditions, and food safety requirements that poor countries cannot meet, as well as high and escalating tariffs for processed commodities, make it difficult for many developing countries to capture benefits from trade liberalization. The developed countries account for 80 percent of the $360 billion world total of annual government payments to farmers—$1 billion per day—many of which are highly trade distorting.

However, developing countries, including the nations of South Asia, may gain leverage in global agricultural trade negotiations through coalitions with certain groups of developed countries. The Cairns Group, consisting of nonsubsidizing agricultural exporters from both the South and the North, has already proved influential.

It is also important for the countries of the region to look at the benefits of intraregional trade liberalization. Because Bangladesh liberalized grain imports in the early 1990s, it was able to maintain stable prices after the 1998 floods devastated local crops with imports from India, thereby reducing the need to seek uncertain supplies of food aid (del Ninno et al., 2001). At present, the region's two largest countries, India and Pakistan, engage in little trade. Expanding economic ties might serve to reduce tensions between neighbors.

Agricultural Input and Output Markets and Rural Infrastructure

Well-functioning domestic markets and good infrastructure are a prerequisite for a wide distribution of benefits from globalization and technological change. However, agricultural input and output market reforms in many developing countries are confronted with severe difficulties, and the progress toward efficient, effective, and competitive private markets has been slow. There are several reasons for the limited success, including

1. structural weaknesses and institutional deficiencies of domestic markets,
2. strong skepticism and lack of commitment on the part of national policymakers,
3. weaknesses in the design and implementation of reform programs,
4. poor infrastructure, and
5. lack of trust by private traders in the level of national governments' commitment to market reform.

The role of the state during and after the transition to private markets is poorly understood. The challenge facing the countries as they pursue successful market reforms is to find the proper balance between the facilitation of private-sector participation and the complementing roles the state needs to play in reducing transaction costs, shaping an appropriate legal environment, promoting effective competition, and easing the transition for low-income producers and consumers.

Rural Capital and Labor Markets

The pace and pattern of adjustments by rural producers to new opportunities provided by technology and globalization will depend on efficiently functioning rural capital and labor markets. We have seen unprecedented progress in the provision of microcredit to the poor in many developing countries, particularly in South Asia. But farmers, traders, and processors in rural areas are still constrained by formidable problems of access to credit. Rural banks have not emerged at the scale warranted or comparable to the growth of financial institutions in urban sectors. Publicly provided credit mechanisms (for example, agricultural banks and cooperatives) have performed very poorly and kept many small- and medium-scale producers out of the loop of public credit systems. It is extremely important to find solutions to this problem for rural financial markets. Without development of rural financial markets, rural sectors are likely to fall behind, particularly in the emerging playing fields evolved from globalization. By the same token, because of these emerging forces, efforts to develop rural banks are more likely to be successful now than in the past.

Rural labor markets are becoming more monetized than in the past. Similarly, participation of women in formal labor markets is increasing. Labor mobility has been increasing because of development or rural infrastructure. These dynamic forces have generated tremendous impact on employment and wage rates in rural areas, which bear direct consequences for the welfare of the rural poor.

Effective rural labor markets are crucial for the generation of income for rural landless workers and for farmers and small enterprises in rural areas, as they produce goods and services in increasingly commercial environments. Because the forces in rural areas are changing so quickly, the adjustments in labor markets often fail to keep pace and thus are confronted with severe disequilibria.

Risk Management and Coping Strategies

Technological change, globalization, and increasing climatic fluctuations are introducing new risks and uncertainties in the food and agriculture area. Economies will be more interconnected, resulting in new risk factors and new opportunities for spreading risks. Individual producing and consuming agents, as well as sectors, will be less protected by government as liberalization and globalization proceed and subsidies are reduced.

Although the trend of global warming is becoming increasingly clear, its effects on food production are still uncertain. Some research suggests that growing conditions will deteriorate in currently tropical areas (where many of the developing countries are located) (Rosenzweig and Parry, 1994; Fischer et al., 1996). However, effects on productivity and production will occur over a long period of time and will be very small in any given year. The associated increase in climatic fluctuations is likely to have much greater effects on food security.

Fortunately, new and innovative approaches to risk management in food production, distribution, and consumption, including new instruments for financial risk control, are beginning to surface. New or improved tools such as better climatic forecasting and the availability of data from geographic information systems are also becoming available. Results from recent work on coping strategies, including social or food security safety nets in food consumption, add to the arsenal of approaches for managing risks and consequences for the poor. However, the application of these opportunities is lagging and appropriate private and public institutions have not been developed.

Rapid Urbanization

Between 2000 and 2030, the urban population of the developing countries is projected to double from about 2 to 4 billion (U.N. Population Division, 2002), while the rural population is expected to increase by only 0.2 percent. Unless policies adapt accordingly, there will be a significant shift in poverty, food security, and child malnutrition from rural to urban areas, even though the prevalence of each of these conditions will continue to be higher in rural regions. South Asia is currently less urbanized than some other developing regions, but the urban population will rise from the current level of 25 to 33 percent to 41 to 49 percent by 2030 in the region's three largest countries: India, Pakistan, and Bangladesh.

Although the essence of urban poverty, food insecurity, and malnutrition will depend on a number of factors such as the speed of rural to urban migration and income-earning opportunities in urban areas, government policy and institutional change can have a major influence. Policies for an efficient future food supply for the rapidly increasing urban populations, including productivity increases and improved infrastructure, markets, and institutions, are needed.

Rural Industrialization

Past research by IFPRI and others have documented the strong linkage effect between productivity increases in small-scale farming and the demand for nonfarm goods and services produced in rural areas (Mellor, 1976; Hazell and Röell, 1983; Hazell and Haggblade, 1989; Delgado et al., 1998). The production and distribution of such goods and services can be important sources of income for the rural poor. Labor-intensive rural industrialization offers new opportunities for employment and income generation for the rural poor. In low-income developing countries, the processing of agricultural commodities provides an important avenue for expanded rural industrialization, thus adding value to the commodities produce while generating employment for the rural poor.

In developing countries, processing food, beverages, and tobacco accounts for one-third of the value added from manufacturing activities. However, in many developing countries, the agricultural processing industry is underdeveloped. Postharvest activities, including processing, storage, and distribution, stimulate backward linkages to farmers through an increase in the demand for agricultural commodities and enhanced off-farm employment. As trade liberalization proceeds, the competitiveness of the domestic food sector becomes an issue of increasing importance, and failure to develop an effective and efficient postharvest sector may relegate low-income countries to mere suppliers of cheap agricultural raw materials, thus forgoing opportunities for added value and employment. However, the development of an efficient postharvest sector in rural areas of poor countries is complex and knowledge intensive. The choice of policies and institutions to facilitate a competitive labor-intensive private sector is critical.

Environmental Concerns and Less-Favored Areas

Population increases, urbanization, income growth, and dietary changes will require continued increases in food production. Almost all of the increase will have to come from increasing productivity on existing agricultural land. Past successes in food production in developing countries have resulted from heavy emphasis on irrigated and high-potential rainfed lands. Large areas of less-favored lands have been neglected, and this is true in South Asia. With rapid population growth, poverty, and food security, natural resource degradation is becoming extremely serious in these areas. If current conditions persist, more than 800 million poor people are expected to live in less-favored areas by 2020 (Hazell and Garrett, 1996).

Past development strategies for those areas have often failed, partly due to lack of serious political commitment and insufficient allocation of public resources to generate public goods and facilitate private investment and partly due to lack of knowledge. Food policy is needed to promote development of these areas, with emphasis on the impact on poverty, food insecurity, and management of natural resources. Recent IFPRI research in India and China has shown that social returns from public and private investments in these areas can be very high.

Growing Water Scarcity

Unless properly managed, fresh water may well emerge as the most important constraint to global food production. Although supplies of water are adequate in the aggregate to meet demand for the foreseeable future, water is poorly distributed across countries, within countries, and between seasons.

Demand for water will continue to grow rapidly. Projections of water demand to 2020 indicate that global water withdrawals will increase by 35 percent between 1995 and 2020. Developed countries are projected to increase their water withdrawals by 22 percent, more than 80 percent of the increase being for industrial uses. Developing countries are projected to increase their withdrawals by 43 percent over the same period and to experience a significant structural change in their demand for water. The share of domestic and industrial uses in their total water demand is projected to double from 13 percent in 1995 to 27 percent in 2020, reducing the share for agricultural uses.

The costs of developing new sources of water are high and rising, and nontraditional sources such as desalination, reuse of waste water, and water harvesting are unlikely to add much to global water availability in the near future, although they may be important in some local or regional ecosystems. The rapidly growing domestic and industrial demand for water will have to be met with reduced use in the agriculture sector, which is by far the largest water user, accounting for 72 percent of global water withdrawals and 87 percent of withdrawals in developing countries in 1995 (Rosegrant, Ringler, and Gerpacio, 1997).

In India and Pakistan, excessive irrigation and poor management of irrigation systems have led to soil salinization, which seriously reduces agricultural productivity. In eastern India and the Terai zone of Nepal, excessive water use and poor management have led to waterlogged soils, which likewise lower farm productivity (Rosegrant and Hazell, 2000).

Reforming policies that have contributed to the wasteful use of water offers considerable opportunity to save water, improve efficiency of water use, and boost crop output per unit of water. Required policy reforms include establishing secure water rights of users; decentralizing and privatizing water management functions; and setting incentives for water conservation, including markets for tradable water rights, pricing reform, reduction and improved targeting of subsidies, and effluent or pollution charges (Rosegrant, 1997).

Declining Soil Fertility

Improved soil fertility is a critical component of low-income countries' drive to increase sustainable agricultural production. Past and current failures to replenish soil nutrients in many countries must be rectified through the balanced and efficient use of organic and inorganic plant nutrients and through improved soil-management practices and crop varieties.

Although some of the plant nutrient requirements can be met through the application of locally available organic materials, such materials are insufficient to replenish the plant nutrients removed from the soils and to further expand crop yields. But the use of chemical fertilizers has decreased worldwide during the past few years, particularly in the developed countries and in parts of Asia. Reduced use of fertilizers is warranted in some locations, including some of the Green Revolution areas of South Asia, because of negative environmental effects. However, it is critical that fertilizer use be expanded where soil fertility is low and a large share of the population is food insecure, as in less-favored areas of South Asia. Fertilizer consumption in these areas is generally low because of high prices, insecure supplies, and the greater risk associated with food production in marginal areas (World Bank, 1997).

In view of the size and seriousness of the soil fertility problem in many low-income countries, a cost-effective fertilizer sector and policies providing incentives for farmers and communities to implement soil fertility programs are needed. Such policies should focus on supporting agricultural research to generate appropriate technology; clear long-term property rights to land; access to credit, improved crop varieties, water, and information about effective and efficient use of plant nutrients in various production systems; efficient and effective markets for plant nutrients; and investments in roads and rural transportation systems.

Impact of HIV/AIDS on Food Security

The devastating effects of HIV/AIDS on the well-being of millions of people and the grim prospects for its rapid expansion, particularly among low-income people who cannot afford the new drug treatments, have serious implications for future food security. Having surpassed both measles and malaria, HIV/AIDS has become the second leading cause of child mortality in Sub-Saharan Africa (Brown, 1996), and life expectancy in several countries of the region has dropped significantly. In addition to the direct health effects on individuals, HIV/AIDS affects food security through reduced earning opportunities by unhealthy adults, by the loss of parents and the rapid increase in the number of orphaned children, and by placing increasing demands on scarce public resources for health.

Although the prevalence is lower in Asia and Latin America, the spread of the disease in South Asia is worrisome. In India, only about 1 percent of adults lived with HIV/AIDS in 2000, but the total population affected, nearly 4 million people, is larger than in any country but South Africa. Prevalences are higher than the national average in some states and major cities, including Bangalore, Hyderabad, Mumbai, and Chennai. Needle sharing among intravenous drug users has transmitted the disease in Manipur, as well as in Nepal and Bangladesh. Low rates of condom use among sex-industry workers in the region is another way in which HIV has spread.

Policy is needed to provide access by poor rural and urban households to AIDS-prevention programs and related health care and to reduce the negative impact of HIV/AIDS on agricultural productivity, income earnings, and child nutrition. Gender-specific aspects should be incorporated in such policy. Efforts should be made to identify feasible policy options to mitigate the negative food security and nutrition effects.

Role of the State and of Good Governance

The roles of the state, the market, private voluntary organizations, and the for-profit private sector have changed markedly in countries exposed to globalization, structural adjustment, and related policy and market reforms. However, lack of knowledge about the proper role of each of these agents in the new socioeconomic and political environments within which the many developing countries find themselves is a major bottleneck to successful transformation. Failure to arrive at proper roles and appropriate institutions is a major reason why the reforms have been disappointing in many developing countries.

The role of the public sector appears to be shrinking in many aspects of food security, while civil society and the private sector have taken on increasing importance. Although such a shift may be appropriate, recent research and experience clearly show the importance of an effective public sector in many areas related to food security such as agricultural research to develop appropriate technology for small farmers, rural infrastructure, health care, education, development and enforcement of a legal system, and the creation of public goods in general. Market liberalization and globalization require different institutions, rules, and regulations. An effective government is needed to facilitate privatization and guide the transformation of the agricultural sector in a direction beneficial for the poor.

The impact of governance (including democracy, adherence to human rights principles, the rule of law, and empowerment of civil society) on transaction costs and efficiency of food systems and poor people's access to food should take high priority and efforts should be made to identify appropriate governance structures. Current efforts in many developing countries to decentralize public-sector decision making and resource allocation is hampered by a lack of understanding of how best to implement local government action.

Market liberalization often assumes that the private sector is capable and willing to take over the roles traditionally managed by the government. Where that assumption has been taken too far, the elimination of inefficient government agencies and institutions have not resulted in appropriate private-sector agents and the poor have been left worse off than before.

CONCLUSION

Addressing the high levels of poverty, food insecurity, and child malnutrition in South Asia is urgent from an ethical point of view. It also makes good economic sense, given the high costs in lost productivity. But the region is unlikely to make rapid progress toward food security for all unless there is a break with business as usual. The needed investments and changes in policies and institutions will require substantial sums of money, but there will be a substantial payoff in terms of economic growth, equity, and more sustainable management of natural resources, to the benefit of both well-fed and food-insecure South Asians.

REFERENCES

ACC/SCN (UN Administrative Committee on Coordination/Subcommittee on Nutrition) and IFPRI. 2000. *Fourth report on the world nutrition situation.* Geneva: ACC/SCN and Washington, DC: IFPRI.

Allen, L. and S. Gillespie. 2001. *What works? A review of the efficacy and effectiveness of nutrition interventions.* Geneva: ACC/SCN and Manila: Asian Development Bank.

Brown, L. 1996. *The potential impact of AIDS on population and economic growth rates.* 2020 Vision for Food, Agriculture, and the Environment Discussion Paper 15. Washington, DC: IFPRI.

Delgado, C. L., J. Hopkins, and V. A. Kelly, with P. Hazell, A. A. McKenna, P. Gruhn, B. Hojjati, J. Sil, and C. Courbois. 1998. *Agricultural growth linkages in sub-Saharan Africa.* Research Report 107. Washington, DC: IFPRI.

del Ninno, C., P. A. Dorosh, L. C. Smith, and D. K. Roy. 2001. *The 1998 floods in Bangladesh.* Research Report 122. Washington, DC: IFPRI.

Fan, S., P. Hazell, and S. Thorat. 1999. *Linkages between government spending, growth, and poverty in rural India.* Research Report 110. Washington, DC: IFPRI.

FAO (Food and Agriculture Organization of the United Nations). 2000. *Agriculture toward 2015/2030: Technical interim report.* Rome: FAO.

———. 2001. *State of food insecurity in the world, 2001.* Rome: FAO.

Fischer, G., K. Frohberg, M. L. Parry, and C. Rosenzweig. 1996. Impacts of potential climate change in global and regional food production and vulnerability. In *Climate change and world food security,* ed. T. E. Downing. NATO ASI Series, pp. 115-160. Berlin: Springer Verlag.

Gillespie, S. and L. Haddad. 2000. *Attacking the double burden of malnutrition in Asia.* Washington, DC: IFPRI for the Asian Development Bank.

Hazell, P. and J. L. Garrett. 1996. *Reducing poverty and protecting the environment: The overlooked potential of less-favored lands.* 2020 Vision for Food, Agriculture, and the Environment Brief 39. Washington, DC: IFPRI.

Hazell, P. B. R. and S. Haggblade. 1989. Farm-nonfarm growth linkages and the welfare of the poor. Paper presented at the World Bank/IFPRI Poverty Research Conference, Airlie House, Virginia, October 2.

Hazell, P. B. R., and A. Röell. 1983. *Rural growth linkages: Household expenditure patterns in Malaysia and Nigeria.* Research Report 41. Washington, DC: IFPRI.

IFPRI. 2000. *Women: The key to food security.* Issue Brief 3. Washington, DC: IFPRI.

————. 2001. *Food for education.* Issue Brief 4. Washington, DC: IFPRI.

Meinzen-Dick, R., L. Brown, H. Feldstein, and A. Quisumbing. 1997. Gender, property rights, and natural resources. *World Development* 25(8): 1303-1316.

Mellor, J. W. 1976. *The new economics of growth: A strategy for India and the developing world.* Ithaca, NY: Cornell University Press.

Pinstrup-Andersen, P. and E. Schioler. 2001. *Seeds of contention: World hunger and the global controversy over GM crops.* Baltimore and London: The Johns Hopkins University Press for IFPRI.

Rosegrant, M. W. 1997. *Water resources in the twenty-first century: Challenges and implications for action.* 2020 Vision for Food, Agriculture, and the Environment Discussion Paper 20. Washington, DC: IFPRI.

Rosegrant, M. W. and P. B. R. Hazell. 2000. *Transforming the rural Asian economy: The unfinished revolution.* Oxford, UK: Oxford University Press for the Asian Development Bank.

Rosegrant, M. W., M. S. Paisner, S. Meijer, and J. Witcover. 2001a. *2020 global food outlook: Trends, alternatives, and choices.* Food Policy Report. Washington, DC: IFPRI.

————. 2001b. *Global food projections to 2020: Emerging trends and alternative futures.* Washington, DC: IFPRI.

Rosegrant, M. W., C. Ringler, and R. V. Gerpacio. 1997. Water and land resources and global food supply. Paper prepared for the Twenty-Third International Conference of Agricultural Economists, Sacramento, California, August 10-16.

Rosenzweig, C. and M. L. Parry. 1994. Potential impact of climate change on world food supply. *Nature* 367 (January 13): 133-138.

Smith, L. C. and L. Haddad. 2000. *Explaining child malnutrition in developing countries: A cross-country analysis.* Research Report 111. Washington, DC: IFPRI.

UN Population Division. 2002. *World urbanization prospects: The 2001 revision.* New York: United Nations.

Watkins, K. 2001. *The Oxfam education report.* London: Blackwell's.

World Bank. 1997. *World development report 1997.* New York: Oxford University Press for the World Bank.

————. 2001. *World development indicators 2001.* CD-ROM.

————. 2002. Povertynet. Available at <http://www.worldbank.org/poverty>.

Chapter 3

Food Security in South Asia: Issues, Options, and Opportunities

Vijay S. Vyas

South Asia today has the largest concentration of poor and food-insecure people. This needs some explanation, given that governments in the region have been committed for a long time to banishing hunger and poverty from their respective countries. Their slogans have not been empty. Vast resources have been expended on poverty alleviation and food security programs. During the past decade GDP growth in these countries has been fairly satisfactory and has been increasing over the past few years. Furthermore, in most of the countries of this region agricultural growth has surpassed population growth. Why then has no major dent been made in resolving food insecurity, which a large part of the population faces, and in undernutrition which an even a larger number of people in this region experience?

This chapter addresses the problem of food insecurity in South Asia, mainly from the angle of public policies, with some allusion to the functioning of markets and civil institutions. The first section clarifies the concept of food security (FS) and spells out the implications of this concept for public policies, market behavior, and action at the household level. In the second section the present status of food security in the selected countries over the past decade or so is presented. The third section reviews the public policies implemented to resolve food insecurity and summarizes the main results. In the fourth section the scope for regional collaborative action is examined. The discussion is centred on five of the countries of the region: Bangladesh, India, Nepal, Pakistan, and Sri Lanka.

At the outset, some of the features of these countries may be noted. In geographical terms, these may be considered medium to large countries (Table 3.1). These countries have high population growth rates, a high proportion of poor households, low per capita income, and a GDP rate of growth that was low but has been rising in recent years. Agriculture is the dominant sector of their economies. Food grains occupy a large though de-

TABLE 3.1. Area, population, and GDP of the selected countries

Country	Geographical area (1,000 sq. km)	Total population (millions)		Annual population growth (1980-1999) (%)	Gross domestic product average annual growth (%) (1965-1999)*		Gross national product per capita	
		1980	1999		Per capita GDP	Per capita in PPQ	Dollars (1999)**	Average annual growth rate (%) 1998-1999
Bangladesh	144	86.7	127.7	2.0	3.8	1.3	370	3.3
India	3,288	687.3	997.5	2.0	4.6	2.4	450	4.9
Nepal	147	14.5	23.4	2.5	3.7	1.2	220	2.2
Pakistan	796	82.7	134.8	2.6	5.6	2.7	470	1.2
Sri Lanka	66	14.7	19.0	1.3	4.6	3.0	820	2.7

Source: World Bank, 2001a,b.
*Pertains for these years.
**Preliminary World Bank Estimates calculated using the World Bank Atlas Method.

clining portion of the cropped area. The agrarian structure is dominated by landless laborers and marginal and small farmers. Yet growth in agriculture, including the food grain sector, outpaced that of population, resulting in a progressively larger per capita supply (Figures 3.1 to 3.5). These countries

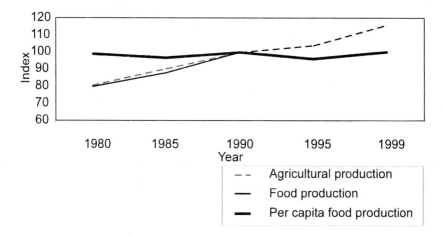

FIGURE 3.1. Index of agricultural production and food production in Bangladesh (Index 1989-1991 = 100).

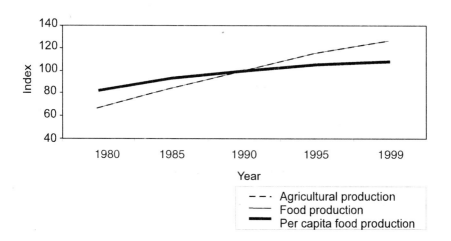

FIGURE 3.2. Index of agricultural production and food production in India (Index 1989-1991 = 100).

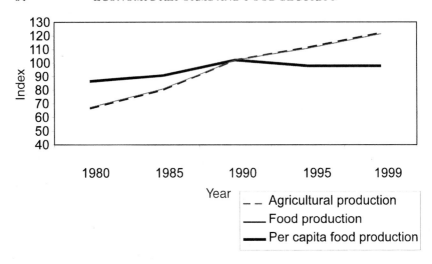

FIGURE 3.3. Index of agricultural and food production in Nepal (Index 1989-1991 = 100).

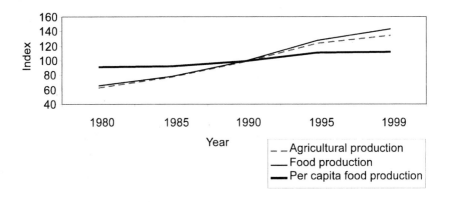

FIGURE 3.4. Index of agricultural and food production in Pakistan (Index 1989-1991 = 100).

have more developed political, legal, and administrative institutions but also more fragmented and divisive societies. Practically all of them experience internal conflicts and civil strife. Given these economic and social features, it would be instructive to learn what these countries could achieve and what they failed to achieve in ensuring food security for their people.

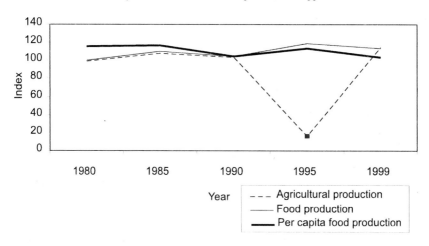

FIGURE 3.5. Index of agricultural and food production in Sri Lanka (Index 1989-1991 = 100).

THE CONCEPT OF FOOD SECURITY

The concept of FS has evolved gradually. It started with an emphasis on national food self-sufficiency but subsequently incorporated households' "entitlement" to food, nutritional adequacy, intrahousehold distributive justice, and other issues. Elaboration of what FS connotes continues. There are several variations on the main theme, expounded by different scholars in different discourses. As Maxwell (1996, p. 155) has remarked, "From its simple beginning, food security has become, it seems, a cornucopia of ideas." However, a common thread runs through various definitions. The following definition, first propounded by the World Bank, succinctly expresses the basic concept and highlights the main components: "access by all people at all times to enough food for an active healthy life" (World Bank, 1986, p. 1). In international forums and national policy pronouncements FS is generally expressed in similar terms.

The definition emphasises four basic elements: availability, stability, accessibility, and nutritional content. At a general level there can be little dispute on any of the four essential ingredients of FS. Only when the implications of these attributes are understood do they give rise to disagreements and difficulties in decision making.

Availability

A precondition for ensuring food security is having an adequate quantity of food within the country.[1] It must be produced domestically or exported from abroad. A doctrinaire approach would suggest that a country can obtain food from the world market and therefore should not insist on achieving food self-sufficiency unless it has a comparative advantage in producing food. For reasons that shall be discussed in the next section, for the large-to-medium developing countries where agriculture is a major sector food consumption claims a large share with the of consumers' budget and there a is productivity gap which can "realistically" be filled, the concept of comparative advantage has to be interpreted in a dynamic sense. In such countries there is a strong case for agricultural development with an emphasis on food self-sufficiency.

Stability

Because the demand for food is fairly stable and continuous throughout the year, even though production is seasonal, its availability at all times can be ensured only by efficient storage and marketing functions. Only a small section of consumers in developing countries can afford to store supplies for the whole year. Although the state can play a role by operating buffer stocks, the market has a much more important role to play. Ensuring a stable supply becomes more important once we take into account annual fluctuations in food-grain production. The challenge, then, is not only to smooth the seasonal flow but also to cope with fluctuations.

Accessibility

The most critical aspect is accessibility, or entitlement, to food. In the countries examined here, the major source of entitlement is private income; the state, or for that matter civil society, can ensure food availability to only a limited number of vulnerable households. For the large majority of the poor, the issue of food security is inextricably linked with that of livelihood. One need not go to the other extreme to equate the provision of livelihood with food security. There are aspects of food security, i.e., availability and stability, that go beyond the questions of livelihood. But livelihood is in most cases the minimum condition for ensuring food security.

Nutritional Content

Food security in the current discussion is not restricted to the availability of a specific number of calories. It consists of more than the avoidance of hunger and starvation, and receiving a daily energy supply (DES). It also means the avoidance of malnutrition, which could occur if the DES is insufficient, excessive, or, more often, imbalanced. Once "healthy life" is included as a component of food security the nutritional dimensions have to be considered. There is a role of the state as well as civil society to ensure this qualitative aspect of nutrition.

In evaluating the status of food security in the countries of South Asia, the four components discussed here and their wider implications must be considered. This task is addressed in the following sections.

THE STATUS OF FOOD SECURITY

Food-Grain Availability

An impressive aspect of the food economies of South Asia has been their ability to increase food-grain availability. What makes this achievement all the more remarkable is the constraint under which this was achieved. The population growth in these countries, though decelerating, was very high by world standards.[2] There was very little scope for expansion of the area under cultivation. Therefore, the growth in output had to be obtained by increasing productivity. Furthermore, the area under nonfood-grain crops, so-called commercial crops, was increasing and that under food grains was declining. This suggests that not only overall productivity but also productivity in food grains was improving. The countries that had not achieved self-sufficiency in food grains imported sufficient quantities to ensure adequate availability (Table 3.2).

Although adequate food-grain availability by itself is not a sufficient condition for the food security of poorer households, it makes equitable distribution easier. Similarly, scarcity will encourage unequal distribution. A disquieting feature of developments in the 1990s was the slackening of the trend in the reduction of food-insecure (malnourished) people in the 1990s compared to the 1980s. In two of the countries, Bangladesh and Nepal, there has been an actual reversal of the trend, and there were proportionately more malnourished people in them in the late 1990s compared to the beginning of the decade (Table 3.3).

A second important feature of the food situation in the region is the faster growth of noncereal foods, such as sugar, oilseeds, fruits, vegetables, and

TABLE 3.2. Index of agricultural and food production in selected countries (1989-1991 = 100)

Years	1980	1985	1990	1995	1999
Bangladesh					
Agriculture production	80.4	89.9	99.5	103.3	115.1
Food production	79.6	87.6	99.6	103.7	115.9
Per capita food production	98.8	96.4	99.6	95.7	99.9
India					
Agriculture production	67.2	84.3	99.4	115.6	126.1
Food production	66.6	83.8	99.6	115.6	126.5
Per capita food production	82.2	92.8	99.6	105.3	107.8
Nepal					
Agriculture production	67.5	80.4	101.9	111.0	121.6
Food production	66.6	80.0	102.0	111.3	122.0
Per capita food production	86.3	91.0	101.9	98.2	97.9
Pakistan					
Agriculture production	62.3	77.8	98.5	123.2	134.0
Food production	65.5	78.7	99.5	127.2	143.0
Per capita food production	91.5	92.6	99.5	111.2	111.9
Sri Lanka					
Agriculture production	99.1	107.9	104.1	116.3	114.4
Food production	100.4	110.0	104.6	119.1	113.3
Per capita food production	115.5	116.9	104.6	113.3	103.6

Source: Time Series for SOFA 2000, FAOSTAT TS Software.

TABLE 3.3. Prevalence of undernourishment

Country	Proportion of total population undernourished		
	Proportion of population 1979-1981 (%)	Proportion of population 1990-1992 (%)	Proportion of population 1996-1998 (%)
Bangladesh	42	35	38
India	38	26	21
Nepal	31	26	28
Pakistan	47	21	20
Sri Lanka	22	28	25

Source: FAO, 2000.

animal and fishery products. This change has been basically demand driven. With increases in income, demand for noncereal food tends to increase. The production of these commodities was encouraged by public policies and concerted research. The entry of large multinational firms as well as producer cooperatives contributed to the production growth of these commodities. Among the cereals, there was a sharper decline in the production of pulses and coarse cereals. These supply and demand factors led to changes in the composition of the "food basket" (Table 3.4). It became more varied but at the same time more expensive.

Stability in Food Supplies

Another noteworthy achievement has been the development of the capacity to cope with fluctuations in food-grain production. Institutions and procedures to cope with periodic shortfalls in production were established. During the past decade or so there have been few instances of starvation deaths even though the region continues to face severe droughts, floods, and cyclones, which result in heavy crop loss. This is all the more notable as the proportion of food aid, including emergency food aid—which was never substantial except for Bangladesh and to an extent Nepal—was even further reduced in recent years. A steady supply at the national level and a quick dispatch of aid to calamity-affected areas was made possible with the better management of available food stocks and the integration of domestic food markets. Thus shortfalls in food-grain production did not result in large-scale shortfalls in availability.

However, the same cannot be said about the seasonal fluctuations in supply. The latter were reflected in sharp seasonal fluctuations in prices. To some extent such fluctuations are inevitable, even desirable, as far as agricultural products are concerned. However, excessive fluctuations are undesirable, especially for the poor who cannot stock food grain and have to meet their requirements on a daily basis. Fluctuations have been particularly severe for small and marginal farmers and agricultural laborers, whose employment and other sources of income are also seasonal. This is one of the most serious aspects of food insecurity in the region and has not been tackled with any degree of success.

Access to Food

The most important aspect of food security is household access to food. It is evident that some progress has been made in this direction. The clearest evidence is the reduction in the percentage of households below the poverty

TABLE 3.4. Food availability, 1996-1998

Country	Dietary energy supply (DES) (Kcal/person/day)	Depth of undernourishment			Diet composition: share of cereals, roots, and tubers in total DES
		DES of under-nourishment (Kcal/person/day)*	Minimum energy requirement (Kcal/person/day)	Food deficit of the under-nourishment (Kcal/person/day)	
Bangladesh	2,060	1,460	1,790	340	84
India	2,470	1,520	1,810	290	64
Nepal	2,190	1,530	1,800	260	80
Pakistan	2,430	1,490	1,760	270	57
Sri Lanka	2,300	1,570	1,830	260	56

Source: FAO, 2000.
*High values of DES indicate low diet diversification and vice versa.

line. In the countries examined here, the poverty line is defined in terms of the amount required to meet expenditure on food plus other essential items. The ratio of households below the poverty line is therefore a sensitive indicator of the state of food security. Different methodologies are used to estimate poverty ratios. In recent years, the World Bank has given estimates of the people earning less than US$1 a day. Since these estimates are based on the norm of purchasing power parity, there is a broad comparability of these figures. Whether one looks at the available national estimates or relies on the World Bank data, it is clear that the poverty ratio has generally declined over the course of the past two decades. But the degree of progress varies from country to country. A hard core of 20 to 25 percent seems to persist in all these countries, although the ratio is as high as 40 percent in Bangladesh and Nepal. These figures can be taken as indicators of the percentage of food-insecure households in these countries (Table 3.5).

In recent years, the FAO has estimated the proportion of undernourished people on the basis of the total availability of food grains and the skewness in distribution (see Box 3.1). These estimates also corroborate the findings reported previously, that there is a decline in the proportion of malnourished households over the decades. Yet in Bangladesh today, 38 percent of the population remains undernourished (Table 3.3).

Information on food intake by different household groups is difficult to find in the absence of regular household surveys. Such surveys are not common in these countries. India is an exception to this rule, as for the past thirty years the National Sample Surveys (NSS) has been carried out. Successive rounds of NSS reveal that the calorie intake in all household groups is increasing and the extent of hunger is declining; it was already reduced to 10 percent by 1990. However, during this time severe hunger did exist in some parts of the country, especially the tribal belt, and in the bottom income deciles. Similar data are available for Nepal and Bangladesh. It is reasonable to infer that the overall incidence of hunger has declined but is still severe in the backward regions and poorer sections.

Nutritional Dimensions

The failure of the South Asian countries lays not so much in their inability to abolish hunger in its starkest form but in the persistence of widespread malnutrition. The incidence of child protein energy malnutrition (PEM), measured by weight for height (wasting) and height for age or weight for age (stunting), although falling, is still shockingly high (FAO, 2000). Data given in the UNDP Human Development Report, revealing the dismal state of nutrition and nutrition-related aspects, can be seen in Table 3.6.

TABLE 3.5. Poverty ratio

Country	National poverty lines								International poverty lines		
	Year	Poverty level (%)			Year	Poverty level (%)			Year	Poverty level $1/day (%)	Poverty gap $1/day (%)
		Rural	Urban	National		Rural	Urban	National			
Bangladesh	1991-1992	46.0	23.3	42.7	1995-1996	39.8	14.3	35.6	1996	29.1	5.9
India	1992	43.5	33.7	40.9	1994	36.7	30.5	35.0	1997	44.2	12.0
Nepal	1995-1996	44.0	23.0	42.0					1995	37.7	9.7
Pakistan	1991	36.9	28.0	34.0					1996	31.0	6.2
Sri Lanka	1985-1986	45.5	26.8	40.6	1990-1991	38.1	28.4	35.3	1995	6.6	1.0

Source: World Bank, 2001b.

Box 3.1. Method for Estimating Prevalence and Depth of Hunger by FAO

The following is is a brief description of the method FAO uses to estimate the prevalence and depth of undernourishment.

- Calculate the total number of calories available from local food production, trades, and stocks.
- Calculate an average minimum calorie requirement for the population, based on the number of calories needed by different age and gender groups and the proportion of the population represented by each group.
- Divide the total number of calories available by the number of people in the country.
- Factor in coefficient for distribution to take account of inequality in access to food.
- Combine the above information to construct the distribution of the food supply within the country and determine the percentage of the population whose food intake falls below the minimum requirement. This is the prevalence of undernourishment.
- Multiply this percentage by the size of the population to obtain the number of undernourished people.
- Divide the total calories to the undernourished by the number of undernourished to obtain the average dietary energy intake per undernourished person.
- Subtract the average dietary energy intake of undernourished people from their minimum energy requirement [expressed in kilocalories per person per day] to get the average dietary energy deficit of the undernourished. This is the depth of hunger.

Source: FAO, 2000.

To sum up, countries of this region did reasonably well in obtaining adequate supplies of food grain, principally from domestic production. Annual fluctuations in food-grain production were muted to a large extent through supply management. Seasonal fluctuations in supply and food-grain prices could not be corrected to a desired extent. This resulted in harm to the poor, who could not store grains and who did not have steady incomes. The head-count poverty ratio suggests that poverty declined in all of these countries, though the pace slowed down in more recent years. The countries could ensure FS to larger number of people, yet severe pockets of hunger remained. Hardly any dent was made in malnutrition, which remains a very serious problem.

TABLE 3.6. State of nutrition

Country	Undernourished people (% of total population) 1996-1998	Underweight (% under age 5) 1995-2000[a]	Stunting (% under age 5) 1995-2000[a]	Low birth weight (%) 1995-1999[a]
Bangladesh	38	56	55	30
India	21	53[b]	52	33[b]
Nepal	28	47	54	—
Pakistan	20	26[b]	23	25[b]
Sri Lanka	25	34	15	25[b]

Source: United Nations Development Programme, 2001.
[a]Data refer to most recent year available during the period specified.
[b]Data refer to a year or period other than that specified differ from the standard definition or refer to only part of country.

PUBLIC POLICIES

A number of public policy measures affected FS in the region. Indirect though substantial impact was made through macropolicies, generally economic reforms or structural changes that the countries had initiated from the late 1980s to the 1990s. At the same time, public policies and programs were implemented explicitly to augment food supplies (availability) and to create purchasing power among the poorer section (access). Public interventions for ensuring interseasonal stability of supplies or enhancing nutritional content were not widespread.

Macropolicies

Success in ensuring FS depends as much, and probably more, on the macroeconomic environment as on the specific measures undertaken to achieve it. Macroeconomic policies in these countries as in several other developing countries in different parts of the world were pursued with two major objectives: *liberalization,* or rather debureaucratization, of the domestic economy and *globalization,* the progressive integration of the domestic economies with the global economy. Careful sequencing of reforms and concern for stable prices were distinguishing features of the economic reforms (for detailed discussion of the economic reforms as they impinged on agricultural sector in India, see Vyas and Reddy, 2000). An aspect of reforms that is relevant to the present discussion is the move toward the removal of special treatment given to domestic industries by protective tariffs, quotas, and licenses which had tilted the terms of trade in favor of industry and against agriculture. Similarly, closer approximation of the value of the domestic currencies to sustainable exchange rate helped agricultural exports and favored agricultural producers. The removal of hurdles on the private sector in domestic as well as foreign trade presumably yielded gains in efficiency. The vigor and effectiveness with which these policies were pursued naturally differed from one country to another, but the main thrust of the reforms was the same.

However, for a variety of reasons, the rate of economic growth in the region was not very high. To that extent, the "spread effect" on the poor was weak. Within the countries there were regions where spectacular growth took place and a remarkable impact on poverty alleviation and presumably on food insecurity occurred. The Punjab region in India witnessed a 7 to 8 percent annual rate of growth in the 1970s and 1980s, with a perceptible impact on poverty levels. The proportion of poor households in rural areas declined to less than 20 percent. Most of these countries had to cope with the problem of FS with an overall growth rate of 4 to 6 percent. When translated in per capita terms, rate of growth was barely 2 to 3 percent per annum.

Another important and pro-poor feature of growth in these countries was the relatively low level of inflation. During the past two decades inflation remained at the two-digit level, coming down to a single digit after a number of years. This distinguishes the growth experience of the countries in this region with that of developing countries in Africa and Latin America.

Public Policies to Stimulate Growth in Agriculture

Every country in this region kept food self-sufficiency as the primary objective of its agricultural policies. By judiciously pursuing a policy of food self-sufficiency the poor were helped both as producers and consumers (Vyas, 2000). It is now well established that growth in agriculture assists the poor and to that extent contributes to enhancing FS (Datt and Ravallion, 1998). Countries in this region took several measures to stimulate growth in agriculture. These can be categorized as nonprice measures and price measures. Among the former the spread of new technology was the prominent one.

As mentioned previously, agriculture in these countries, with the possible exception of Pakistan, is characterized by the high proportion of small and marginal holdings. A radical measure to address this problem would be redistributive land reforms. However, by the 1970s, structural reforms, especially land distribution, were already over. In the preceding two decades all the countries had enacted various land-reform measures with varying degrees of seriousness. But in none of the countries was the agrarian structure significantly affected, although "ceiling" legislation indirectly contributed to stopping the process of large holdings from expanding (Vyas, 1979).

High-yielding varieties swept the irrigated wheat-growing areas of this region. This was followed with a less dramatic sweep of improved varieties of rice. These technologies not only yielded high output but, because of high total factor productivity, also enhanced farmers' incomes. Two important conditions for the spread of this technology were irrigated land and efficient delivery systems for improved seeds, fertilizers, and credit for all who adopted this technology. Because of these reasons the HYV strategy could not penetrate in the large dry areas. Nor it was availed of by large sections of small and marginal farmers who could not be served efficiently by the existing credit and marketing arrangements. It is no wonder that this technology has not reached its barrier and the tempo of agricultural growth has slowed down.

An equally important contribution of public policies was in the form of public investments for improved rural infrastructure, particularly for extending irrigation. Public investment also attracted private capital formation in agriculture, which was, by and large, of supplementary character. Investment in rural infrastructure, led by public investment, played a major role in agricultural growth in this region. Unfortunately, with misplaced emphasis on curtailing government expenditure, public spending in rural infrastructure has slackened in the 1990s. This explains to a large extent the faltering agriculture growth during this period.

The type of agricultural growth that may benefit small, marginal farmers and agricultural laborers requires strong infrastructure support, research and development, and the enhancement of institutional capability. In the countries of this region, the state cannot abdicate the responsibility of providing these public goods. In general, private resources will not be invested in activities requiring lump payments and long gestation periods and that bring low returns to private actors.

Lessons from the Green Revolution are very pertinent to the present situation. The declining role of public investment in agriculture and excessive faith in market institutions today will negate the gains that rapid agricultural growth bestowed on the poor. Thus, public investment in irrigation and land development, focused research on dryland crops and supplementary enterprises, and arrangements to reduce transaction costs of the delivery system stand out as important preconditions for rejuvenating the stalled Green Revolution in disadvantaged areas and to make an impact on the incomes of small and marginal producers.

Most of the countries in this region also introduced price incentives for agricultural producers by subsidizing inputs and guaranteeing remunerative minimum support prices. These measures did create a favorable environment for the surplus-generating agricultural producers. However, limited public resources subsidizing inputs and output meant lower investment for agricultural development.

In any event, the price policy pursued in these countries proved to be a double-edged sword. Although it led to a marketable surplus, those who depend on the market wholly or partially to meet their food requirement, namely the rural poor, were net losers. The adverse impact of high prices on this group could be neutralized if higher agricultural production also leads to higher employment in agriculture or stimulates demand and consequently employment in the nonfarm sector. But in both these respects developments were slow and inadequate.

It should be noted that the main function of prices is to act as signals for resources allocation. To this important function it is legitimate to add an "insurance" function by declaration and implementation of minimum support prices to protect the producers, to some extent, from market-induced uncertainties. Price policies are weak and inefficient instruments for income transfer. The main policy objective should be to improve income terms of trade, rather than barter terms of trade. For the former, the development and extension of appropriate technologies, improvements of delivery systems, and investment in rural infrastructure are more important. A strategy of agricultural growth relying on these measures will not only be cost-effective and sustainable but also contribute to the enhancements of FS.

Entitlement and Access

Important instruments for ensuring FS in South Asia were the poverty alleviation programs (PAP) implemented. Countries of this region have a long tradition of carrying out such programs. These PAPs have had three goals: (1) to provide social safety nets, (2) to create assets in poor households, and (3) to generate wage-paid employment. The successes achieved in these programs, though they vary from country to country, have not, in any way, been spectacular (Vyas, 1995; South Asian Association for Regional Cooperation, 1992).

Programs that could provide a safety net to indigent households were not very prominent. It is only in recent years, and in a few countries, that programs such as help for widows and old and disabled persons have been instituted. One program in this category, the midday meal program for school children, is gaining in popularity. All of the countries have by now accepted the responsibility of coping with food scarcities during natural calamities such as floods, drought, and cyclones. During such emergencies free food is distributed to the needy and/or food-for-work schemes are organized on a massive scale. The record of disaster management in South Asia has been noteworthy. (Countries in this part of the world had a long tradition of religious, civil society, and philanthropic institutions coming to the aid of the hungry and destitute. These traditions are unfortunately dying out.)

More common programs for poverty alleviation and FS are those that involve the distribution of assets among poor households or the provision of gainful employment. As noted earlier, land redistribution among the landless and marginal farmers has not achieved the desired objective in any of the countries to any remarkable extent. The emphasis is now on the distribution of reproducible assets (livestock, tools and implements, credit) among poor households. Evaluation of these programs suggests that only those poor households with a minimum of wherewithal have benefited, though there are several exceptions. The Grameen Bank of Bangladesh and the self-help group movement in India, though, have been able to reach the "unreachable."

For the assetless poor households, a variety of employment programs have been sponsored. Most of these are different versions of the food-for-work programs. Employment generated by these programs is not very substantial (except in Maharashtra state, India, where a scheme has been in existence for a number of years). Nor are many of the works created under these programs sustainable, mainly because the complementary input of capital is inadequate. The scope of such programs and their geographical coverage is widening, and the effectiveness in sustaining employment is improving. Measures such as decentralization of planning and implementation as it is occurring in Kerala state, India, Nepal, and Sri Lanka, and integrating these works with natural resource development (as in India) have improved their effectiveness.

The countries of South Asia have used different modes to enable poor households better access to food and other necessities. Efforts are in the right direction. However, the major criticism to which these initiatives can be subjected are that: (1) programs are not cost-effective, (2) they generally benefit the "near poor" but not the abject poor, and (3) they do not encourage genuine participation of the poor but instead encourage a "dependency syndrome."

Public Distribution of Food Grain

The countries of South Asia have a long tradition, dating back to World War II, of distributing food grain at below-market prices to poorer households. The earlier arrangements were in the form of rationing of food grains for a large, generally urban population. In recent years, the common mode has been a dual price system, with food made available to the poor at below-market prices. The government purchases food grain from the domestic market at preannounced procurement prices and sells it through a chain of public distribution system (PDS), or "fair price," shops. The difference between the procurement price and the issue price was too small to bear the cost of procurement, storage, and distribution. Previously this cost was fully subsidized. In later years with the rise in procurement prices, due to the pressure from the large farmers' lobby, this gap started to widen. With the pressure to minimize subsidies the governments took the easy option of raising the issue price, thus defeating the very purpose of the PDS. Apart from ever-increasing support prices, the rising cost of procurement and distribution management contributed to the increase in the procurement and issue prices. At the same time, the progressive widening in coverage of PDS made it a general food subsidy scheme rather than one targeted to the poor. According to several evaluation studies, PDS has deviated sharply from its avowed objective. Certain reforms have been introduced in recent years, mainly in the area of better targeting. The largest PDS system, in India, has introduced two categories of beneficiaries, below poverty line (BPL) and above poverty line (APL) households, with discriminating prices in favor of BPL families. In spite of these deficiencies, no superior alternatives to PDS can benefit the poor. The alternative is to make the operations more efficient. Various suggestions have been made in this regard (Vyas, 1996; Krishanaji and Krishan, 2000).

In the end one could summarize that with national FS ensured, the environment for similarly ensuring household FS is quite conducive. The problems that have been highlighted can be resolved in the given institutional setup of the countries of this region. There is also scope for collaborative regional activities. In the next section these are briefly introduced.

REGIONAL COLLABORATION

As noted earlier, all of the countries hold poverty alleviation as an important goal. Progressively greater emphasis is being placed on FS for the poor households in their policies and programs. There are several common points in the development strategy pursued by the South Asian countries for achieving these goals. This in itself is a promising start for collaborative action among the countries of this region. Added to that is the fact of geographical proximity that can facilitate coordinated activities in a variety of fields. The more promising among the areas of collaborative action, from the point of food security, are the following:

- Instituting early warning systems and facing natural calamities with coordinated action
- Conservation and proper utilization of the natural resources, especially water
- Mutual learning and collaborative action in the area of agricultural research
- Sharing the experience on poverty alleviation and food security
- Coordinating pricing and trade policies
- Establishing common reserves for emergency relief
- Taking a common stand in the international forums

Instituting Early Warning Systems

South Asia is a disaster-prone area. Large parts of the region are subject to floods and droughts, sometimes both. Volcanic eruptions, fire, and frost are also common occurrences. There is a view that because of deforestation and extension of cultivation on marginal lands the incidence of natural disasters has increased. The countries of the region have succeeded in coping with natural disasters to a great extent, but at the present each country has to fend itself more or less on its own, even when a calamity is spread over more than one country. The region can benefit if an early warning system is instituted to alert all of the countries about an impending calamity. With modern advances in techniques and instruments of surveillance this would be a manageable task.

Conservation and Proper Utilization of Natural Resources

In this resource-poor region, efficient and sustainable use of natural resources should be given high priority. One of the most important areas for collaborative action from this perspective is the integrated use of international rivers. Basins of major rivers in the region straddle national borders. Up- and downstream states can make best use of these waters, for irrigation, hydropower generation, navigation, fishery development, and drinking purposes if they can agree on comprehensive water use planning. The Indo-Pak treaty on the Sindh river system, mediated through the World Bank, provides an excellent example. Serious attempts are underway to arrive at such an understanding between India, Nepal, and Bangladesh on the use of international rivers. An encouraging beginning has been made in the case of the waters of the river Ganga at Farraka. Such arrangements will need political sagacity and far-sightedness on the part of all concerned.

Even if comprehensive understanding on all aspects of river waters cannot be arrived at, a beginning can be made by exchanging information on vital aspects of rainfall, river flows, evapotranspiration, groundwater regimes, and other aspects. High-level commissions on integrated water use have been established, or are in the process of establishment, in a number of countries of the region. There could be mutual consultation and exchange of information among these bodies right from the start.

Collaborative Activities and Mutual Learning in Agricultural Research

The national agricultural research systems (NARS) in some of these countries are fairly well developed and have reached international standards. India's agricultural research system is considered one of the best in the developing countries. Pakistan had made significant advances in the use of water in agriculture. Sri Lanka has fairly developed system for plantation crops. Other countries have also specialized in some crops or resource uses. All countries of the region can benefit from the exchange of information and collaboration in organizing relevant research activities.

Such collaboration had great potential in the region because of the numerous large, cross-border, agroecological tracts; The areas in each set, Punjab, India and Punjab, Pakistan, West Bengal State, India and Bangladesh, Sindh, Pakistan and Western Rajasthan, India, the Tarai region, Nepal, and eastern Uttar Pradesh and Bihar, India, share more or less the same agroecological features. Research findings applicable to one part could be of use to other parts. Second, cropping patterns in these countries are dominated by rice and wheat, which makes the generic research on these crops useful for large areas of the different countries growing these crops. It should be remembered that the Green Revolution came to this part of the world on the strength of adapting the results of the generic research conducted in different parts of the world, for wheat in Mexico and rice in the Philippines. Third, some countries in the region, India in particular, have made significant advances in what might be called "frontier" research in biotechnology, tissue culture, and plant genetics. Other countries in the region should be enabled to take advantage of these advances in crop and animal sciences, rather than have to "rediscover the wheel."

Sharing Experiences in Rural Development

All the countries in this region have implemented a variety of rural development and poverty alleviation programs. They have made several innovations in their asset distribution, employment generation, social security, and public distribution programs. Many of these innovations have succeeded, although quite a few have failed. Other countries can learn valuable lessons from these successes and failures. Experiences in targeting programs and involving people in their identification and management, and making the projects financially viable, are of relevance to those countries considering similar objectives.

Such exchange of experiences should not be restricted to only the government-sponsored activities. Many NGOs and civic organizations have done exemplary work with the poor and food insecure. The Aga Khan Foundation in Pakistan, the Self-Employed Women's Association (SEWA) in India, and Grameen Bank in Bangladesh are the better known, but by no means only, examples of such NGOs. All such organizations have valuable lessons to offer in the most difficult area of poverty alleviation, developing people-friendly, economically viable, and sustainable delivery systems. Periodic meetings and, wherever feasible, exchange of personnel will be mutually beneficial.

Coordinating Price Policies and Liberalizing Border Trade

Due to the long and porous frontiers, particularly between India, Bangladesh, and Nepal, any effort to isolate the agricultural economies of these countries will lead only to distortions and defeat the intent of policies. The example of Nepal's fertilizer price policy is illustrative. Both India and Nepal will gain if they have a common understanding of fertilizer prices. If the fertilizer prices in these countries are not coordinated, the result will be large-scale smuggling from the country where the prices are low to the country where prices are high. Policing such activities will be difficult. What is true in the context of India and Nepal also applies, though to a lesser extent, to other countries sharing common borders. Machinery for regular and comprehensive consultation on price policies should be created. In the absence of such comprehensive consultation, even ad hoc and limited discussion would be of mutual benefit.

The same applies to trade policy. For a variety of reasons—diversity and complementarity of their nation's agricultural sector, more or less similar stage of development, and the similar growth strategies pursued by these countries—they can mutually benefit by promoting trade on preferential basis. In this respect they are lagging behind other regional groups. To start, they should legitimize, in fact encourage, the border trade, which is largely informal and brisk. This could well be a cautious first step in the direction of better economic coordination.

Establishment of Common Buffer Stock

As argued earlier, FS becomes a reality only at the household level. Food-insecure households must be supported by the national policies and programs. The latter can be more effective with coordinated actions at the international, more particularly at the regional, level. An important plank of the national-level programs to ensure FS is the creation and operation of a food-grain buffer stock to counter annual fluctuations in agricultural production. On the principal that the "risk shared is risk reduced," the ideal solutions would be to lay more emphasis on global reserves than on national reserves. Though greater emphasis will be placed on national buffer stocks, the regional food reserves could also have a role to play, supplementing national efforts.

Several attempts have been made in different parts of the world to institute regional programs for food security. The most important initiative in Asia is the ASEAN Food Security Reserves. There is a formal agreement among the member countries of ASEAN on these reserves and the modalities of the operations are agreed upon. A food security board has been constituted for this purpose. The relative success of this mechanism in the ASEAN region can be ascribed to two important conditions. First, there is a growing political cohesion and economic coordination among the member countries. Second, there is a much greater complementarity among the food economies of the ASEAN countries. There is a major food-surplus country (Thailand) and major food-importing country (Malaysia). These countries provide strong ground for regional cooperation (Tygai and Vyas, 1990). In South Asia both these conditions are more or less absent. In the absence of an established, surplus food economy in the region, the stocks earmarked for regional emergencies cannot be relied upon in the event of a food deficit in a participating country. Such schemes, in the present circumstances, are likely to be unreliable. Second, difficulty arises due to the paucity of resources to operate such schemes that require large investment for the procurement of grain and the infrastructure to store and distribute them.

In the present circumstances the goods will have to be more realistic. The possibility, and even justification, of regional stocks are limited. Other forms of regional cooperation are more useful as well as more practical. Coordinated action in (1) promotion of a buying organization to procure food on economic terms, (2) arranging investment for transnational infrastructure for agricultural development, and (3) cooperation in the development and exchange of technology for food production, processing, and storage is likely to make greater contribution to the emergence of an integrated regional food economy.

Taking a Common Stand in International Forums

Momentous changes are taking place at the international level that have serious implications for FS in the developing countries. Most important among these are the Uruguay Round (UR) agreements and the establishment of the World Trade Organization. UR agreements have, for the first time, brought agricultural trade under international discipline. Developed countries have made firm commitments to bring down the level of subsidies and give larger access to their markets. Also, some provisions have been made to help the developing countries, especially the poor food-importing ones, cope with the transition period and gradually adjust their trade regimes to the changed economic environment. As the countries in this region are either large food-importing countries, such as Sri Lanka, or marginally surplus ones, such as India and Nepal, they have to watch carefully that the concessions allowed to them under the WTO agreements are not nullified for one reason or other. Similarly, all these countries are large exporters of nonfood agricultural crops: plantation crops, cotton, sugar, and others. They have to see to it that their products are not denied access to developed-countries' markets. So far developed countries have not yielded much on the provisions because they may hurt their agricultural producers. Only with collective bargaining do the developing countries of this region have some hope to secure just conditions.

In the changed economic environment the WTO and its specialized committees, for example, the committee on provisions on the technical barriers to trade (TBT) and the committee on the provision of application of sanitary and phytosanitary (SPS) measures, have acquired great importance. The countries of South Asia whose interests converge on these issues should act with solidarity. They have to watch that the developed countries do not take away the gains of tariff liberalization by imposing nontariff barriers, or by bringing in extraneous matters, such as environmental concerns or child labor, in the trade negotiations. WTO is one forum, an important one no doubt, where the countries of this region have to take united stand. There are also other international organizations in which these countries have to watch their interest collectively.

There is an equally urgent need, in fact, an opportunity to organize coordinated action to take maximum advantages from UN agencies and other international organizations. This would also include the Consultative Group on International Agricultural Research (CGIAR) institutions, as some of these institutions are located in Asia (two of them, the International Crops Research Institute for the Semi-Arid Tropics [ICRISAT] and the International Water Management Institute [IMI] are located in India and Sri Lanka respectively). Many other international research centers have outreach programs in this region. NARS in these countries should take full advantage of the research capabilities of the international centers.

There is a worldwide move toward regional economic blocs with varying degrees of economic integration. The South Asia region is rather behind in this respect, not withstanding initiatives such as SAARC. A meaningful regional cooperation could be forged around the desire to fulfil the basic and pervasive need of FS for poor people of the region. The international agencies can help these countries in coming together in some of the areas identified in this chapter and make tangible progress in achieving the noble objective of ensuring FS.

NOTES

1. It may be stated that much of the discussion in this chapter is on dietary energy supply, i.e., food available for human consumption, expressed in kilocalories per person per day. At this stage food grains are a major source of calories. Hence "food" and "food grains" are used interchangeably.

2. A rising proportion of adults in these countries was a favorable factor.

REFERENCES

Datt G. and M. Ravallion (1998). Farm productivity and rural poverty in India. *Journal of Development Studies*, 34(4), 62-85.

Food and Agriculture Organization of the United Nations (2000). *The State of Food Security in the World 2000*. Rome: FAO.

Food and Agriculture Organization of the United Nations (2001). *The State of Food and Agriculture 2001*. Rome: FAO.

Krishnaji, N. and T.N. Krishnan (eds.) (2000). *Public Support for Food Security: The Public Distribution System in India*. UNDP, Strategies for Human Development in India, Volume 1. New Delhi: Sage Publications India PVt. Ltd.

Maxwell, S. (1996). Food security: A post modern perspective. *Food Policy*, 21(2), 155-170.

South Asian Association for Regional Cooperation (1992). Meeting the challenge. Report of Independent South Asian Commission on Poverty Alleviation. South Commission. Kathmandu, Nepal: SAARC Secretariat.

Tyagi, D.S. and V.S. Vyas (1990). *Increasing Access to Food: The Asian Experience.* New Delhi: Sage Publication.

United Nations Development Programme (2001). *Human Development Report 2001.* Oxford, UK: Oxford University Press.

Vyas, V.S. (1979). Some aspects of structural change in agriculture. Presidential Address Delivered at Thirty-eighth Annual Conference of Indian Society of Agricultural Economists, held under the Auspices of the Assam Agricultural University.

Vyas, V.S. (1996). Diversification in agriculture: Concept, rationale, and approaches. *Indian Journal of Agricultural Economics,* 51(4), 636-643.

Vyas, V.S. (2000). Ensuring food security—Role of the state, market and civil society. Presidential Address to the Third Conference of Asian Society of Agriculture Economists. *Economic and Political Weekly,* 35(50), 4402-4407.

Vyas, V.S. and P. Bhargava. (1995). Public intervention for poverty alleviation: An overview. *Economic and Political Weekly,* 30(41/42), 2559-2572.

Vyas, V.S. and V.R. Reddy (2000). Economic reforms and Indian agriculture: Long-term issues and recent experience. In R. Kapila and U. Kapila (eds.), *Economic Developments in India: Analysis, Reports, Policy Documents* (Volume 25, pp. 41-68). New Delhi, India: Academic Foundation.

World Bank (1986). Poverty and hunger: Issue and options for food security in developing countries. World Bank Policy Study. Washington, DC: Author.

World Bank (2001a). *World Development Indicators.* Washington, DC: Author.

World Bank (2001b). *World Development Report, 2000/2001.* Washington, DC: Author.

PART II:
TRADE LIBERALIZATION
AND FOOD SECURITY
IN SOUTH ASIA

Chapter 4

WTO Agricultural Negotiations and Food Security in Developing Countries

Eugenio Díaz-Bonilla
Marcelle Thomas
Sherman Robinson

INTRODUCTION

The Doha Declaration (November 2001) renewed the commitment by the World Trade Organization (WTO) Agreement on Agriculture (AOA) to take into account food security concerns in the coming negotiations. To properly address those concerns, two questions must be considered:

1. What is the relevance of the current WTO classification of countries with respect to their food security status?
2. Are the current legal texts that define WTO commitments on the basis of those categories of countries adequately handling food security concerns through AOA special and differential treatment?

Both questions are related; if the categories are badly defined to capture food security concerns, then it is unlikely that the different treatment under WTO rules will deal with those concerns in a meaningful way. But even if these categories capture the variety in the situations of food (in)security, WTO rules and commitments may still be inadequate to take into account the problems of food-insecure countries.

Assuming that a classification of countries can adequately identify food-insecure WTO members, the negotiation strategy of these nations should focus on two main aspects. First is the elimination or substantial reduction of subsidies and protectionism in industrialized countries, which are spending one billion dollars per day to support their own agriculture (OECD, 2001). Those subsidies and protection undermine agricultural production in developing countries. The second aspect is the need for some changes in the language of the AOA to address domestic food security concerns in developing countries. A group of developing countries have included some of these changes in their argument for adding a "development box" and/or a "food security box" in the AOA, but the substance of the changes is more important than how they are packaged.

The chapter is organized as follows. The next section briefly presents the notion of food security utilized, followed by a section on a short background of food security trends. The rest of the chapter discusses in greater detail three main issues related to WTO negotiations: country classification according to their food security situation; protectionism and subsidies in industrialized countries; and domestic policies for food security in developing countries. The chapter closes with some remarks for the WTO negotiations from the perspective of developing countries.

FOOD AND NUTRITION SECURITY

Food security can be analyzed at the global, national, regional, household, and individual levels (Figure 4.1). Since the World Food Conference of 1974, food security has been analyzed not only at the global and national levels but also at the household and individual levels, where issues of food security emerge in a more concrete way (Maxwell, 1996). In addition to food supply, poverty, and lack of income opportunities (Sen, 1981), variability around the trend of both food supply and access, and their sustainability over time were becoming the main obstacles to food access (Maxwell, 1990). Food intakes are required to provide more than what is needed for survival; they also have to support an active and healthy life (Maxwell and Frankenberger, 1992). The 1996 World Food Summit included several of those components when it asserted that "food security exists when all people, at all times, have physical and economic access to sufficient, safe, and nutritious food to meet their dietary needs and food preferences for an active and healthy life" (FAO, 1996, p. 7).

But availability and access are only preconditions for adequate utilization of food. They do not determine the substantive issue of malnutrition or nutrition insecurity at the individual level (Smith, 1998; Smith and Haddad, 2000). The Food and Agriculture Organization's (FAO) recent report on the state of food insecurity in the world distinguishes between malnourishment linked to food intake and malnutrition, a physiological condition also related to food intake but affected by other determinants as well. In the report, malnourishment in 99 developing countries is measured using an indicator of food availability at the national level, doubly corrected by the gender and age structure of the population, and by the consumption or income distribution profile of the country (FAO, 1999).

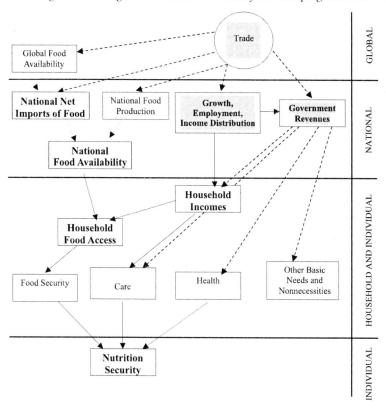

FIGURE 4.1. Conceptual framework for food security. (*Source:* Adapted from Smith, 1998.)

National indicators of malnutrition, although showing an almost perfect and highly significant correlation with national food availability measured by national consumption of calories per capita, are far more weakly correlated with "deeper" measures of malnutrition, such as the percentage of child malnutrition based on anthropometrical measures (Smith, 1998). Analyzing nutrition insecurity at the individual level (utilizing child malnutrition as the indicator) requires the consideration of household and individual food access, as well as other determinants such as the health environment, women's education, and women's relative status in the society (Smith and Haddad, 2000).

Acknowledging the relevance of nutrition indicators for analyzing food insecurity at the household and individual levels, this chapter nonetheless takes a national perspective. This is the level at which WTO negotiations take place. It focuses mainly on food availability issues, utilizing consumption, production, and trade measures (Figure 4.1).

TRENDS IN FOOD SECURITY

Food security appears to have improved over the past four decades. Total food availability in developing countries, measured in daily calories and grams of protein per capita, was about 30 percent higher at the end of the 1990s than in the 1960s, even though the world population almost doubled during that time (Tables 4.1 and 4.2). The number of malnourished children

TABLE 4.1. Average caloric consumption figures

	Calories per capita per day					% Change in 1995-1999		
	1960s	1970s	1980s	1990s	1995-1999	Since 1960s	Since 1970s	Since 1980s
World	2,347	2,453	2,636	2,750	2,790	18.9	13.7	5.8
Industrialized countries	2,956	3,079	3,201	3,337	3,359	13.6	9.1	4.9
Developing countries	2,036	2,173	2,424	2,607	2,667	31.0	22.7	10.0
Least-developed countries	2,016	2,018	2,078	2,067	2,073	2.8	2.7	−0.2
Africa south of Sahara	2,070	2,077	2,075	2,160	2,189	5.7	5.4	5.5
Transition markets	3,236	3,366	3,383	2,992	2,906	−10.2	−13.7	−14.1

Source: Díaz-Bonilla, Thomas, and Robinson, 2002.

TABLE 4.2. Average protein consumption figures

	Proteins per capita per day (grams)					% Change in 1995-1999		
	1960s	1970s	1980s	1990s	1995-1999	Since 1960s	Since 1970s	Since 1980s
World	64	65	70	73	75	17.2	15.4	7.1
Industrialized countries	90	94	99	103	104	15.6	10.6	4.8
Developing countries	51	53	59	66	68	33.3	28.3	15.3
Least-developed countries	50	51	51	51	51	2.0	0.0	0.0
Africa south of Sahara	53	52	51	52	53	0.0	1.9	3.9
Transition markets	97	102	103	90	86	−11.3	−15.7	−16.5

Source: Díaz-Bonilla, Thomas, and Robinson, 2002.

under age five (a better indicator of food problems than average food availability) declined between the 1970s and the mid-1990s by about 37 million (see Table 4.3), and the incidence of malnutrition dropped from 47 to 31 percent (Smith and Haddad, 2000).

Other points are worth noting:

- Food availability in developing countries comes mostly from domestic production: imports were about 15 percent of total food production in the 1990s (up from 10 percent in the 1960s and 1970s).
- Food trade, along with stocks, contributed to reduce the variability of food consumption in developing countries to about one-third to one-fifth of that of food production.
- The burden of the total food bill (measured by food imports as a percentage of total exports) declined on average for developing countries from almost 20 percent in the 1960s to about 6 percent in the 1990s (Figure 4.2). This was caused by the expansion of total trade, which has grown faster than food imports, along with a decline in real food prices.
- Volatility of agricultural prices in world markets in the last half of the 1990s—since the implementation of the World Trade Organization (WTO) agricultural agreements—does not seem to be higher than for the whole period since the 1960s (Table 4.4). It is less clear what has happened to the volatility of agricultural prices within developing countries, which also depends on domestic policies.

TABLE 4.3. Number of malnourished children since 1970

Regions	Millions of children under age five						
	1970	1975	1980	1985	1990	1995	1997
Latin America and the Caribbean	9.5	8.2	6.2	5.7	6.2	5.2	5.1
Sub-Saharan Africa	18.5	18.5	19.9	24.1	25.7	31.4	32.7
West Asia/ North Africa	5.9	5.2	5.0	5.0	—	6.3	5.9
South Asia	92.2	90.6	89.9	100.1	95.4	86.0	85.0
East Asia	77.6	45.1	43.3	42.8	42.5	38.2	37.6
All regions	203.7	167.6	164.3	177.7	169.8	167.1	166.3

Source: Smith and Haddad (2000) from 1970 through 1995; 1997 data are the IMPACT base-year values extrapolated from 1995 values using the IMPACT model (Rosegrant et al., 2001).

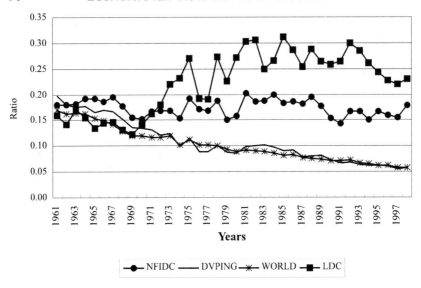

FIGURE 4.2. Ratio of food imports over total exports.

Although food security has improved in general, some regions and countries are at risk, and some have become more food insecure:

- Average food availability is still low for regions such as sub-Saharan Africa (SSA). For more than one-fourth of all developing countries, per capita indicators have decreased since the 1960s (Tables 4.5 and 4.6 and Box 4.1). In most cases those declines appear to be associated with war.
- The number of malnourished children under the age of five has actually increased in SSA, and the incidence of malnutrition is still very high there and in South Asia (Table 4.3).
- For the 49 least-developed countries (LDCs), the total food bill has remained high, at 20 percent, and several developing countries with large external debts face additional constraints in financing their food imports (Figure 4.2).

In summary, although aggregate trends of food security indicators for developing countries seem positive, the situation is deteriorating in specific cases. A more disaggregated analysis of individual country situations is needed. The issue of the heterogeneity of developing countries is discussed next.

TABLE 4.4. Coefficient of variability for price: Constant value

	1960-1999	1990s	1995-1999
Cocoa (cents/kg)	0.54	0.14	0.13
Coffee, mild (cents/kg)	0.40	0.29	0.21
Coffee, robusta (cents/kg)	0.55	0.26	0.14
Tea (cents/kg)	0.20	0.19	0.21
Sugar (cents/kg)	0.81	0.16	0.17
Orange ($/mt)	0.11	0.08	0.01
Banana ($/mt)	0.11	0.12	0.11
Beef (cents/kg)	0.21	0.13	0.06
Wheat ($/mt)	0.22	0.14	0.16
Rice ($/mt)	0.34	0.13	0.07
Maize ($/mt)	0.21	0.16	0.17
Sorghum ($/mt)	0.21	0.13	0.15
Coconut oil ($/mt)	0.36	0.29	0.15
Soybean oil ($/mt)	0.30	0.18	0.13
Groundnut oil ($/mt)	0.28	0.15	0.08
Palm oil ($/mt)	0.30	0.29	0.19
Soybean ($/mt)	0.22	0.11	0.12
Soybean meal ($/mt)	0.27	0.16	0.21
Cotton (cents/kg)	0.19	0.14	0.12

Source: Díaz-Bonilla, Thomas, and Robinson, 2002.

TABLE 4.5. Developing countries with worsening indicators for calories and proteins

	Countries with lower indicators in the 1990s than in the 1960s as well as the group average in the 1990s	Countries with lower indicators in the 1990s than in the 1980s as well as the group average in 1990s
Level calories	26 (20 %)	37 (28 %)
Level proteins	33 (25 %)	42 (32 %)
	Countries with higher indicators in the 1990s than in the 1960s, as well as the group average in the 1990s	Countries with higher indicators in the 1990s than in the 1980s, as well as the group average in 1990s
Volatility calories	23 (17 %)	16 (12 %)
Volatility proteins	23 (17 %)	21 (16 %)

Source: Díaz-Bonilla, Thomas, and Robinson, 2002.
Note: Based on data for 132 developing countries.

BOX 4.1. Developing Countries with Lower Indicators in the 1990s Than in the 1960s and Than the Group Average in the 1990s

Level of Calories Only

Central African Republic
Comoros
Congo, Republic of

Mongolia
Mozambique

Level of Proteins Only

Botswana
Cameroon
Democratic Republic of Congo
Cote d'Ivoire
Gambia
Guatemala
Iraq

Korea, Democratic People's
　Republic
Ruwanda
Sao Tome and Principe
Swaziland
Tajikistan
Vanuatu

Levels of Both Calories and Proteins

Afghanistan
Angola
Burundi
Cambodia
Chad
Congo, Democratic Republic of
Cuba
Guinea
Haiti
Kenya
Liberia

Madagascar
Malawi
Namibia
Nicaragua
Senegal
Solomon Islands
Somalia
Uganda
Zambia
Zimbabwe

Source: Author's calculation from FAOSTAT database (FAO, 2000).

TABLE 4.6. Countries with worsening indicators of malnutrition of children under five years old in the late 1990s

	Compared to the 1960s	Compared to the 1980s
Worse off	0	8 (5 %)
Improved by one-third or less and were below the average at the end of 1990s	17 (11 %)	39 (26 %)

Source: Author's calculation based on World Development Indicators data (World Bank, 2001).
Note: 149 developing countries with data for 1960s and 1990s; 150 developing countries with data for 1980s and 1990s.

VARIETY OF FOOD SECURITY SITUATIONS AND IMPLICATIONS FOR WTO NEGOTIATIONS

In addition to the obvious distinction between developed and developing countries, which are both self-identified groups, the WTO recognizes two other groups within developing countries: Least-developed countries (LDC), a United Nations (UN) classification; and net food importing developing countries (NFIDC), which are selected through the Committee on Agriculture of the WTO. LDCs have several legal implications under the WTO framework, and both types of countries were considered in a special ministerial decision approved at the end of the Uruguay Round.[1]

The question is how well those categories capture the heterogeneity of developing countries. Díaz-Bonilla and colleagues (2000) use various methods of cluster analysis and data for 167 countries to identify groups of countries categorized according to five measures of food security: food production per capita, the ratio of total exports to food imports, calories per capita, protein per capita, and the share of the nonagricultural population.[2] The results identify twelve clusters of countries according to their similarities in their food security profiles (measured by the variables listed earlier) from very food insecure, cluster 1, to very food secure, cluster 12 (Table 4.7).

Clusters with centers (in z-score values) falling below –0.5 (minus half a standard deviation from zero) are defined as "food insecure." Clusters 1, 2, 3, and 4 fall in that category. Clusters 5, 6, 7, and 8 have most of their variables in the –0.5 to +0.5 range (plus or minus half a standard deviation around zero). They are considered to be in the "food neutral" category. Fi-

TABLE 4.7. Classification of countries in twelve clusters: Mean values of the food security variables (per capita)

	Calories	Protein (grams)	Food production (US %)	Export to food import ratio	Share of food import to total export	Share of nonagricultural population
Cluster 1	1,982.9	48.6	81.8	4.9	20.4	0.23
Cluster 2	2,229.2	58.8	117.6	5.3	19.0	0.71
Cluster 3	2,244.6	52.6	120.3	14.1	7.1	0.41
Cluster 4	2,581.5	70.8	157.2	4.8	20.8	0.39
Cluster 5	2,602.3	66.5	210.4	11.3	8.8	0.75
Cluster 6	2,672.9	72.8	124.1	19.8	5.0	0.41
Cluster 7	2,976.1	82.7	135.1	9.1	11.0	0.82
Cluster 8	2,827.7	78.4	233.3	25.6	3.9	0.83
Cluster 9	3,231.3	100.1	254.2	18.6	5.4	0.88
Cluster 10	3,271.8	97.7	304.2	35.9	2.8	0.93
Cluster 11	3,303.7	103.3	520.6	17.7	5.7	0.93
Cluster 12	3,374.1	107.5	923.9	32.7	3.1	0.93

Source: Díaz-Bonilla and colleagues (2000).

nally, clusters 9, 10, 11, and 12, with most of the variables above +0.5, are considered "food secure."

Figure 4.3 illustrates the relative position of the twelve clusters in a diagram where the average value of the z-score variables for the combined consumption of calories and proteins is plotted against the trade indicator showing the burden of the food bill (also in z-score values). India (a member of cluster 3) is also plotted. The solid lines at the values of –0.5 across both axis of the chart divide the space into four main quadrants separating the food-insecure clusters from the rest (the dotted lines at the +0.5 values add other quadrants differentiating among clusters that are food neutral or food secure): clusters 1 and 2 appear in the quadrant that is consumption vulnerable and trade stressed (southwest quadrant), with the values below –0.5 on both dimensions; cluster 3 is in the quadrant that identifies consumption vulnerability but not trade stress (southeast quadrant); cluster 4 is in the trade-stressed quadrant but is above the level of –0.5 for consumption (northwest quadrant). The rest of the clusters appear in the intermediate or high levels of consumption and trade security (northeast quadrant), with both dimensions above the –0.5 value.

Cluster 1 includes the most food-insecure countries. They show the lowest levels of availability of calories and proteins per capita, and of food production per capita. Their food imports require over 20 percent of their total export earnings, compared to the world weighted average of 6 percent, and they are predominantly rural (only about 23 percent of the population is ur-

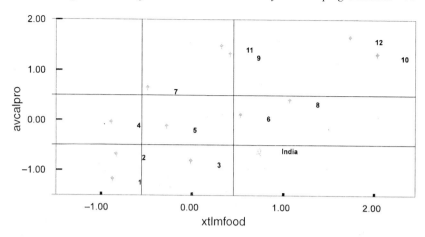

FIGURE 4.3. Scatter plot of consumption per capita (average of calories and proteins indicators) versus trade indicators.

ban; Table 4.7). This group includes thirty countries, all of them LDCs, except Kenya, a country classified as NFIDC by the WTO. They are mostly from Africa (23 out of the total 30). They include 21 WTO members and four WTO observers (Table 4.8).

Cluster 2 includes food-insecure countries with an urban profile. Those countries show somewhat higher levels of consumption and production than cluster 1, but they are still "consumption vulnerable" and trade stressed. The main difference is that these countries are far less rural than those in other food-vulnerable clusters; in fact, on average, more than 70 percent of the population is classified as urban. This raises the issue of urban food insecurity, which has its own special characteristics (see Garrett and Ruel, 2000). While countries in the previous cluster, being mostly rural, may be more concerned about food insecurity in the countryside and the impact of agricultural imports on poor agricultural producers, in countries with larger urban populations (such as those in cluster 2), and where conceivably an important percentage of poor and food-insecure groups may be urban dwellers, there is a clear trade-off for policies aimed at agricultural trade protection: they may maintain higher incomes for poor producers, but they may also act as a tax on poor consumers (both effects depending on other policies and the interaction of markets and institutions).[3] Among the 14 members of this cluster, two are LDCs from Africa and five are NFIDCs (mostly from Latin America). The other seven members are basically former republics of the ex-Soviet Union and Latin American countries. Except

TABLE 4.8. Country membership in clusters 1 to 12

Food-Insecure Groups

Cluster		LDC	NFIDC	Others
Cluster 1 (30 countries)	WTO members	Angola, Bangladesh, Burkina Faso, Burundi, Central African Republic, Chad, Democratic Republic of Congo, Gambia, Guinea, Guinea-Bissau, Haiti, Madagascar, Malawi, Mali, Mozambique, Niger, Rwanda, Sierra Leone, United Republic of Tanzania, Uganda	Kenya	
	WTO observers	Cambodia[1], Ethiopia, Nepal, Yemen		
	Others	Afghanistan, Comoros, Eritrea, Liberia, Somalia		
Cluster 2 (22 countries)	WTO members	Djibouti, Lesotho	Botswana, Cuba, Dominican Republic, Honduras, Peru	El Salvador, Georgia, Mongolia, Nicaragua
	WTO observers			Armenia, Azerbaijan
	Others			Tajikistan
Cluster 3 (17 countries)	WTO members	Solomon Islands, Togo, Zambia	Côte d'Ivoire, Sri Lanka	Bolivia, Cameroon, Republic of Congo, Ghana, Guatemala, India, Namibia, Papua New Guinea, Philippines, Zimbabwe
	WTO observers	Laos[1]		Vietnam
	Others			
Cluster 4 (13 countries)	WTO members	Benin, Mauritania, Senegal	Pakistan, Saint Lucia	Albania, Grenada, Saint Kitts and Nevis, Saint Vincent/Grenadines
	WTO observers	Sudan[1], Vanuatu[1]		Seychelles
	Others	Kiribati		

Cluster		LDC	NFIDC	Others
Food-Neutral Groups				
Cluster 5 (18 countries)	WTO members		Jamaica, Trinidad and Tobago, Venezuela	Belize, Brazil, Colombia, Costa Rica, Ecuador, Fiji Islands, Guyana, Kyrgyzstan, Nigeria, Paraguay, Suriname, Swaziland
	WTO observers			Croatia, Macedonia (The Former Yugolav Republic), Uzbekistan
	Others			
Cluster 6 (5 countries)	WTO members	Myanmar		Antigua and Barbuda, Gabon, Indonesia
	WTO observers			China
	Others			
Cluster 7 (14 countries)	WTO members	Maldives	Barbados, Egypt, Mauritius, Morocco, Tunisia	Brunei Darussalam, Dominica, Estonia, Jordan, Kuwait, Macau, Mexico
	WTO observers	Cape Verde		Algeria, Lebanon, Russian Federation, Saudi Arabia
	Others			Bahamas, Islamic Republic of Iran, Libyan Arab Jamahiriya, Syrian Arab Republic
Food-Secure Groups				
Cluster 8 (9 countries)	WTO members			Bulgaria, Chile, Republic of Korea, Latvia, Malaysia, Republic of Moldova, Panama, Slovakia, South Africa
	WTO observers			
	Others			
Cluster 9 (16 countries)	WTO members			Czech Republic, Germany, Iceland, Israel, Japan, Lithuania, Malta, Poland, Portugal, Romania, Slovenia, Turkey, United Arab Emirates, United Kingdom

TABLE 4.8 (continued)

Cluster		LDC	NFIDC	Others
	WTO observers			Belarus, Kazakhstan
	Others			
Cluster 10 (9 countries)	WTO members			Austria, China—Hong Kong SAR, Finland, Hungary, Norway, Sweden, Switzerland, United States
	WTO observers			Ukraine
	Others			
Cluster 11 (9 countries)	WTO members			Argentina, Belgium-Luxembourg, Canada, France, Greece, Italy, Netherlands, Spain, Uruguay
	WTO observers			
	Others			
Cluster 12 (3 countries)	WTO members			Australia, Denmark, Ireland
Outliers (2 countries)	WTO members			New Zealand, Thailand

Source: Díaz-Bonilla and colleagues (2000).

Notes: WTO members not included because of data unavailability: Bahrain, Cyprus, Liechtenstein, Qatar, and Singapore.

LDC: Least developing countries. LDCs not included because of data unavailability: Bhutan, Equatorial Guinea, Samoa, Sao Tome Principe, and Tuvalu.

NFIDC: Net food importing developing countries.

The majority of countries have been classified in the same group by all three clustering methods; the countries in bold have been classified in the same group by two out the three clustering methods.

¹Countries in the process of accession to the WTO.

for Tajikistan, all of the countries are either WTO members (11) or observers (2) (Table 4.8).

Cluster 3 includes food-insecure countries with consumption vulnerability. This cluster has availability of proteins and calories below cluster 2, but it is better off than cluster 1. It is also slightly below cluster 2 in production (but above cluster 1), and it is as rural as cluster 4. The main characteristic is that the burden of the food bill is at an intermediate level. This cluster can be characterized as consumption vulnerable but trade neutral, the mirror image of cluster 4 (Figure 4.3).

Cluster 3 includes 17 countries, four of which are LDCs and two are NFIDCs. All belong to the WTO as members or observers, and are developing countries in Africa, Asia, and Latin America (Table 4.8). Three countries from the Cairns Group appear in this group (Bolivia, Guatemala, and the Philippines).[4] India is in this group. Along with Namibia, the Philippines, and Vietnam (also in cluster 3), these countries have a low incidence of the food bill on total exports ("trade stress"): 4.5 percent for India, 5.3 percent for Vietnam, and about 6 percent for Namibia and the Philippines. Except for the Philippines, these countries are all net food exporters. Some of them may also exemplify a possible policy dilemma: because they are not trade stressed, they could expand food imports to improve their low levels of consumption; but at the same time, because they have large poor agricultural populations, there is concern regarding the impact of additional food imports on those rural groups.

Cluster 4 is composed of food-insecure countries with trade vulnerability. While the previous cluster had low consumption but intermediate levels of trade burden, cluster 4 shows the opposite profile: it has intermediate levels of consumption but it is very trade stressed; in fact, this group has the heaviest trade burden, with a food bill of almost 21 percent of total exports. Figure 4.3 shows cluster 4 in the trade-stressed quadrant but with an average consumption of calories and proteins above not only clusters 1, 2, and 3, the other food-insecure groups, but also the food-neutral cluster 5.

Cluster 4 has 13 members, including six LDCs and two NFIDCs.[5] All of them except one are WTO members or observers. Although the inclusion of some bigger countries in this group (such as Pakistan, Sudan, and Senegal) conform to the notion of having intermediate consumption but being trade stressed, the classification of some small islands from the Caribbean and the Pacific in this group is less clear. This may simply reflect lack of data regarding exports of services (such as tourism) and/or the fact that the urban/rural distinction does not have the same meaning in small islands as in bigger continental countries.

The rest of the countries are classified as food neutral or food secure. The conclusion is that some of the categories utilized by the WTO appear inadequate to capture food security concerns. The most obvious case is the category of "developing countries." These countries, which are self-identified, are subject to special and differential treatment first in the GATT and now in the WTO. In this analysis developing countries appear scattered across all levels of food (in)security, except cluster 12, the very high food-secure group (Table 4.8).

In the category of NFIDCs, ten out of the eighteen countries are in food-insecure groups (clusters 1 to 4); the remaining eight are in food-neutral groups (clusters 5 and 7), which have intermediate levels of food security. Being a net food importer appears to be only a weak indicator of food vulnerability. Some countries may be net food exporters but still have a larger percentage of their total exports allocated to buy food, and vice versa (for example, Mali is a net food exporter but its food bill is about 15 percent of total exports, while Venezuela, a NFIDC, spends about 5 percent of total exports on imported food). In addition, some countries may be net food importers just because of a dominant tourist industry (such as Barbados, which also has the highest income per capita of the NFIDCs, about US$7,000). Other NFIDCs have important levels of oil exports (such as the case of Venezuela, and Trinidad and Tobago) and therefore imports of food reflect only the comparative advantages of their production structure. With the exception of Egypt, food imports of the NFIDCs in the food-neutral group represent about 9 percent of total exports; for the food-insecure NFIDCs (including Egypt), the average is about 16 percent.

The category of LDCs, on the other hand, does correspond broadly to countries suffering from food insecurity, even though food security criteria were not explicit in their definition. Only three out of the 44 LDCs covered in this study are not among the first four clusters (most vulnerable countries).[6] At the same time, some countries, such as Kenya, have a food security profile similar to the more vulnerable LDCs but are not included in this category. Others, which have somewhat better profiles but are still in the food-insecure categories, are neither LDCs nor NFIDCs, such as El Salvador, Georgia, Mongolia, and Nicaragua (all WTO members).

In terms of the WTO negotiations, this analysis suggests that to define specific rights and obligations in the WTO using the category of LDCs appears an appropriate starting point, even though this group is not defined by food security criteria. Yet some countries that are neither LDC nor NFIDC and are therefore excluded from WTO special treatment are food insecure. One approach would be to give countries classified as food insecure by some objective criteria the same special rights and obligations in domestic support and market access as the ones received by LDCs. In addition, they could be considered for the food aid, financial support, and technical assistance envisaged in the ministerial decision on possible negative effects of the agricultural reform program on LDCs and NFIDCs. The issue of special access to other countries' markets for LDCs and the additional benefits conferred upon LDCs because of reasons other than food security would still be limited only to the countries specified by the United Nations.

The current category of NFIDCs, a classification negotiated during the Uruguay Round, has some implications as defined in the ministerial decision, and constitutes an acquired right. The implementation of that decision, as discussed in the meetings of the Committee on Agriculture of the WTO, appears to have been limited mostly to exchanges of information among multilateral organizations and bilateral donors about programs already under execution. In particular, no special action was taken during the 1995-1996 increase in agricultural prices, because the aid agencies considered that the rise was not related to the implementation of the Uruguay Round agricultural agreements. For that reason, many LDCs and NFIDCs have been calling for objective criteria to "operationalize" the ministerial decision (UNCTAD, 2000).

Defining more precisely the group of countries that appear vulnerable to food security problems would help accomplish such operationalization. It can be argued that the perception that the category of NFIDCs is not adequate (because it leaves vulnerable countries out, while including countries that are relatively better off) may have contributed to the lack of implementation of the decision. In any case, the current category of NFIDCs has been defined for reasons beyond food security, and this analysis does not suggest that it be changed because it is less useful in addressing food security issues. Rather, the operationalization of the ministerial decision of food security concerns should also include other countries not currently considered within the NFIDCs.

Any quantitative measure of food security would provide a cutoff point, which would help differentiate developing countries that may need special treatment in terms of food security from those that do not. It is also relevant to ask about the food security situation of the developed countries. Several developed countries have raised the issue of food security in the debate on the "multifunctionality" of agriculture, or, more generally, in the nontrade concerns. The cluster analysis classification, however, shows that developed countries are unanimously concentrated in the food-secure groups (clusters 9 to 12). Therefore, the term "food security" must have a different meaning in developed than in developing countries. In terms of policy implications and the agricultural negotiations, maintaining the same label for two altogether different situations only obscures the issues being negotiated. The discussion of food security should be limited to the vulnerability of developing countries, using a different terminology for developed countries. In summary:

- All developed countries are classified into food-secure groups. Food security does not seem to be a rationale for some of the exceptions claimed by developed countries.
- Developing countries are classified across all levels of food security (except the most-food-secure group), attesting to the heterogeneity of this category.
- Only ten of the eighteen NFIDCs appear in food-insecure groups. Being a net food importer seems to be a weak indicator of food insecurity.
- With a few exceptions, all the LDCs covered by the study are classified among the food-insecure clusters. The category of LDCs appears to be an appropriate starting point, but it may not be enough.
- Some WTO members that are neither LDC nor NFIDC are clearly classified into food-insecure groups.

The implication is that there is a need for a better definition of food-insecure countries, based on objective quantitative indicators. Based on the new classification, food-insecure countries would be granted specific exceptions. This special and differential treatment should be limited to food security considerations and should not change the balance of rights and obligations for LDCs and NFIDCs for reasons other than food security.

TRADE LIBERALIZATION, FOOD SECURITY, AND INDUSTRIALIZED COUNTRIES

The combination of domestic support, market protection, and export subsidies in industrialized countries has displaced agricultural production and exports from developing countries. This is an especially important issue for the poorer countries, where two-thirds of the population lives in rural areas, agriculture generates over one-third of the gross domestic product (GDP), and a substantial percentage of exports depend on agriculture. A key objective for developing countries, therefore, is the elimination or the substantial reduction of subsidies and protectionism in industrialized countries during the current WTO negotiations. This general proposition has raised three concerns:

- Will the liberalization of agricultural policies in industrialized countries increase the food bill of net-importing countries?
- Will the liberalization of trade erode the trade preferences of developing countries that have preferential access to these protected markets of rich countries?
- Will export expansion have harmful effects on poverty and food security?

In the first two cases, a welfare-enhancing approach would be to offer cash grants or other financial schemes to compensate developing countries for higher prices and lost preferences caused by the liberalization of markets in rich countries, rather than maintaining protection in the latter.

The third question is linked to earlier criticisms of the Green Revolution, later extended to commercialization and international trade: it has been argued, first, that the limited resources of small farmers could prevent them from participating in expanding markets and lead to worsening income distribution; second, and more worrisome, if relative prices shift against the poor, or if the power of already dominant actors (large landowners, big commercial enterprises) is reinforced to allow them to extract income from the poor or to appropriate their assets, the poor could become worse off in absolute terms. It has also been argued that food security could decrease if cash or export production displaces staple crops and if these changes result in women having less decision-making power and fewer resources.

Yet several studies have shown that the Green Revolution—and domestic and international commercialization—can yield benefits for the poor because of its effect on production, employment, and food prices, although any uniform attainment of benefits is by no means guaranteed (see Hazell and Ramasamy, 1991; Von Braun and Kennedy, 1994). Trade expansion that creates income opportunities for women may also give them greater control over expenditures, with positive impact on child nutrition and development, as well as greater incentives to invest in the education and health of girls. But there may be a tradeoff between income-generating activities and the time allocated for child care (Paolisso et al., 2001). In general, complementary policies are needed to increase the physical and human capital owned by the poor and by women, to build general infrastructures and services, to ensure that markets operate competitively, and to eliminate institutional, political, and social biases that discriminate against these groups.

FOOD SECURITY PROBLEMS
IN DEVELOPING COUNTRIES

During the current WTO negotiations, several developing countries have indicated concerns that liberalizing their agricultural trade may affect negatively those countries' large, poor agricultural populations. Developing countries have argued for a slower pace in reducing their tariffs (or maintaining and even increasing current levels) on the premise that industrialized countries should first eliminate their higher levels of protection and subsidies. Another concern is how to avoid any sudden negative impact on poor producers, whose vulnerable livelihoods may be irreparably damaged by drastic shocks (for instance, by forcing poor families to sell productive assets or to take children from school).

This discussion reflects a permanent tension between maintaining high prices for producers versus assuring low prices for consumers. While industrialized countries have used transfers from consumers and taxpayers to maintain high prices for producers, developing countries have enforced low agricultural prices to further the process of industrialization. Several studies have shown that poverty alleviation in developing countries was impaired by policies that protected capital-intensive industrialization and discriminated against agriculture. Post-1980s policy reforms in developing countries appear to have reduced or eliminated general policy biases against agriculture, but in some cases they may have contributed to the decline of the infrastructure and institutions needed for agricultural production and commercialization. Further correction of market distortions may still be needed in some countries, but now the emphasis should be on policies for investing in the rural economy, focusing on the poor.

Out of concern for small farmers, some have argued that developing countries should move even further toward protection of the agricultural sector. However, considering that poor households may spend as much as 50 percent of their income on food, these recommendations could have a negative impact on the poverty and food security of not only the increasing number of poor urban households and landless rural workers, but also poor small farmers, who tend to be net buyers of food. Trade protection for food products is equivalent to a very regressive implicit tax on food consumption, mostly captured by large agricultural producers, with a greater impact on poor consumers. Also, trade protection for any sector usually implies negative employment and production effects in other sectors, and the general effect of widespread trade protection is a reduction in exports.

The best approach for developing countries to support agriculture is to eliminate biases against that sector in the general policy framework and to increase investments in human capital, land tenure, water access, technology, infrastructure, nonagricultural rural enterprises, organizations of small farmers, and other forms of expansion of social capital and political participation for the poor and vulnerable. At the same time, developing countries may legitimately insist that industrialized countries reduce their higher levels of subsidization and protection (as argued in the previous section), and ask for policy instruments to protect the livelihoods of the rural poor from import shocks that could cause irreparable damage (which is discussed next). Although most of the policies suggested here are not restricted under the AOA, some adjustments to the current language may be needed to make sure that food security concerns are properly addressed.

Emergency Food Security Stocks (Green Box Annex)

The AOA establishes the conditions for emergency stocks, which must be built on clearly defined targets, for instance as a percentage of total consumption. The AOA also requires transparent financial arrangements to avoid waste and corruption. This is different from using stocks to stabilize domestic grain prices, an expensive and relatively ineffective policy. A key point is that those stocks must be bought and sold at market prices. The language of the AOA is clear on sales from the stock but less so in the case of buying food products. For poor countries, administered prices, which tend to generate losses, buying high to support farmers and selling low to subsidize consumers, add to the costs of the food security program. To avoid any confusion on applying this measure, the legal texts could indicate that food-insecure countries are in compliance with the AOA when they build food security stocks by buying at domestic market prices a small number of pre-specified products in volumes not exceeding some limited percentage of domestic consumption (i.e., stocks for not more than 10 percent of domestic consumption for up to three products).

Domestic Food Aid (Green Box Annex)

The AOA allows food security interventions subject to well-defined nutritional criteria, so for many developing countries, the issue is not the legal constraints under the AOA but rather the means to design and finance adequate interventions.

Support to Poor Producers and Production for Food Security

Agricultural and trade policies may affect developing countries' large rural populations, where poverty is still concentrated. These concerns are related to issues of domestic support (how to provide meaningful support to agricultural producers, especially small farmers), market access (particularly how to manage import surges), and export subsidies (that may displace local producers). The AOA allows countries a great latitude in domestic support policies: green and blue box measures, the de minimis exemptions, and the change from product specific to the Aggregate Measure of Support (AMS). Developing countries, in addition, are allowed smaller reductions and longer implementation periods than developed countries. LDCs are completely exempt from any reduction in domestic support. In addition, developing countries are legally entitled under WTO to provide additional investment support to their agricultural producers provided the measures are an integral part of their development programs, or in the case of input subsidies (from credit to fertilizers or water) that they are given to "low-income or resource-poor producers" (Article 6.2). From an equity standpoint, these measures have the advantage of encouraging developing countries to design specific programs for rural development or alleviation of rural poverty, instead of resorting to general and nontransparent subsidy schemes that may benefit richer farmers or be wasted in corruption. The only constraint is that these measures may be subject to complaint by a WTO member if those subsidies exceed the budgeted level in 1992 by commodity, and the complaining country can prove either serious prejudice, or nullification/impairment of benefits from the subsidies (Article 13). Although such complaints appear unlikely in the case of most, if not all, poor developing countries, it might be advisable to include language in the AOA to exempt food-insecure countries from the eventuality of such actions. In addition, the definition of "low-income or resource poor producers" could be more specific, either by using a measure of the poverty line used in international comparisons (below one dollar or two dollars a day) or a relative measure within the country (for instance, producers with less than 40 percent of national income per capita).

Special Safeguard for Food Security and Other Trade Remedies

Some developing countries have requested that the special safeguard (available only to countries, mostly developed ones, which have changed nontariff barriers into tariffs) be eliminated and a new special safeguard created for food security reasons. Conceivably, the latter can be done by adding some modifications to the common safeguard such as (1) streamlined and faster procedures for a limited number of designated crops for food security reasons and (2) exemptions from the need to offer compensations, linked to the temporary use of the safeguard. Similarly, streamlined procedures can be defined to counter export subsidies and dumping activities that may affect food-insecure countries when they affect a limited group of designated crops with food security implications. The most straightforward option would be to apply the common WTO rules for export subsidies, instead of using, as is now the case, the stricter requirements of Article 13 of the AOA to apply countervailing duties.

Food Aid, Access to Food, and Development Funds

Food aid, which has declined in recent years, should be made available in adequate levels and be counter-, not pro-cyclical. Food aid should be made available in grant form, focused toward poor countries and social groups, and delivered in ways that do not displace domestic production in the receiving countries. A special aspect is to make sure that export controls and export bans on food items are tightly disciplined so as not to hamper access to food by importing countries. It is also necessary to maintain and expand financial facilities (both multilateral and bilateral) to help with short-term difficulties in financing food imports.

CONCLUSION

Can the WTO legal framework protect the interests of the world's poor and hungry? Obviously, the WTO cannot make sure that everyone on the planet gets enough to eat. But it can help to prevent unfair competition that hurts the poor. In the negotiations developing countries are legitimately insisting that industrialized countries reduce their higher levels of subsidies and protection. Also, WTO country members need to set up a new category of "food-insecure" countries, those for whom famine is just one catastrophic harvest away, and make sure that they have the policy instruments to protect the livelihoods of the rural poor. Most of those changes can be accommodated with some changes in the current language of the AOA, whether they are called a food security box or not. While actively seeking the changes suggested (or some similar ones), food-insecure developing countries may want to consider not making the issue of the food security box something for which they pay a heavy negotiating price, when a well-selected set of changes in language should suffice without the need of labels that suggest that whole new concessions have been obtained. Also, in the negotiations developing countries must be careful not to ask for legal room that they will not be able to utilize later because they do not have the financial resources to implement the suggested policies. The problems facing developing countries in ensuring food security are not legal constraints under the WTO but, rather, a lack of financial, human, and institutional resources. This should be recognized by ensuring that agricultural trade negotiations proceed in tandem with increased funding by international and bilateral organizations for agricultural and rural development, food security, and rural poverty alleviation.

NOTES

1. See Díaz-Bonilla, Piñeiro, and Thomas (1999) for a more detailed discussion of these groups. The Food and Agriculture Organization also defines low income food deficit countries (LIFCDs), but they are not subject to any special treatment or legal consideration under the WTO.

2. The indicators utilized in the study are considered proxies for three elements of food security at the national level: food availability, access, and utilization.

3. The case of vulnerable rural groups that are net consumers of food must also be considered.

4. The Cairns Group is a negotiating block of agricultural exporting countries that has argued for greater liberalization in world agricultural markets. The current 17 members are Argentina, Australia, Bolivia, Brazil, Canada, Chile, Colombia, Costa Rica, Guatemala, Indonesia, Malaysia, New Zealand, Paraguay, the Philippines, South Africa, Thailand, and Uruguay.

5. Senegal became an LDC in 2002. The two groups, LDC and NFIDC, being mutually exclusive, Senegal was removed from the NFIDC group.

6. Cape Verde, Maldives, and Myanmar are in clusters 6 and 7.

REFERENCES

Díaz-Bonilla, E., V. Piñeiro, and M. Thomas. 1999. *Getting Ready for the millennium round trade negotiations: Least-developed countries' perspective.* Brief 8 in IFPRI 2020 Focus 1. Washington, DC: International Food Policy Research Institute.

Díaz-Bonilla, E., M. Thomas, A. Cattaneo, and S. Robinson. 2000. *Food security and trade negotiations in the World Trade Organization: A cluster analysis of country groups.* Trade and Macroeconomics Discussion Paper 59. Washington, DC: IFPRI.

Díaz-Bonilla, E., M. Thomas, and S. Robinson. 2002. *Trade liberalization, WTO, and food security.* Trade and Macroeconomics Discussion Paper 82. Washington, DC: International Food Policy Research Institute.

FAOSTAT (FAO Statistical Database). 2000. Database from the Food and Agricultural Organization, available at <http://faostat.fao.org/faostat/collections>.

Food and Agriculture Organization of the United Nations (FAO). 1996. Rome declaration on world food security and World Food Summit plan of action. Rome, Italy: Author.

———. 1999. *Food insecurity: When people must live with hunger and fear starvation. The state of the food insecurity in the world 1999.* Rome, Italy: Author.

Garrett, J. and M. Ruel, eds. 2000. *Achieving urban food and nutrition security in the developing world.* 2020 Vision Focus. Washington, DC: International Food Policy Research Institute.

Hazell, P.B. and C. Ramasamy. 1991. *The Green Revolution reconsidered: The impact of high-yielding rice varieties in South India.* Baltimore, MD: The Johns Hopkins University Press.

Maxwell, S. 1990. Food security in developing countries: Issues and options for the 1990s. *Institute of Development Studies Bulletin* 21(3): 2-13.

———. 1996. Food security: A post-modern perspective. *Food Policy* 21(6): 155-170.

Maxwell, S. and T.R. Frankenberger. 1992. *Household food security: Concepts, indicators, measurements: A technical overview.* New York/Rome: UNICEF/FAO.

Organisation for Economic Co-operation and Development (OECD). 2001. Agricultural policies in OECD countries: Monitoring and evaluation.

Paolisso, M.J., K. Hallman, L. Haddad, and S. Regmi. 2001. *Does cash crop adoption detract from childcare provision? Evidence from rural Nepal.* Food Consumption and Nutrition Division Paper 109. Washington, DC: International Food Policy Research Institute.

Rosegrant, M., M. Paisner, S. Meijer, and J. Witcover. 2001. *2020 global food outook: Trends, alternatives, and choices.* Food Policy Report. Washington, DC: International Food Policy Research Institute.

Sen, A. 1981. *Poverty and famines: An essay on entitlement and deprivation.* Oxford, UK: Clarendon Press.

Smith, L.C. 1998. Can FAO's measure of chronic undernourishment be strengthened? *Food Policy* 23(5): 425-445.

Smith, L.C. and L. Haddad. 2000. *Explaining child malnutrition in developing countries: A cross-country analysis.* IFPRI Research Report 111. Washington, DC: International Food Policy Research Institute.

United Nations Conference on Trade and Development (UNCTAD). 2000. Impact of the reform process in agriculture on LDCs and net food-importing developing countries and ways to address their concerns in multilateral trade negotiations. Background note by the UNCTAD Secretariat. TD/B/COM.1/EM.11/2. Geneva.

von Braun, J. and Eileen Kennedy (Eds.) 1994. *Agricultural commercialization, economic development, and nutrition.* Baltimore, MD: The Johns Hopkins University Press.

World Bank. 2001. World development indicators 2001 [CD-ROM]. Washington, DC: Author.

Chapter 5

Indian Agriculture, Food Security, and the WTO-AOA

Anwarul Hoda
Ashok Gulati

January 1, 1995, marked the beginning of a new era for agricultural commodities with the implementation of the WTO Agreement on Agriculture (AOA). It was lauded for attempting to rein in high levels of distortionary support in agricultural sectors across the world, but in many developing countries, including India, it was perceived as a threat to food security. There has been strong opposition to agricultural trade liberalization in some quarters in India, stemming from food-security concerns. What specifically are these concerns? How does the AOA impinge on India's food policies? And what can India do to ensure food security? This chapter seeks to shed some light on these questions, focusing on India's food security in the context of the WTO-AOA implementation. Toward this end, the chapter is organized into seven sections. The second section outlines the main concerns that have surfaced in India regarding food security. The next section discusses India's commitments under the WTO and its experience with implementation. Following this, the chapter examines whether the AOA really does represent a threat to India's food security. The fifth and sixth sections identify the problem areas and briefly chart out a course that would ensure India's food security in a liberalized environment. The final section concludes the discussion.

THE EMERGENCE OF FOOD SECURITY CONCERNS IN INDIA

Since India became a signatory to the AOA, a concern that has been raised repeatedly is whether agricultural trade liberalization would destroy India's food security. This is a critical issue because a large section of the population is dependent on agriculture for their livelihood, and poor consumers already spend an overwhelming share of their income on food.[1] In this context, three issues are often raised in political and bureaucratic circles related to policy making.

First, there is a lurking fear that trade liberalization under the AOA and consequently greater integration of Indian agriculture into world markets would lead to a deluge of cheap imports. The effect would be of wiping out the production base, creating unemployment, and deepening poverty, thus eroding food security. The surging import of edible oils during 1998/1999 and 1999/2000, which amounted to more than 40 percent of domestic consumption, is often cited as an example. This was not considered good from the standpoint of food security.

Second, liberalization of domestic agricultural markets would mean that price volatility in world markets would be further transmitted to domestic markets (adversely affecting cultivators when prices decline or consumers when prices increase). For poor producers, this could mean being pushed to the very brink of destitution given their limited capacity to bear risks. The crash in domestic prices of wheat in the postharvest period of 2000 was considered to be a direct result of this policy. It is feared that in time this could adversely affect food security.

Third, agreeing to contain domestic subsidies to agriculture within a prescribed limit was not considered to be in India's interest from the viewpoint of ensuring food security for its people.

There is a fear in some circles in India that importing agricultural commodities means importing unemployment. Based on such apprehensions, many political leaders have at times expressed extreme views. They have gone to the extent of saying that India should rethink whether it would like to remain a member of the WTO if its food security was being compromised under any agreements relating to agriculture. To evaluate these concerns, it is essential to examine what the AOA entails for India in terms of specific commitments, how India has implemented these commitments, and in what ways these impinge on India's food security.

INDIA AND THE AOA

The Uruguay Round of the AOA is built on three pillars: market access, domestic support, and export competition. Reduction commitments had to be undertaken by different countries, with developing countries being accorded special and differential treatment.

India has no reduction commitments, either with respect to domestic support or export competition. As a developing country, India also benefited from additional exemptions on generally available investment subsidies and input subsidies to resource-poor farmers. The permissible limit on distortionary domestic support, both product specific and non–product specific, was 10 percent of the value of production for developing countries such as India against the lower figure of 5 percent for developed countries. Furthermore, as far as export competition was concerned, subsidies for market promotion, international freight, and internal transport of export consignments were all exempt from reduction commitments.

Even in the area of market access India was not required to make any substantial adjustments. Because quantitative restrictions (QRs) maintained for balance-of-payments reasons were exempted from the tariffication requirement (AOA, Article 4.2), India continued with these restrictions even after the WTO agreement entered into force. Free of the tariff rule, India was also exempted from the requirement of maintaining access to its market and providing minimum access in cases where current imports are negligible. In making its tariff commitments, India availed of the flexibility provided to developing countries in the modalities to bind agricultural tariffs at ceiling levels in respect of all tariff lines in which tariffs were not bound in earlier negotiations. In India's tariff schedule at the Uruguay Round, all 686 lines at the six-digit level (subheadings under the Harmonized System) falling within the purview of the Agreement on Agriculture were subject to tariff commitments.[2] Most of the tariff commitments did not entail any reduction in the applied levels of tariffs, as they were far below the ceiling rates at which bindings were offered. Only in respect to a handful of products such as alcoholic beverages did India make substantial cuts in tariffs. The tariff commitments made in earlier rounds of negotiations were continued without any change.

An examination of India's implementation experience since 1995 reveals that the AOA has played only a limited role in shaping agricultural policies. For instance, in domestic support, the twin policies of price support through the Minimum Support Prices policy and subsidies on agricultural inputs continue as before. Support prices are announced for 24 commodities (with procurement taking place mainly for rice and wheat). Product-specific support remains negative for most products and below the de minimis for others. It is only with the crash of international prices to extraordinarily low levels that for some products in 2000-2001 the product specific support may have turned positive. But this too is unlikely to have exceeded de minimis levels. Non-product-specific support, or input subsidies, have actually been on the rise but also remain below the de minimis limit, even if calculations are made without taking into account the exemptions for input subsidies for low-income and resource-poor farmers. This has been true of the 1980s and has continued well into the 1990s. India used no export subsidies until 2001-2002, when an unprecedented accumulation of food-grain stocks coupled with a steep fall in international prices made its use inevitable.

In the area of market access, the AOA has resulted in changes with respect to QRs but not so much in tariffs. During the Uruguay Round, India bound its tariffs at levels that were among the highest in the world, at 100 percent for raw commodities, 150 percent for processed agrocommodities, and 300 percent for most edible oils (with a notable exception of soya oil at 45 percent). In offering such high rates, the government took into account the possibility in the future of phasing out import restrictions. The high bound rates have given India considerable flexibility to raise tariffs on agricultural products whenever a need is felt for such action. In fact, India has also been able to revise tariffs upward after renegotiations on products that had been bound at low levels in past negotiations.[3] These have included milk powder, rice, maize, and other commodities (see Table 5.1). Only on a few products, such as alcoholic beverages, has the tariff level actually been reduced in light of the commitments. Recent declines in tariff rates have meant that the applied rate of customs tariffs in India, in recent years, has generally been far below the high rates of ceiling bindings for almost all agricultural products. As depicted in Figure 5.1, this is true for all commodity groups. Only alcoholic beverages have duties that are close to their bound rates.

The only significant change in India's import policy has been the removal of QRs. Since these were put in place following an adverse balance-of-payments situation they had to be phased out when the balance-of-payments situation improved considerably in the second half of the 1990s. In 1997, India agreed in the WTO to phase out these restrictions and offered to do so over a nine-year period. Disagreements over the length of the phase-out period led to a number of WTO members raising a dispute against India on the matter. Subsequently, India reached an agreement with the European Union (EU), Canada, Australia, New Zealand, and Switzerland to phase out the restrictions over the period 1997-2003, but the United States pursued the dispute and obtained the final verdict in its favor. The Appellate Body Report of August 23, 1999, recommended that "India bring its balance-of-payments restrictions, which the Panel found to be inconsistent with Article XI: 1 and XVIII: 11 of GATT 1994, and with Article 4.2 of the Agreement on Agriculture, into conformity with its obligations under these agreements" (WTO, 1999). Pursuant to this recommendation, India agreed to eliminate the balance-of-payments restrictions on 1,429 six-digit tariff lines in two installments in 2000 and 2001. On April 1, 2001, India eliminated the last quantitative restrictions on 715 six-digit tariff lines, including 147 tariff lines pertaining to agriculture. After this date, quantitative restrictions on imports are maintained only for reasons of protection of health and mor-

TABLE 5.1. List of goods for which the bound rates were renegotiated

Code	Description	Old bound rate	New bound rates	Note
0402.10	Milk, concentrated or containing added sugar. In powder, granules or other solid forms, of a fat content, by weight, not exceeding 1.5 percent	0	60	A global tariff rate quota of 10,000 tonnes at an in-quota tariff rate of 15 percent is applicable cumulatively to both the tariff lines 0402.10 and 0402.21.
0402.21	Not containing added sugar or other sweetening matter	0	60	
806.1	Grapes, fresh	30	40	
Ex1001.90	Spelt	0	80	
1005.10	Maize, seed	0	70	
1005.90	Maize, other	0	60	A global tariff rate quota (350,000 tonnes in 2000-2001, rising to 500,000 tonnes in 2004-2005) at an in-quota rate of 15 percent is applicable.
1006.10	Rice in the husk (paddy or rough)	0	80	
1006.20	Husked (brown) rice	0	80	
1006.30	Semi-milled or wholly milled rice, whether or not polished or glazed	0	70	
1006.40	Broken rice	0	80	
1007.00	Grain sorghum	0	80	
1008.20	Millet	0	70	
1514.10	Rape, colza and mustard oil, crude	45	75	
1514.90	Rape, colza or mustard oil and fractions thereof, other	45	75	A global tariff rate quota of 150,000 MT at in-quota tariff rate of 45 percent is applicable.
1901.10	Preparations for infant use, put up for retail sale	17.5	50	

Source: Based on data from Ministry of Commerce, Government of India; available at <commerce.nic.in/indsched.htm>.

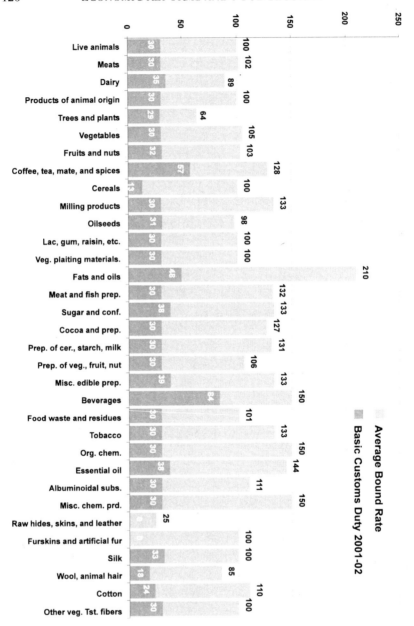

FIGURE 5.1. Bound rate versus basic customs duty 2001-2002 (chapterwise) in India.

als as permitted under Article XX of GATT 1994. The list of such restrictions includes certain animal fats, which are prohibited from imports for religious reasons. When the QRs were removed in April 2001, it was feared that there would be an import flood. In response to such fears, a "war room" was set up to track imports of about 300 sensitive commodities, most of which were agricultural products.

IS THE AOA A THREAT TO INDIA'S FOOD SECURITY?

Against this background of implementation experience, the question of whether the AOA has impinged India's food security can now be addressed. One way of doing so is to examine the manner in which the current food policies may conflict with the AOA. In this context, India's agricultural policy can be seen as consisting of six interrelated elements:

- Price support through minimum support prices
- Buffer stocking
- Food subsidies through the public distribution system
- Subsidies on agricultural inputs (power, fertilizers, irrigation water, seeds, etc.)
- Public investment in rural infrastructure
- Public support to R&D, extension, etc.

Of these, the first three can be regarded as directly pertaining to food security and represent the three functions carried out by the Food Corporation of India (FCI).

First, minimum support prices protect farmers from price falls and fluctuations (seasonal and annual), thereby ensuring stability in domestic food production. Second, the buffer stocks that the FCI maintains help in the short term to make food available in the event of sudden shortages. Third, the supply of essential foodstuffs at lower prices improves the food security of the weaker sections of society. The FCI carries this out through one of the world's largest public distribution systems (PDS) with a network of retail outlets called fair price shops (FPS) numbering more than 460,000 and spread across the country. The PDS has been an important constituent of India's strategy for poverty alleviation and seeks to improve the situation of the economically weaker sections of society by enhancing their food security. The PDS was revitalized in June 1997 when it was introduced as the targeted public distribution system (TPDS), which targets families below the poverty line (BPL) at highly subsidized rates. The quantity of food grain allocated to an estimated 65.2 million BPL families is 18.52 million tons

per annum, at 50 percent of the economic cost. Within this overall allocation the *Antyodaya Anna Yojana* scheme launched in December 2000 focuses on 10 million of the poorest families of the BPL population, providing 25 kgs of food grains to families at a token price of 2 rupees per kg for
wheat and 3 rupees per kg for rice. This scheme involves a cost of 23.15 billion rupees. Among the other special schemes aimed at enhancing food security are the *Annapurna scheme* of 10 kg of food grain per person per
month for indigent senior citizens free of cost, and the Mid-Day Meal
scheme.[4]

The other three elements of policy are of a more general nature but nevertheless have important implications for food security. Provision of input
subsidies is an ingredient of government policy in a country in which the
vast majority of farmers are low income and resource poor. By making these
inputs more affordable, it was hoped that this would encourage the adoption
of new technologies that would in turn augment food production. Similarly,
the government's provision of general services for research, extension and
advice, pest and disease control, and inspection and training is needed to
help farmers strive constantly for technological advancement and productivity increase. Finally, public investment in irrigation, power, roads, and
market complexes provides the bedrock for agricultural development in the
country.

Of these six elements of India's food policy, it is interesting to note that
as many as four—public investment in infrastructure projects (capital cost
only), general services, buffer stocks for food security purposes, and domestic food aid—are enumerated in the list of domestic support programs
exempted from AOA reduction commitments not only for developing countries but also for the developed. In other words, these are fully permissible
under the AOA (Box 5.1).

With regard to the other two—input subsidies and minimum price
support—there are limits under the AOA. In respect of input subsidies, for
India, as for other developing countries, it is not to exceed 10 percent of the
value of production. However, for developing countries, there is an exemption for the programs covering low-income and resource-poor farmers. A
substantial portion of India's input subsidies would in fact qualify as exempt, given the number of farmers that are low income and resource poor.
Yet India has so far not availed of this exemption. In non-product-specific
support, calculations made on the basis of the least favorable interpretation
of some of the provisions still show that the percentage, while creeping up
remains well below the 10 percent de minimis limit.

Likewise, minimum price support operations are affected by the obligation to phase down subsidies. However, here too product-specific support
up to 10 percent is exempt from reduction commitments for developing

---

Box 5.1. Six elements in India's policy: What does the AOA Allow?

Buffer stocking	✓
Public distribution system	✓
Public investment	✓
Government support to R&D, etc.	✓
Miniumum support prices	within de minimis limits
Input subsidies	within de minimis limits to low-income, and resource-poor farmers exempt

---

countries and may not be a cause for concern. In domestic support the high level of the fixed external reference price incorporated in India's commitments has ensured that the product-specific support is either negative or well within the de minimis limit of 10 percent. Thus, there are no existing measures for food security and rural development in India that the AOA prohibits.

The potential threat that agricultural trade liberalization poses can be evaluated against India's own competitiveness in agriculture. An extensive overview of several crops over the past three decades suggests that India is competitive in several major agricultural commodities (Figure 5.2) and that trade liberalization per se would not necessarily have an adverse impact. Commodities such as rice and wheat, India's major staples, are not only efficient import substitutes but also have been export competitive for several years (Figure 5.3). Most other commodities belonging to the pulses group or coarse cereals appear to be efficient import substitutes but are not export competitive. Therefore, under the circumstances, there is little threat of a deluge of cheap imports on a regular basis. The only major uncompetitive commodities—even as import substitutes—seem to be oilseeds and edible oils, which are currently produced at high cost in India. Milk too may be of some concern here, mainly because of the high level of export subsidization of skimmed milk powder (SMP) by the developed countries, although the

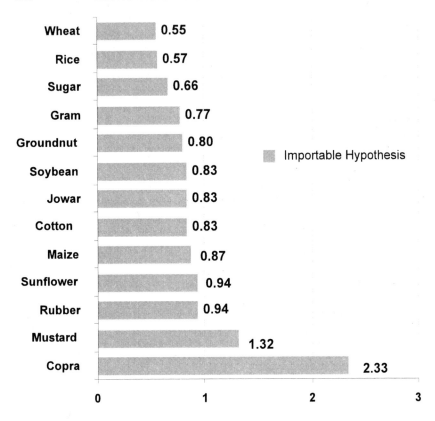

FIGURE 5.2. Nominal protection coefficients for selected agricultural commodities, 1992-1997.

trends in the 1990s are encouraging. Overall, Indian agriculture seems reasonably competitive.

What is important to note is that the commodities that are critical for India's food security have bound rates that are significantly high. Where the need has been felt for raising these duties above the bound level, this has been accomplished through the process of negotiations. Given that the bound level of duties has in most cases proved to be much higher than the levels applied, there is scope to increase tariffs for protection of agriculture if the situation so demands. Figure 5.4, for instance indicates that for as many as 565 lines, the applied rate is more than 50 percentage points below the Uruguay Round bound rate during 2001-2002. This is leaving aside the fact that there is enough of a cushion even in the applied rates.

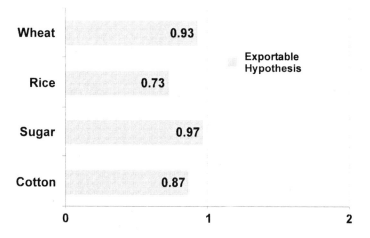

FIGURE 5.3. Nominal protection coefficients for selected agricultural commodities, 1992-1997.

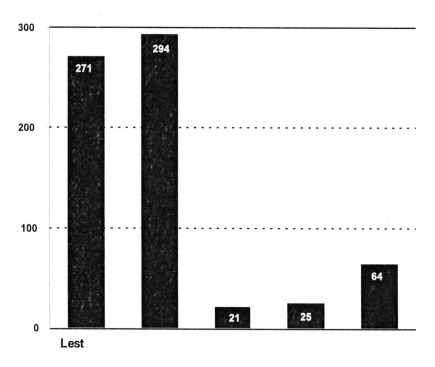

FIGURE 5.4. Tariffs below UR bound rate, 2001-2002.

Due to the adequate protection that tariffs offer, combined with India's general competitiveness, the withdrawal of quantitative restrictions by India (more generally, import liberalization) has not led to any large increase of imports of agricultural products. The only commodity group that has seen a dramatic rise in imports is edible oils, in which India is highly uncompetitive. Overall, it is evident that the AOA does not present any significant threat to India's food security.

THE REAL TRADE CHALLENGES TO INDIA'S FOOD SECURITY

The main problem for India is the fact that several developed-country members of the AOA continue to both subsidize their exports and heavily support their domestic agricultural sectors. From the beginning, it was widely felt, in the developing countries and in the academic community within the OECD countries, that the liberalization the AOA actually achieved was meager. Since then further analysis has led to greater skepticism on whether there was any liberalization at all for most of the heavily protected products. Many OECD countries, for instance, had protection levels in 1997-1999 that were not far different from pre–Uruguay Round levels (Figure 5.5).

Several developed countries of the EU, for example, continue to rely heavily on export subsidies. Regarding the 25 countries that have export subsidy commitments, it is well known that the EU accounts for almost 90 percent of the export subsidies. The subsidy is mostly on livestock products such as meat and milk, followed by grains and sugar (Figure 5.6). Under the circumstances, further trade liberalization, particularly for the heavily distorted products (dairy, for instance), could be detrimental to India's agricultural production base.

Similarly, with greater volatility in prices, huge declines in international prices can wipe out sections of the farm population. India, unlike the United States, which has the resources to compensate its farmers in such an eventuality, would have to look at other avenues, such as special safeguards to protect domestic markets from world markets when they are highly distorted.[5]

Equally important, domestic marketing reforms continue to be the weakest link in Indian agriculture. There is no other explanation for the peculiar paradox that India confronts today. While large sections of the more than one billion Indians suffer from poverty and malnutrition and lack easy access to food, the government holds mammoth food stocks—estimated at about 60 million tonnes—some of which are rotting in the warehouses. The FCI, which is responsible for procurement, stocking, and supplying the

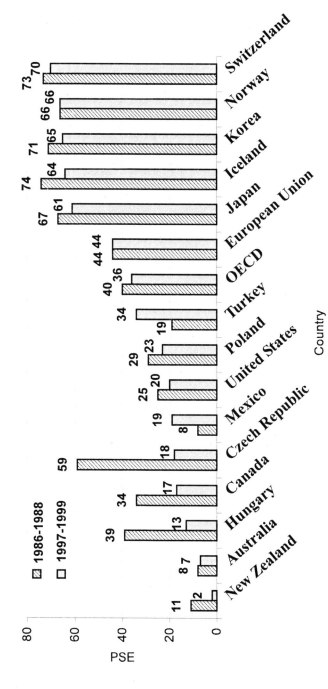

FIGURE 5.5. Producer support estimates (PSEs) of OECD countries, 1986-1988 and 1997-1999. (*Source:* OECD, 2001.)

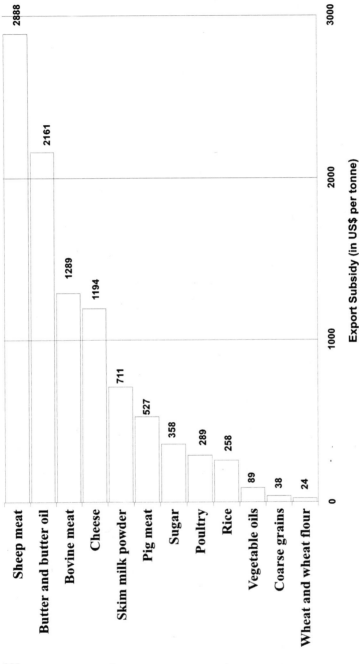

FIGURE 5.6. (Weighted) average export subsidy per tonne of notified subsidized exports selected commodities 1995-1999. (*Source:* Gulati and Narayanan, 2003.)

128

PDS, is highly inefficient. Leakages from the PDS are huge. As Dev and Ranade (1999) have shown, given the leakages in the system and lack of proper targeting, it takes 5.37 rupees to transfer one rupee of income to poor people through the PDS. Meanwhile, food subsidies have reached unsustainable proportions estimated at 136.7 billion rupees in 2001-2002. With a low off-take and increasing procurement, the food subsidy consists largely of a subsidy for the maintenance of buffer stocks, which for 2001-2002 is estimated to be 41.6 percent of the food subsidy (as compared with 12.5 percent in 1997-1998). Although there has been a steady increase in procurement due to large increases in the minimum support price (MSP), the take off has been low due to a small differential between the market price and that of the PDS. Also, the geographical or spatial coverage of the PDS has not been satisfactory. It is imperative to address these issues first in the interest of food security.

There is also the very important question of domestic marketing restrictions. At present, state governments are constraining wholesale trade in food grains through various controls on trading and interstate movement under the Essential Commodities Act of 1955. Restrictions on private stockholding (through limits), movement of grains, and levies (on rice and sugar) all fragment the Indian market. As a result, grain flow from surplus to deficit areas is hindered and with the inefficacy of the PDS, particularly in areas where it is most needed, it is natural that food security becomes a major cause for concern.

ENSURING FOOD SECURITY: THE WAY FORWARD

Given these rather diverse issues that determine food security in a liberalized context, how can India ensure that its food security is safeguarded? What is required is a multipronged approach that works simultaneously on both fronts—international negotiations and domestic marketing reforms.

At the Negotiating Table

India has often voiced its food security concerns in international forums. For instance, in a statement that India submitted to the second special session of the Committee on Agriculture, June 2000, it said:

> For India, and this will come as no surprise to this Body, the concept of food security includes the ability and flexibility to provide support to agriculture in order to achieve self-sufficiency in food grains and in order to ensure that agriculture remains a viable proposition for mil-

lions of farmers. We believe this to be critical not only from our national interest but also that of other developing countries, for whom net-food imports are a matter of necessity. (WTO, 2000b, p. 4)

Similarly, a joint proposal that Cuba, the Dominican Republic, Honduras, Pakistan, Haiti, Nicaragua, Kenya, Uganda, Zimbabwe, Sri Lanka, and El Salvador submitted to the Committee on Agriculture, states:

> For reasons of national security, economic and political stability, S&D [special and differential treatment] provisions giving more flexibility in agricultural trade policies must therefore be allowed to developing countries. Key products, especially food staples, should be exempted from liberalization, and the domestic production capacity of developing countries must be encouraged and helped along to become more competitive, rather than destroyed on the basis of noncompetitiveness. Under GATT Article XXI, national security issues may be exempted from WTO trade disciplines. Food security is also inextricably connected to national security and political sovereignty. Chronic food insecurity puts national security in jeopardy by placing at risk, the health of a large number of people, and also because it incites internal turmoil and instability. (WTO, 2000a, p. 1)

Although these concerns are justified, it is difficult to find a rationale for India's demand for a "food security box." The fact is that a review of the AOA commitments and India's agricultural policy shows that so far, the AOA has imposed no constraint on India to meet its food security concerns. Public stockholding, domestic food aid, and other policies are all exempt as green box measures. Similarly, investment and input subsidies to low-income and resource-poor farmers have also been exempted from any reductions for developing countries. Moreover, bound rates have been fixed at adequately high levels and there is enough of a cushion in our tariffs for protection against import surges.

Under the circumstance, demanding a "food security box" as India has done, or a "development box," as some other developing countries have demanded, seems quite unnecessary. In fact, such proposals could well be counterproductive. They could reinforce demands from the highly protected developed countries, in Europe and East Asia, for the exemption of domestic support measures adopted for preserving the multifunctional role of agriculture. That would work against developing countries such as India since it would not only prevent them from gaining access to these markets but also place them in danger of being at the receiving end of highly subsidized exports from these countries and having their production base wiped out.

There has to be a fundamental shift in the approach that India and other developing countries adopt in the negotiations. They need to emphasize the need for special and differential treatment less and the need for equal treatment more. There was little sense in allowing the developing countries in the Uruguay Round to make reductions in tariffs at a lower rate and over a longer period while some developed countries retained the possibility of maintaining tariffs in multiples of 100 percent on key products. Similarly, while many developing countries were pinned down to the de minimis level of 10 percent of product-specific and non-product-specific support, in 1998 the EEC had a Total Aggregate Measurement of Support (Total AMS) of more than 20 percent. If direct payments under production-limiting programs, commonly known as blue box measures, are included in the calculation, as they should be on account of their high trade- and production-distorting effect, the Total AMS of the EEC in that year comes to more than 30 percent. With the blue box taken into account the product-specific support in the EEC was 88.7 percent for common wheat and 134.2 percent for rice in that year.

As for export subsidies, the significance of the exemption from reduction of certain practices pales in significance when compared to the flexibility that some developed countries have retained for subsidizing exports. During the period 1995-1997, the rate of per unit subsidization of the EEC was 145 percent for rice, 154 percent for sugar, 118 percent for butter, and 378 percent for pig meat. While the benefits given to the developing countries under the guise of special and differential treatment were cosmetic, the flexibility that some developed countries retained are of titanic proportions. In light of this, a more effective approach would be to design the rules for uniform application in such a way that they secure reductions commensurate with the level of distortions caused. This would lead to a worldwide production shift to relatively more efficient economies, benefiting countries such as India.

India must adopt a bold approach in the negotiations on agriculture not merely in seeking reductions in support and protection in the industrialized countries but in offering concessions itself as well. Toward this end, taking a proactive stance and aligning itself with countries such as those in the Cairns Group or China is essential. Such an approach is conceivable because the analysis of the AOA and its likely implications for food security clearly reveal that there is no danger to India's food security from the commitments India has undertaken. Some proposals India could put forward in the areas of market access, domestic support, and export subsidies include the following.

Market Access

- The adoption of a harmonization formula is the best way to obtain an across-the-board reduction in a manner that ensures that higher tariffs are subjected to deeper cuts and that tariff escalation is reduced. The main elements of the harmonization formula must be the reduction of bound levels of tariffs that are above 100 percent down to 100 percent, and the reduction of rates between 40 and 100 percent by one-third. Below 40 percent, reduction should be optional, subject not to a formula but based on a reciprocal arrangement worked out through negotiations. India must make its proposal credible by not at all seeking special and differential treatment on market access.
- Because the period of transition is likely to be long, seeking agreement on various methods of administration of tariff rate quotas (e.g., first come first served, allocation on historical shares, auctioning quotas, etc.) must remain an important objective of India. There would be justification also for seeking increases in the TRQs. Even more important from India's perspective would be to ensure that the TRQs are not reserved for exporting countries with a preferential relationship with importing countries, as is the case at present in the EEC. India must press for increases in TRQs, if not TRQs as a whole, to be allocated on a most-favored-nation basis. The objective of simplifying tariffs for achieving greater transparency also deserves to be pursued vigorously. If the demand that all tariffs must be expressed in only ad valorem terms looks unattainable, India must at least press for the elimination of compound and mixed tariffs.
- From the point of view of food security, India should also negotiate for the right to use special safeguards (SSGs), under which a country can use higher duties than even bound rates when there is a case of "import surge." Under the Uruguay Round on Agriculture, it has mainly been developed countries that have claimed the right to use SSGs. The proposal for eliminating special agricultural safeguards has rationale, but it has to be remembered that there is a trade-off between reductions in developed-country tariffs and maintaining this provision. It may be a better strategy for India to seek that all developing countries have access to this provision, particularly for the products for which international markets are highly distorted. In India special agricultural safeguards with the provision for an automatic increase in tariffs on the basis of certain triggers will be far simpler to administer than general safeguard measures that can be taken only on the basis of serious injury to domestic industry. In the situation in

which the developing countries take recourse to special safeguards, they must also have the ability to impose countervailing duties without the need to establish material injury. India should also ask for an automatic right to impose countervailing duties on subsidized imports.

- If the willingness to open our own market is not evident fairly early in the negotiating process neither the developed nor the developing countries will respond to our requests. The concept of nonreciprocity notwithstanding, in the real world of trade negotiations, concessions cannot be won without reciprocity. One of the first steps for India in reducing developed-country tariffs for products of interest to it is for it to make a conditional offer for reduction of the bound level of tariffs or at least give a broad indication of our willingness to do so. We have seen that there is a wide gap between the bound and applied rates for a large number of tariff lines. For many cereals and pulses a norm of 40 percent can be considered to be safe in the situation in which the developed countries agree to eliminate or reduce subsidies substantially. We have to remember that a major concern for India pursuant to the food security objective is to make at least the key staples in the diet of the population available to the people at affordable prices. In fact, the Chelliah Committee had recommended the elimination of import tariffs on cereals (Government of India, 1992-1993). Tariffs on these products must not therefore be set at a high level. In other products also there is considerable scope for narrowing the gap between the bound and applied levels. The conditional offer can be made definitive only if India's objectives in all areas of the negotiations are fulfilled. In some products, such as milk, milk products, and edible oils, India could retain higher tariffs to safeguard the industry when world prices decline drastically. Access to special agricultural safeguards would facilitate reduction of tariffs to moderate levels.

Domestic Support

- There should be only two categories of domestic support measures, those that cause no or minimal trade distortions and those that do not fall in this category. The exemption for the blue box measures must be withdrawn and also made subject to reduction commitments. In other words, these measures should also be brought within the purview of the amber box, as it has come to be known. Blue box measures are the source of very substantial economic distortions, and they need to be taken into account in the calculation of Total AMS and subjected to progressive reduction as well.

- The modality of making reduction commitments on the basis of a percentage cut in the bound level of Total AMS must be replaced by the requirement that the Total AMS must be reduced to a fixed percentage of the value of agricultural production. We suggest a figure of 10 percent.

- Although some flexibility can be retained for modulating the product-specific support, a ceiling for such support must be fixed for each product. We propose that the product-specific support must not exceed 30 percent of the value of production of the product concerned. In this scheme there should be no place for exemption for de minimis support. We propose that the final target of 10 percent for the reduced Total AMS and the ceiling of 30 percent on product-specific support apply to developing and developed countries alike. Developed countries with very high levels of Total AMS or product-specific support may have to be given a longer implementation period (special and differential treatment).

- A two-pronged approach is necessary to limit recourse to measures in the green box that could distort trade or hamper production. First, support through measures under the green box (excluding general services, public stockholding for food security purposes, and domestic food aid, which are widely regarded as non–trade distorting) must be capped at the monetary level existing at the commencement of negotiations in 2000. No reduction may need to be made, but no increase must be permitted either. Second, the criteria of these green box measures already provided for in the Agreement on Agriculture must be reviewed to see if additional criteria need to be stipulated to ensure that the distorting potential of these measures is reduced even further.

- Considering the way the exemption on input subsidies is formulated in Article 6 of the Agreement on Agriculture and the restrictive interpretation of exemptions that is generally made by the dispute settlement panels, it appears necessary to propose a clarification of the intent of the article. It should be proposed that where all farmers are eligible for input subsidies the proportion of the subsidies given to low-income or resource-poor farmers must be deducted from the computation of non-product-specific support. Such a clarification will give considerably greater flexibility to India on non-product-specific support.

Export Subsidies

India must be steadfast in supporting the elimination of export subsidies that are scheduled in the specific commitments and ask that no rollover facility be permitted. Other forms of export subsidization also need attention, as they provide a ready means for circumventing commitments made on explicit export subsidies. These include officially supported export credit, export credit guarantees, and insurance programs, which must ideally be finalized and multilaterally approved before the conclusion of the negotiations. These programs should be accessible only for exports to the net food-importing developing countries and the least-developed countries. It is also imperative to stop the misuse of food aid as an instrument of surplus disposal without hindering its legitimate use for humanitarian ends. For this purpose, it is suggested that a multilateral institution be given complete charge of food aid operations. The institution must receive cash donations from donor countries and make purchases of food from the international market to meet the demands that needy countries place on it. Important too is the cessation of export monopolies for state trading enterprises so that the private sector may also compete. Stricter rules also need to be framed for export taxes and restrictions to alleviate the food security concerns of countries that rely on imports. Seeking an indefinite extension of the provision that allowed developing countries to exclude certain export promotion measures from reduction commitments during the initial implementation period would undercut the proposal for eliminating export subsidies and is therefore not advisable. What India should support is a time-limited extension of the existing provision as the Cairns Group has proposed.

Behind-the-Border Reforms

Equally important are "behind-the-border reforms" of considerable scale and magnitude. Notwithstanding the consistency of the domestic support programs with the requirements of the AOA, many of these policies have been criticized for a number of reasons (see Gulati and Hoda, 2002). These are three of the elements mentioned previously: input subsidies, minimum support price, and domestic food aid. Domestic marketing constraints are far too many to provide the right kind of incentives to agricultural producers in India. All this requires broad-based domestic reform, which must entail the following fundamental changes:

- With regard to domestic marketing, the MSP must be divorced from the procurement price. The latter should essentially be market determined, while the former serves as insurance for farmers by covering their paid-out costs.

- The FCI's procurement operations need to be limited to maintaining minimum buffer stock requirements. This could be done by directly going to the market or by inviting tenders from grain companies. The idea is to encourage private-sector participation in grain trade rather than replacing it.
- The FCI should not be allowed to sell below cost in its open market operations, as this displaces private sector initiative and burdens the treasury with a considerable cost.
- Subsidized sales by the PDS need to be gradually replaced with food coupons to poor consumers, i.e., a switch from a price policy (in the form of subsidized food) to direct income support to the poor and needy. However, this would have to be accomplished over a period of time, and institutions would have to be created to make such a scheme operational and free of vested interests.
- In the meantime, the removal of other controls could be achieved swiftly. These measures, preferably, should aim to enable greater transparency and efficiency in the functioning of domestic grain markets, and can be realized through the abolition of the Essential Commodities Act and the levies and restrictions under it.
- Simultaneously, there must be massive investments in rural infrastructure (power, roads, irrigation canals, etc.) that would enable small and poor producers to gain access to inputs and markets.
- Regarding input subsidies, they have largely failed to achieve efficiency, equity, and financial and environmental sustainability. Major pricing and institutional reforms need to be undertaken.

Domestic marketing reforms should be undertaken to create one integrated food market in which interregional flows are not impeded. For this objective, apart from price reform, a crucial element is institutional reform in both procurement and distribution, which must be included to ensure better household food security. Furthermore, institutional reform that brings about transparency, financial autonomy, and accountability in these activities are necessary. The AOA imposes no restriction on these, and they would have high payoffs for Indian agriculture.

CONCLUSION

This chapter has sought to evaluate India's food security concerns in the context of the implementation of the WTO AOA. When India became a member, there was widespread fear that the integration of Indian agriculture into world markets would make it more vulnerable and erode the food security of the large number of poor. A review of the AOA, however, shows that it does not impinge per se on India's food policy or food security in any significant way. The fact is that of the six interrelated elements of India's food policy (namely, price support, buffer stocking, food subsidy, input subsidies, public investment, and government support to R&D and extension), four are fully permissible under the AOA. With respect to the other two, price support and input subsidies, the AOA allows the government enough room to provide the poor with an adequate level of protection. As far as trade policies are concerned, India has applied tariffs that are much lower than the bound rates. These bound rates themselves have been set at very high levels. Moreover, even the current tariff levels have enough cushion in them to protect against import surges. Given India's reasonable competitiveness and the adequate protection that tariffs offer, a deluge of imports of critical commodities on a regular basis seems unlikely. The problem areas as far as food security is concerned may well be that international markets for major commodities are so highly distorted, particularly through domestic and export subsidies, that opening up India's borders in haste could have an adverse impact. India must therefore focus on the multilateral negotiations in the ongoing Doha Round to secure a fair trading environment in agriculture. Besides, ensuring food security has much to do with domestic policy reform. So far, domestic marketing policies have been too restrictive and the institutions tackling procurement and distribution ineffective, if not dysfunctional. Substantial behind-the-border reforms are essential. Domestic marketing reforms should be undertaken so that there is one integrated market for food within India and restrictions do not prevent interregional flows. In this context, apart from price reform, a crucial element is institutional reform, in both procurement and distribution, which must be developed and undertaken to ensure better household food security. Proper targeting of food subsidies and the eventual move to a food stamp system should also be pursued.

NOTES

1. In India, for the year 1993-1994 for which comprehensive information on consumption patters is available, rural people spent 63 percent of their monthly per capita expenditure (MPCE) on food (24 percent on cereals alone), while urban people spend only 55 percent of ther MPCE on food (14 percent on cereals). For people in the lowest income bracket, however, the share of expenditure going to food can go up to as high as 75 percent in rural areas. These ratios, of course, have come down significantly over the past two decades. In 1972-1973, for example, an average In-

dian in rural areas spent 73 percent of MPCE on food (40.6 percent on cereals) compared to an average urbanite that spent 64.5 percent of MPCE on food (23 percent on cereals) (Bansil, 2000).

2. There were only 673 when the URA was signed in 1994. Some lines were split and some others combined since then, and 686 was the figure as of April 2001.

3. Zero-tariff bindings for some commodities, including rice, plums, fresh grapes, and dried skimmed milk, were committed in 1947 with the Geneva Protocol; maize, millet, and spelt were bound at zero at the Torquay Protocol, 1951, and sorghum at the Geneva Protocol in the Dillon Round, 1962.

4. Along with the PDS and food aid programs, in an attempt to spur employment in rural areas and to ensure food security, the central government has extended a nuiversal Food for Work program in August 2001, called the *Sampoorna Gramin Rozgar Yojna* (SGRY). Under this umbrella program (which envelops the ongoing *Employment Assurance Scheme* and *Jawahar Gram Samridhi Yojana*), 5 million tons of food grain worth INR 5 billion (approximately US$108.5 million) will be distributed to the states annually, free of cost. The plan is to implement the scheme on a cost-sharing ratio of 75:25 between the central government and the states. However, due to extant lack of necessary infrastructure, financial resources (and political will), as of September 2001, just over 10 percent of the total allocated food grain had been off-taken, by a few states.

5. The U.S. Congress authorized four consecutive emergency packages of market loss assistance payments during the years 1998-2001, amounting to a total of US$28 billion (Gulati and Hoda, 2002).

REFERENCES

Bansil, P.C. (2000). "Demand for Foodgrains by 2020 A.D." Paper presented at the National Seminar on Food Security in India: The Emerging Challenges in the Context of Economic Liberalization, organized by Centre for Economic and Social Studies, Hyderabad, India, March 25-27, 2000.

Dev, Mahendra and Ajit Ranade (1999). Persisting Poverty and Social Insecurity: A Selective Assessment. In Kirit Parikh (Ed.), *India Development Report 1999-2000* (pp. 49-67). New Delhi: Oxford University Press.

Government of India (1992-1993). *Tax Reform Committee*, Final Report (Chairman: Raja J. Chelliah). New Delhi: Ministry of Finance.

Gulati, Ashok and Anwarul Hoda (in press). *Negotiating Beyond Doha*. Delhi: Oxford University Press.

Gulati, Ashok and Sudha Narayanan (2003). *Subsidy Syndrome in Indian Agriculture*. New Delhi: Oxford University Press.

Organisation for Economic Co-operation and Development (OECD) (2001). *Agricultural Policies in OECD Countries: Monitoring and Evaluation 2000*. Paris: Author.

World Trade Organization (WTO) (1999). India: Quantitative Restrictions on Imports of Agricultural, Textile, and Industrial Products. AB-1999-3. WT/

DS90/AB/R. 23 August (99-3500). Available at <www.worldtradelaw.net/ reports/wtoab/india-qrs(ab).pdf>.

World Trade Organization (WTO) (2000a). Agreement on Agriculture: Special and Differential Treatment and a Development Box. Committee on Agriculture Special Session. G/AG/NG/W/13. 23 June (00-2616). Available at <www.wto.org/ english/tratop_e/agric_e/ngw13_e.doc>.

World Trade Organization (WTO) (2000b). *Second Special Session of the Committee on Agriculture, 29-30 June, 2000. Statement by India,* G/AG/NG/W/33, July 13.

Chapter 6

Trade Liberalization and Food Security in Bangladesh

Paul Dorosh
Quazi Shahabuddin

INTRODUCTION

In the past three decades, Bangladesh, as is true of other countries of South Asia, has succeeded in substantially increasing production of rice and wheat. A food-grain-deficit country from the early 1970s to the end of the 1990s, Bangladesh's net food-grain production exceeded the target level of 454 grams/ person/day in both 1999/2000 and 2000/2001 (Figure 6.1). Yet as recently as 1998/1999 the country imported more than 3 million tons of rice and wheat (nearly 15 percent of total consumption) as widespread floods caused substantial damage to rice crops. Moreover, with approximately half the population living below the poverty line, food security at the household level remains a major concern.

Agricultural research and extension, investments in irrigation, and liberalization of input markets facilitated wide adoption of Green Revolution technology (high-yielding variety seeds, fertilizer, and irrigation) that doubled food-grain production in Bangladesh since the early 1970s. A large expansion in the number of participants, the size of the market, investments in infrastructure (roads, bridges, electricity, and telecommunications), and a gradual easing of restrictions on private-sector trade (including lifting a ban on commercial bank credit for the food-grain trade) have resulted in a well-functioning private market (Ahmed, Haggblade, and Chowdhury, 2000).

We would like to thank Raisuddin Ahmed, Ruhul Amin, M. Abdul Aziz, Carlo del Ninno, Eugenio Díaz-Bonilla, Naser Farid, Nurul Islam, Shahjahan Miah, K. A. S. Murshid, S. R. Osmani, and Mahfoozur Rahman for their insights on Bangladesh rice markets and helpful comments on this and earlier manuscripts. Thanks also to USAID/ Dhaka who provided funding for food policy analysis under the Bangladesh Food Management and Research Support Project from 1997 to 2001.

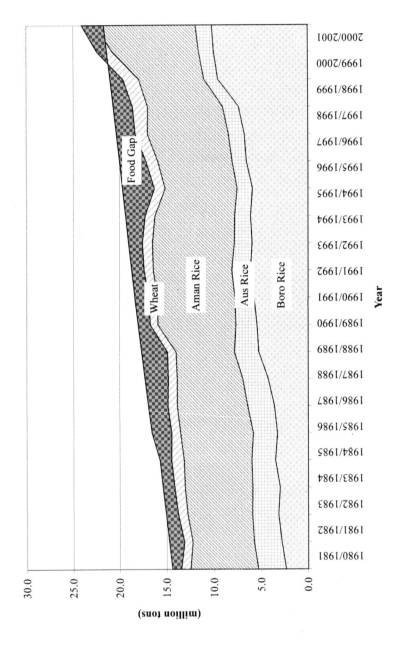

FIGURE 6.1. Net food-grain production and the food gap in Bangladesh, 1980/1981-2000/2001. (*Source:* Calculated from data from the Bangladesh Food Planning and Monitoring Unit [FPMU].)

Liberalization of international trade in food grains has also contributed to national food security, adding to domestic supplies and stabilizing markets after production shortfalls. Private-sector rice imports from India, made possible by broad trade liberalization in India and Bangladesh in the early 1990s, were especially important following major rice production shortfalls in Bangladesh in late 1997 and again in late 1998 (Dorosh, 2001). By mid-2001, however, after several excellent rice harvests, low producer prices rather than food shortages were the major concern, and Bangladesh sharply increased its import tariffs on rice to protect domestic producers.

This chapter examines the positive contribution of trade liberalization to short-run food security in Bangladesh in recent years, as well as the role of trade policy for protection of farmers from low-cost imports in years of good harvests. Section two presents an overview of the rice economy of Bangladesh, highlighting production patterns, the role of the public sector in rice markets, trade policy reforms, and trade flows. It also describes the surges in cross-border trade of rice between India and Bangladesh in 1998 and 1999, following production shortfalls in Bangladesh, and changes in import tariffs following good harvests in 2000 and 2001. The third section discusses Bangladesh's agricultural policy in the context of the World Trade Organization (WTO). The final section summarizes and discusses the implications for Bangladeshi food policy.

THE RICE ECONOMY OF BANGLADESH

Rice dominates food production and consumption in Bangladesh, accounting for 75.6 percent of the 2,123 calories/person/day consumed in Bangladesh from 1997 to 1999. Wheat accounted for 7.6 percent of calories in the same years. By comparison, rice and wheat accounted for 30.9 and 20.1 percent of the 2,434 calories/person/day consumed in India in the same period (Table 6.1) (FAO, 2002). Per capita rice consumption in Bangladesh was thus more than double that of India (161.1 and 75.8 kgs/person/year, respectively), though, since India's population was 7.4 times larger than that of Bangladesh, total rice consumption of Bangladesh was less than one-third that of India.

Bangladesh produces three rice crops per year: monsoon-season *aman* rice, typically transplanted in July or August and harvested in November or December; winter-season *boro* rice, transplanted in January and harvested in April or May, and a smaller *aus* season crop, harvested in July and August. Wheat is cultivated in the winter season. Most of the rapid expansion in Bangladeshi rice production over the past two decades is due to the rapid increase in irrigated winter-season (boro) rice production. The monsoon-

TABLE 6.1. The Bangladeshi and Indian food economies, 1997-1999 (average)

Comparison factor	Bangladesh	India
Population (million)	131.8	976.3
Rice production (1,000 tons)	20,518	85,990
Aman (Kharif) (1,000 tons) (1997/1998)	8,850	72,500
Boro/Aus (Rabi) (1,000 tons) (1997/1998)	10,012	11,000
Imports (1,000 metric tons)	1,169	41
Exports (1,000 metric tons)	0	3,315
Net imports (1,000 metric tons)	1,169	−3,274
Net imports/production (percent)	5.7	−3.8
Government rice stocks (1,000 tons)	537	12,027
Government rice stocks/production (percent)	2.6	14.0
Rice consumption (kg/person/year)	161.1	75.8
Calorie share (percent)	75.6	30.9
Wheat consumption (kg/person/year)	19.0	57.3
Calorie share (percent)	7.6	20.1

season (aman) rice crop now accounts for less than half of annual rice production in Bangladesh.

Food grain for the public food-grain distribution system in Bangladesh is procured through domestic purchases at a fixed procurement price, international commercial tenders, or food aid.[1] Prior to major reforms in the early 1990s, subsidized sale of grain through ration shops was the major distribution channel. Between 1988/1989 and 1990/1991, on average 612,000 tons of rice and wheat were sold through the rural rationing and the urban statutory rationing channels, 26.7 percent of total food-grain distribution (which averaged 2.294 million tons). Total sales channels, including open-market sales and other programs, accounted for 63.5 percent of distribution, with relief and food-for-work channels accounting for the other 36.5 percent of distribution in these years (Dorosh, 2001). Reforms in 1991/1992 and 1992/1993 closed the rural rationing and statutory rationing channels, in an effort to improve the targeting of food-grain distribution, as well as to reduce fiscal costs (Ahmed, Haggblade, and Chowdhury, 2000). As a result, both the percentage and total amount of food grain distributed through targeted and relief channels increased in the mid- to late 1990s, averaging 1.166 million tons per year from 1995/1996 to 1997/1998, 72.8 percent of the 1.603 million ton total annual average distribution during these three years.

Bangladesh has been a consistent net importer of rice throughout the past two decades, though substantial increases in rice production have reduced net imports over time. In the 1980s, rice imports (exclusively by the public sector) averaged 266,000 tons per year (Table 6.2). During the 1990s, rice imports fell to an average of 133,000 tons, though substantial year-to-year fluctuations have been noted.

Throughout the 1980s and early 1990s, Thailand was the major source of Bangladeshi rice imports. However, the 1994 liberalization that permitted private-sector imports coincided with India's rice trade liberalization and buildup of public rice stocks, dramatically changing the rice import trade. India, which enjoys the advantages of lower transport costs, reduced time of delivery (for private-sector imports), and the possibility of smaller import contracts delivered by truck, quickly replaced Thailand as the major source of imports of Bangladesh through 1999. In 1996/1997 and 1997/1998, 91.6 percent of Bangladeshi rice imports came from India, with the next largest import sources, Pakistan, Vietnam, and Thailand, each accounting for only 1 to 3 percent of the trade (Government of Bangladesh, 1996-2000).[2]

Private Sector Rice Trade Between India and Bangladesh

Soon after the liberalization of international trade of rice in 1994, Bangladesh imported substantial quantities of rice from India during a period of three successive poor Bangladeshi rice harvests beginning in December 1994. During this 17-month period, the private sector imported 1.127 mil-

TABLE 6.2. Bangladeshi food-grain trade, in thousands of tons, 1980/1981-1999/2000

Crop	Aid/Grant	Commercial	Private	Total
Rice				
1980/1981-1991/1992	73	153	—	226
1992/1993-1999/2000	12	145	676	833
1980/1981-1999/2000	49	150	270	469
Wheat				
1980/1981-1991/1992	1,191	376	—	1,567
1992/1993-1999/2000	780	190	400	1,370
1980/1981-1999/2000	1,027	302	160	1,489
Total				
1980/1981-1991/1992	1,264	529	—	1,793
1992/1993-1999/2000	792	335	1,075	2,202
1980/1981-1999/2000	1,075	451	430	1,956

Source: Based on data from Bangladesh Food Planning and Monitoring Unit.

lion metric tons (an average of 66,000 metric tons per month), in addition to 704,000 metric tons imported by the government. Bangladeshi rice import demand fell to near zero in 1996 and 1997, however, as food rice harvests led to a drop in domestic market prices below import parity levels (Figure 6.2). Following another poor aman rice harvest in Bangladesh in November and December 1997, however, rice prices rose sharply to import parity levels, providing incentives for large-scale private-sector imports.

In early 1998, the government of Bangladesh took deliberate steps to encourage private-sector imports of rice to stabilize domestic markets. A 2.5 percent tariff on rice imports was removed, government open-market rice sales were minimized, instructions were given to expedite clearance of rice imports through customs, and, despite pressure from some groups for more direct market interventions, antihoarding laws were not reimposed. As a result, during the first five months of 1998, the private sector imported 894,000 tons of rice from India (according to government of Bangladesh customs figures), mainly by truck and rail across land borders.

With the boro rice harvest in May 1998, the national average wholesale price of coarse HYV rice fell from a peak of 14.2 taka/kg in April (approximately US$0.31/kg in 1998) to 12.0 taka/kg in June ($\approx$ US$0.26/kg in 1998), and private imports slowed to 59,000 tons in June. Soon thereafter, however, major floods spread across much of the country and rice prices again rose to import parity. By continuing its policy of encouraging private-sector imports, the government enabled the private sector to import substantial quantities of rice and keep the domestic market price from rising above import parity levels. According to official government of Bangladesh estimates, more than 2000,000 tons of rice per month were imported from August 1998 to March 1999, with private rice imports reaching 288,000 tons in January and 345,000 tons in February 1999.[3]

The close correlation between estimated import parity prices and domestic wholesale prices shows that average profit margins for rice imports were in line with normal expected margins. Letters of credit data provide further evidence of the competitive nature of the rice import trade. From January through mid-September 1998, 793 traders opened letters of credit for rice imports, with the import market share of the ten largest traders equal to only 142,369 tons, 16 percent of the total (Dorosh, 2001).

In comparison with private-sector rice imports, government interventions in the domestic rice market were small, only 399,000 tons from July 1998 through April 1999. Private-sector rice imports, equal to 2.42 million tons in this period, were thus 6.1 times larger than government rice distribution. A total of 57.7 percent of rice distribution was targeted to flood-affected households through Vulnerable Group Feeding (VGF) (41.5 percent) and Gratuitous Relief (16.2 percent). Total rice distribution during these

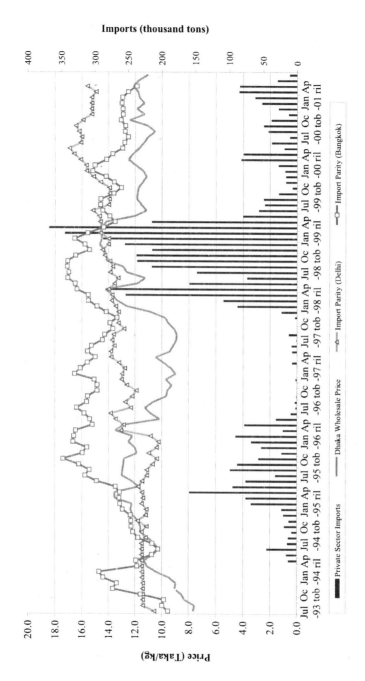

FIGURE 6.2. Rice prices and quantity of private rice imports in Bangladesh, 1993-2001. *Note*: US$1 ≈ 61 taka in 2004. (*Source*: Dorosh [2001] and authors' calculations.)

months, however was only slightly above the original target, in part because the ministry of food faced substantial difficulties in procuring rice through either domestic or international tenders.[4]

The private sector also imported substantial volumes of wheat following the 1998 flood, even though large amounts of wheat food aid flowed into Bangladesh and distribution through VGF and Food for Work was expanded. Private-sector wheat imports from July 1998 through February 1999 reached 624,000 tons, 435,000 tons more than in the same period in 1997/1998. Given the significant private-sector imports, it appears that food aid inflows did not provide a disincentive for domestic wheat producers.[5]

Alternate Estimates of the Volume of Rice Trade

The behavior of market prices in Bangladesh suggests that rice imports from India were a major source of supply. The volume of this rice trade remains somewhat uncertain, however, for two major reasons. First, Bangladeshi import data differ substantially from Indian export data. Bangladeshi customs data indicate that 3.172 million tons of rice were imported from India from April 1998 through March 1999, 2.827 million tons (89.1 percent) by the private sector. Indian data on the quantity of rice exports (calculated from customs data on value of exports and prices) show only 2.215 million tons, 958,000 tons (30.2 percent) less than the Bangladeshi customs figures.[6]

Second, calculations of total availability of rice in Bangladesh using official production and trade data are not consistent with market price movements and estimated rice demand. Using data from 1996/1997 as a base (a year in which private rice imports were negligible), domestic demand can be calculated using an estimated own-price elasticity of demand for rice (−0.15) and observed changes in market prices of rice. Calculations for the December 1997 through April 1999 period (consisting of a full year of harvests, plus the flood-damaged December 1998 aman crop) suggest that in response to the sharp increase in average real prices of rice in Bangladesh, per capita demand was 3.85 to 4.83 percent less in 1996/1997. Subtracting demand from apparent availability (the sum of net production, net public distribution, and private imports) gives an implicit stock change for these seventeen months of 3.546 million tons, 1.953 million tons greater than the typical stock change, estimated for the December 1996 to April 1997 period.

These calculations strongly suggest that total supply has been overestimated (or demand underestimated) in the December 1997 through May 1999 period. A large change in private stocks seems highly unlikely, given that the periods are defined to end just before major harvests. An overestimate of imports equal to the 1.083 million ton total discrepancy between Bangladeshi import and Indian export data could be a major cause of an overestimate of supply, accounting for about half (55.5 percent) of the calculated gap between supply and demand. Other factors may also explain part or all of the mismatch between calculated supply and demand, including an overestimate of production and an underestimate of consumption.[7]

Market Prices in the Absence of Private-Sector Imports

Though the quantity of private-sector imports from India is uncertain, it is clear that this trade substantially augmented Bangladeshi rice supplies in 1997/1998 and 1998/1999. One measure of the impact of this trade on national food security in Bangladesh is to compare actual prices and imports with estimates of prices and imports in the absence of private-sector imports from India. Given the average wholesale price of coarse rice in Dhaka of 13.3 taka/kg in 1998/1999, rice imports from December 1997 through November 1998 were 2.043 million tons (according to the Bangladeshi customs data). Had rice imports from India not been available, the next lowest-cost source for private importers would have been Thailand, for which the import parity price of 15 percent broken rice in Dhaka in the same period was 16.1 taka/kg. Given the 20.9 percent increase in import parity price, estimated rice demand would fall by between 4.2 and 6.3 percent, assuming an own-price elasticity of rice demand of –0.2 to –0.3. In this case, rice imports would decline by approximately 0.7 million to 1 million tons.[8]

Such an increase in rice prices would have had a major impact on rice consumption of poor households. Average daily per capita calorie consumption of a sample of poor flood-exposed households in rural Bangladesh in December 1998 was only 1,638 calories/day. Based on econometric estimates of calorie demand equations, with rice prices 21 percent higher, per capita consumption among the rural poor in 1998/1999 could have been 44 to 109 calories/day less than this very low consumption level (del Ninno, Dorosh, and Smith, 2003).

If private-sector imports were unavailable (or banned) from any source, then, with no change in government imports, total supply would have been 12.1 percent less (apart from private stock changes) and rice prices could have risen by 40 to 60 percent, to an average of between 18.7 taka/kg and 21.3 taka/kg.[9] Such an increase in the rice price level would likely have been unacceptable to the government of Bangladesh and public-sector imports would have been increased. But public-sector imports of a magnitude equal to private-sector flows would not have been feasible.

During the 1998 calendar year alone, private-sector imports, mainly from India, reached 2.26 million tons. Had the government of Bangladesh imported this grain itself, the average cost of the imported rice delivered to local delivery points would have been approximately 14.9 to 15.9 taka/kg, 1.0 to 2.0 taka/kg above the private-sector import costs, due to additional marketing costs totaling 50 to 100 million dollars. Furthermore, if the government received a net price of 11.5 taka/kg (equal to the Open Market sales [OMS] price of 12.0 taka/kg, less 0.5 taka/kg OMS dealer's commission), the total unit subsidy would have been 3.4 to 4.4 taka/kg, and the total fiscal cost would have been 160 to 210 million U.S. dollars.

The potential for price stabilization through import trade in times of rice shortages does not depend on imports from India. Alternative sources of rice imports are possible, as well, though imports by sea might involve fewer importers given economies of scale in shipping. Moreover, international market prices vary over time and other sources of supply can be less expensive than India, in spite of India's transport cost advantage. For example, export prices from Thailand and Vietnam were lower than those for similar grades of rice from India in 1999/2000 and 2000/2001 (Figure 6.2).[10]

Rice Prices and Imports, 1999/2000-2000/2001

In May 1999, the eighteen-month period of high rice prices (broken only by a brief two-month respite following the May 1998 boro rice harvest) finally came to an end with a record boro rice crop. Thereafter, excellent aman and boro harvests in both 1999/2000 and 2000/2001 kept domestic prices low; as Bangladeshi prices fell below import parity, private-sector rice imports essentially stopped.[11]

While Bangladesh enjoyed good harvests, India increasingly faced problems of plenty. Good weather and high procurement prices in the second half of the 1990s led to a sharp increase in government procurement, from an average of 17.2 million tons per year from 1980 to 1992 to an average of 26.6 million tons per year from 1993 to 2000 (Table 6.3). However, domes-

TABLE 6.3. India's average food-grain procurement, 1980/1981-1999/2000

	Average procurement	Average distribution	Average net procurement
1980-1992	17.2	16.5	0.7
1993-2000	26.6	16.2	10.4
1980-2000	20.8	16.4	4.4

Source: Government of India, Ministry of Finance, 2001, Table 1.19.

tic distribution remained at approximately the same levels, so that average net procurement rose from 0.7 million tons per year to 10.4 million tons per year. As a result, public food-grain stocks grew rapidly, from 11.8 million tons at the start of 1993 to 45.7 million tons at the start of 2001 (Figure 6.3). As stocks increased, the government of India took increasingly aggressive measures to promote exports. For example, in 2000/2001, state trading parastatals were permitted to buy wheat at the below-poverty-line (BPL) price for export (USDA, 2001).

In spite of good harvests and low prices in Bangladesh, imports of Indian subsidized food grain from their national stocks were potentially profitable, provided the quality of the food grain was approximately equivalent to Bangladeshi market standards. Relatively little rice flowed to Bangladesh from India through official channels in 2000/2001, as import parity prices of coarse rice from private markets remained above Bangladeshi domestic prices. However, when India lowered its APL sales price for "fine" rice in July 2001 from 11.3 rupees/kg to only 8.3 rupees/kg, the import parity price of this rice in Dhaka fell from 18.0 to 12.1 tk/kg, only 9 percent above the wholesale price of coarse rice in Dhaka. Soon thereafter, Bangladesh increased the rice import tariffs and taxes from 5 percent to 37.5 percent,[12] raising the import parity price 33 percent above domestic price levels (Figure 6.4, Table 6.4). As discussed in the next section, these sharp changes in tariffs were legal under current WTO rules. The threat of subsidized exports for domestic producer prices in Bangladesh raises important issues for the upcoming WTO negotiations, however.

BANGLADESHI AGRICULTURAL POLICY IN THE CONTEXT OF THE WTO

Under the 1995 Agreement on Agriculture (AOA) of the Uruguay Round GATT negotiations, least-developed countries (LDCs) such as Bangladesh face relatively few restrictions. Developed and developing countries are required to reduce tariffs, export subsidies, and domestic support to agriculture according to fixed time schedules; LDCs are exempt from these mandated reductions. Nonetheless, some parts of the agreement—the obligations to freeze domestic support to agriculture at the 1986-1988 level and to bind all tariffs—are binding on Bangladesh and other LDCs. In practice, though, these very loose restrictions have little impact on current agricultural policies.

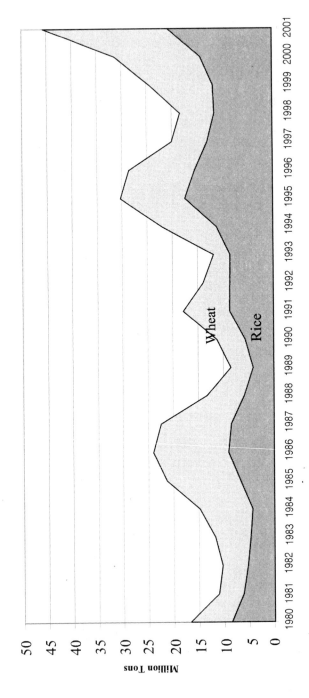

FIGURE 6.3. Central pool food stocks in India, 1980 to 2001, on January 1 of each year.
Note: Minimum January 1 stock norms: wheat 8.4 million tons; rice 8.4 million tons.

152

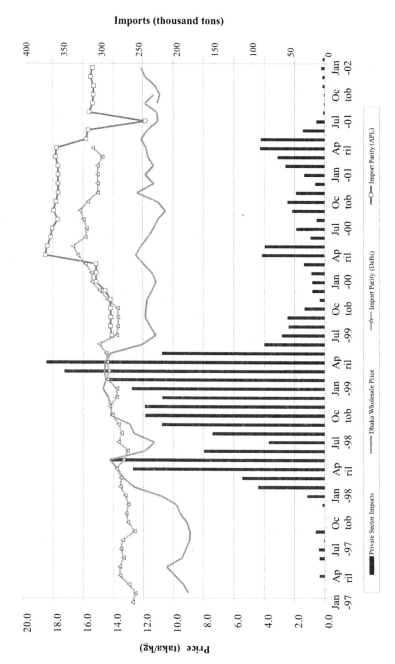

FIGURE 6.4. Rice prices and quantity of private rice imports in Bangladesh, 1997-2001. (*Source:* Dorosh [2001] and authors' calculations.)

TABLE 6.4. Import parity price of APL rice in Dhaka, 2001-2002

	APL (fine rice) price (Rs/kg)	APL import parity (without tax) Dhaka (tk/kg)	Dhaka wholesale coarse rice price (tk/kg)	Import duties[a] (percent)	APL import parity (including tax) Dhaka (tk/kg)	APL import parity/ wholesale price
April 2001	11.30	17.28	12.13	5.0	18.01	1.48
July 2001	8.30	11.50	11.07	5.0	12.08	1.09
August 2001	8.30	11.51	11.86	37.5	15.82	1.33
January 2002	8.30	11.34	12.35	37.5	15.60	1.26

Source: Bangladesh Food Planning and Monitoring Unit (FPMU) and authors' calculations.
[a]Not including advanced income tax of 3 percent and a license fee of 2.5 percent.

Import Tariffs and Restrictions

Bangladesh reported bound tariff rates at 200 percent for all agricultural products, except for 13 six-digit HS code items for which the bound rate is 50 percent.[13] Actual tariff rates are much lower, due in large part to major trade reforms undertaken as part of structural adjustment programs in the early 1990s. In 1995, when Bangladesh declared its bound tariff at 200 percent, the highest unweighted tariff was 66 percent and the import-weighted average tariff rate on all agricultural products was 25.9 percent, compared to 20.8 percent for nonagricultural products.

Economic reforms beginning in the late 1980s liberalized the external trade and foreign exchange regime, increased participation of the private sector, and improved export competitiveness. Tariff reforms, in particular, were wide-ranging, covering both tariff and nontariff barriers. Quantitative restrictions (QRs) on imports were dismantled and import procedures deregulated. As a result, Bangladesh economy became more open in terms of volume of trade, tariff and quantitative restrictions, import procedures, and foreign exchange regime.

As shown in Table 6.5, the import-weighted average nominal protection rate for all sectors declined from about 29 percent in 1991/1992 to around 22 percent in 1995/1996, and further to about 20 percent in 1998/1999. The average unweighted nominal protection rate dropped even more steeply, from about 67 percent in 1991/1992 to 24 percent in 1999/2000. In the case of agriculture, the average unweighted nominal protection rates fell from about 76 percent in 1991/1992 to around 31 percent in 1999/2000 (Table 6.6). For the agricultural products included under the AOA, the average effective rate of protection declined from about 70 percent in 1992/1993 to about 21 percent in 1999/2000. Also, the number of four-digit HS codes

TABLE 6.5. Nominal protection rates in Bangladesh (%)

	Protection rate: All sectors		Protection rate: Agriculture	
	Unweighted	Weighted	Unweighted	Weighted
1991/1992	67.4	28.7	76.5	33.6
1992/1993	55.4	28.2	61.9	31.6
1993/1994	42.4	29.3	45.8	33.5
1994/1995	31.3	26.1	41.0	19.2
1995/1996	27.1	22.3	35.1	15.3
1996/1997	26.9	24.0	34.2	20.4
1997/1998	28.2	23.9	36.2	18.0
1998/1999	27.2	20.3	33.9	12.4
1999/2000	24.1	—	30.9	—

Source: Dowlah (2000).

TABLE 6.6. Effective rate of protection for agricultural commodities under AOA (%)

	1992/ 1993	1993/ 1994	1994/ 1995	1995/ 1996	1996/ 1997	1997/ 1998	1998/ 1999	1999/ 2000
Rice	–8.00	0.90	2.40	–5.80	–5.70	–5.40	–5.30	–4.70
Wheat	–3.90	8.50	11.00	2.00	2.20	2.60	2.80	0.20
Average ERP	69.58	51.93	32.22	27.74	27.52	24.52	22.75	20.66

Source: Dowlah (2000).
Note: The average ERP includes non–food-grain agricultural products included under the Agreement on Agriculture, such as sugarcane, cotton, tobacco, potato, other vegetables, pulses, oilseeds, fruits, tea livestock, forestry, other fruits, edible oil, and sugar/gur.

banned and/or restricted due to trade or nontrade reasons was reduced from 315 in 1989/1990 to 124 during 1997-2002.

Because the current tariff rates of all agricultural commodities are quite below the bound tariff rate of 200 percent, Bangladesh would still retain flexibility in its tariff policies if it reduced its bound tariffs in the next round of WTO negotiations. The 37.5 percent tariff on rice (43 percent, including the license fee and advance income tax) would still be a viable policy even if bound rates were cut in half to 100 percent. In any case, if export subsidies are affecting domestic production, they can be subject to countervailing duties. Also, Bangladesh could apply antidumping measures to private-sector unfair trade practices or common safeguards measures to unusually low export prices, even if export subsidies and/or dumping practices are not involved.

Domestic Support Measures

Potentially more restrictive for Bangladeshi agricultural policy is the commitment to freeze domestic support to agriculture at the 1986-1988 level. The Agreement on Agriculture groups support policies into three boxes: amber, green, and blue. The amber box contains policies that substantially distort production and trade (input subsidies and price support). Policies with little or no impact on production and trade (expenditures on research, extension, pest and disease control, training, marketing, and infrastructure services) fall within the green box. The blue box includes policies that are not included in the other two boxes.

The total monetary value of trade distortions through domestic support policies to agriculture is quantified in the Aggregate Measure of Support (AMS). In the Agreement on Agriculture, the AMS is to be reduced over time (though LDCs are exempt). Only policies with substantial production and trade-distorting effects (amber box) are included in the AMS, though; policies included in the green and blue boxes are not included. Small subsidies are allowed: for developing countries and LDCs, support less than 10 percent of the farmgate value of production is not counted, according to the de minimis clause. Low-income or resource-poor countries such as Bangladesh also have access to a special category of production support policies under Special and Differential Treatment (SDT), including general investment subsidies and agricultural input subsidies.

The AOA stipulates that the developing countries will lower the base-level AMS by 13.3 percent by 2005. In its required schedule LXX submission, Bangladesh reported no import quotas, export subsidies, or domestic support (AMS) to agriculture in 1986-1988. Reported AMS was zero because various support measures were exempt from the calculation because of the de minimis rule or other provisions of the AOA. Bangladesh claimed exemptions for food security stocks, food aid, and natural disaster relief. Investment aid and input subsidies to resource-poor farmers, and administrative costs of seed distribution at government fixed prices were classified as non-trade-distorting green box measures. Using the criterion of government budgetary support to agricultural inputs, the fertilizer subsidy was considered to be zero.[14] The government subsidy on seed production and distribution was also minimal and exceeded the de minimis limit only for maize. The budgetary support measures for food-grain price support (through the public procurement program) were also insignificant.

Thus, Bangladesh's support to agriculture largely falls under the green box, which includes research, extension, pest and disease control, training, marketing, infrastructure, and other services. For 1995/1996, the total support is estimated at tk 9.8 billion, equal to 3.3 percent of the total value of agricultural output (Table 6.7). With liberalization of the irrigation market,

TABLE 6.7. Green box support to Bangladeshi agriculture (million taka)

	Support services		Water resources development	Total	Total as a percent of agricultural output
	Crop specific	Non-specific			
1986/1987	421	2,731	776	3,928	1.8
1987/1988	307	2,606	1,369	4,282	2.0
1988/1989	539	3,972	2,413	6,924	3.1
1993/1994	776	5,304	2,195	8,275	3.4
1994/1995	756	5,420	2,150	8,326	2.9
1995/1996	714	5,709	3,387	9,810	3.3

Source: Asaduzzaman (1999).

the budgetary support to irrigation equipment has declined to about 1.5 percent of agricultural output, largely through subsidies to limited number deep tubewells (shallow tubewells do not receive any subsidy) and major surface water irrigation projects. Thus, under WTO rules, there is ample scope for Bangladesh to increase its domestic support to agriculture.

CONCLUSIONS AND POLICY IMPLICATIONS

The trade liberalization that permitted private rice and wheat imports in the early 1990s has enhanced national food security in Bangladesh by making possible timely private-sector imports following major production shortfalls. In late 1998 and again in the second half of 1999, Bangladesh domestic rice prices rose rapidly to levels equal to import parity with India, providing the financial incentive for several million metric tons of rice imports. By encouraging this trade, the government of Bangladesh was able to augment domestic rice supplies quickly and stabilize market prices.

The positive contribution of trade liberalization to short-run food security in Bangladesh in times of domestic production shortfalls, however, does not minimize the importance of increased agricultural productivity and rural economic growth to provide rural poor households with sufficient incomes to acquire food. Moreover, the same private-sector import trade that so greatly benefited consumers in 1998/1999 posed a threat to medium-term food security in 2001 by reducing already low producer prices in a year of good rice harvests. To protect Bangladeshi producers, the government of Bangladesh raised import tariffs on rice from 5 percent to 37.5 percent (43 percent, including advanced income taxes and license fees) in August 2001. This policy was successful in that private-sector rice imports since the import tariff increase essentially ceased for the rest of 2001. Nonetheless, the danger exists that this temporary measure could become a permanent policy, leading to major distortions in Bangladeshi agriculture even after rice import prices return to more normal levels.

Bangladesh–India rice trade in recent years highlights several important issues for trade policy and WTO negotiations. Under current WTO agreements, a country cannot go over the bound tariff, but it can change the level of applied tariffs as long as the ceiling is not exceeded. With rice import tariffs bound at 200 percent in 1994, the new applied rate of 43 percent was well within the bound rate. If under the new set of WTO agreements, maximum Bangladeshi import tariffs are reduced, however, then the option of raising import duties to protect farmers may be restricted, depending on the negotiated cuts in bound rates. In any case, if export subsidies are affecting domestic production, they can be subject to countervailing duties. Also, Bangladesh could apply antidumping measures to private-sector unfair trade practices or common safeguards measures to unusually low export prices, even if export subsidies and/or dumping practices are not involved.

Thus, the Bangladeshi experience shows that trade liberalization offers potential benefits for national food security by enabling a low-cost, rapid increase of food supplies to stabilize prices following domestic production shortfalls. A flexible trade policy may be needed, however, to protect producer interests and long-term food security, particularly in the face of export subsidies or steep declines in world prices in years of good domestic harvests. Appropriate trade policy, in itself, cannot solve the problems of dietary imbalance or inadequate access to food by the poor, but it can have major impacts on food availability and farmer incomes.

More generally, as South Asian nations liberalize their markets, each country's agricultural and trade policies (particularly those of India) will have significant impact on its neighbors. Given the potential impact on producers and consumers, policy analysis in each of these countries will thus require more attention to intra-regional trade. Erratic weather and frequent changes in neighboring country policies may require periodic adjustments in policy, but intraregional trade has the potential to increase food security and benefit all countries in the region.

NOTES

1. Local tenders have also been used in recent years, particularly when fixed-price procurement has failed to meet government targets.

2. Other factors contributed to India's increase in rice exports, including a 27 percent depreciation of the rupee in real terms (see Dorosh, 2001).

3. The extremely high figures for recorded rice imports in early 1999 may overstate actual rice imports. Other commodities may have been imported in place of rice using false documents to avoid import tariffs and other surcharges. Government of India data and analysis of rice demand in Bangladesh suggest that actual level of imports may have been only about two-thirds the official figures. A lower import figure does not alter the main conclusion that private-sector trade stabilized domestic rice prices at import parity levels (see Dorosh, 2001).

4. Relatively low rice stocks limited rice distribution, as problems related to the instability of prices and unreliability of suppliers constrained actual procurement of rice through commercial local and international tenders (del Ninno et al., 2001).

5. Domestic prices in this period were slightly below import parity of U.S. hard red winter wheat, but most wheat imported by the private sector in this period came from lower-cost suppliers (see del Ninno et al., 2001).

6. One possible explanation for the discrepancies in the data is that other commodities that faced import duties were falsely declared as rice, for which the import duty was zero.

7. Sensitivity analysis using other values of the own-price elasticity of demand result in estimates of the implicit stock change ranging from 920,000 to 2.270 million tons (see Dorosh, 2001).

8. This calculation assumes no problems with supply of imports from Thailand and a competitive import market involving fewer importers and larger shipments. See Dorosh (2001) for a discussion of implications of importing rice from Thailand.

9. In the absence of private-sector imports, domestic supply would have been 14.839 million tons, a 12.1 percent reduction in per capita supplies relative to the actual estimated levels. Assuming an elasticity of demand of -0.2 to -0.3, prices would need to rise by $12.1/0.3$ (40 percent) to $12.1/0.2$ (60 percent) to equalize market supply and demand.

10. Bangladeshi and Indian rice production are not highly correlated (the correlation coefficient of the error terms of linear time-trend regressions of rice production is 0.30 for Bangladesh and India total annual rice production), suggesting that India may be a fairly reliable source of rice supply (see Dorosh, 2001).

11. Small amounts of non-parboiled rice were imported in 2000/2001, mainly from Vietnam. Higher-quality (non-coarse) rice imports likely accounted for much of the rest (see Dorosh, 2001).

12. Including advanced income tax of 3 percent and a license fee of 2.5 percent, the total tariffs and fees were increased from 10.5 percent to 43.0 percent.

13. Bangladesh declared 200 percent bound rate of tariff for all agricultural goods except the following: live horse, live sheep, live fowls, frog legs, human hair, sweet potatoes, green tea, black tea, rice in the husk, canary seed, soyabean seeds, cotton seeds, and molasses.

14. Article 13 in Annex 3 of the Legal Text of UR Agreement stipulates that the value of such measures shall be estimated using the government budgetary outlays.

REFERENCES

Ahmed, Raisuddin, Steven Haggblade, and Tawfiq-e-Elahi Chowdhury. 2000. *Out of the shadow of famine: Evolving food markets and food policy in Bangladesh.* Washington, DC: International Food Policy Research Institute.

Asaduzzaman, M. 1999. The Uruguay Round, WTO rules and Bangladesh agriculture. Paper presented at the Round Table on the Consequences of the Uruguay Round Agreements for Bangladesh Agriculture, jointly sponsored by the Government of Bangladesh, UNDP, and FAO in Dhaka, July 1999.

del Ninno, Carlo, Paul Dorosh, and Lisa Smith. 2001. *Public policy, markets and household coping strategies in Bangladesh: Avoiding a food security crisis following the 1998 flood.* Washington, DC: International Food Policy Research Institute.

del Ninno, Carlo, Paul A. Dorash, and Lisa C. Smith. 2003. *Public policy, food markets, and household coping strategies in Bangladesh: Lessons from the 1998 floods.* Food Consumption and Nutrition Division Discussion Paper 156. Washington, DC: International Food Policy Research Institute.

del Ninno, Carlo, Paul Dorosh, Lisa Smith, and Dilip K. Roy. 2001. *The 1998 floods in Bangladesh: Disaster impact, household coping strategies and response.* IFPRI Research Report 122. Washington, DC: IFPRI.

Dorosh, Paul. 2001. Trade liberalization and national food security: Rice trade between Bangladesh and India. *World Development* 29:4, 673-689.

Dowlah, C.A.F. 2000. Agriculture and the new WTO Round: Economic analysis of interests and options for Bangladesh. Workshop on "A New WTO Round on Agriculture, SPS and the Environment: Capturing the Benefits for South Asia," jointly organized by the World Bank, UNCTAD, and the SAARC Secretariat. New Delhi, January 11-13, 2001.

Food and Agricultural Organization (FAO). 2002. *Food balance sheets,* available at <http://apps.fao.org>.

Government of Bangladesh, Bangladesh Bureau of Statistics. 1996-2000. *Foreign trade statistics of Bangladesh.* Dhaka, Bangladesh: Author.

Government of India, Ministry of Finance. 2001. Economic survey. Delhi.

USDA. (2001). India: Grain and feed annual. Global Agriculture Information Network, available at <http://www.fas.usda.gov>.

PART III:
TECHNOLOGY FOR FOOD SECURITY
IN SOUTH ASIA

Chapter 7

Technology Options for Achieving Food Security in South Asia

Panjab Singh

INTRODUCTION

A technological initiative for identifying strategies for ensuring food security in South Asia is very important to develop. The South Asian region is one of the most populous in the world and faces the problem of supporting the largest number of absolute poor and malnourished people in the world. Nearly half of the "absolute poor" of the world live in this region. In the 1980s and 1990s, there was a perceptible improvement in the eradication of absolute poverty in East Asia and Latin America. In China, the number of absolute poor declined from 360 million in 1990 to 213 in 1998. However, in South Asia, the number of absolute poor increased from 495 to 522 million in the same period. The problems are still unfolding. South Asian countries are experiencing a peculiar demographic trend. From high birth rates and mortality rates in the decades of the 1950s and 1960s, they are now faced with high birth rates and low mortality rates. As a result, the menace of population explosion in this region continues unabated. Ensuring food security is the major challenge before the national governments of the countries in this region.

The concern for ensuring food security is not new. In fact, ensuring adequate food supplies has been the cornerstone of the policies of the governments in this region during the past five decades. Particularly in India, enhanced food-grain production has been the focus of the Indian government since its independence. Efforts were made to enhance food-grain production to attain self-sufficiency in food. These efforts yielded results, leading to a remarkable increase in the production of food grains. However, it was soon realized that self-sufficiency in food grains was not enough to ensure food security. Poverty and food insecurity could coexist with surpluses in food grains. Thus the conceptual framework for ensuring food security

163

needs to be examined in detail, and deliberation among researchers from this standpoint will be extremely important for policymakers.

Technology has played a key role in augmenting production in agriculture. Mechanization has ensured timeliness of operations in agriculture and thus enhanced the capability of farmers to bring large areas under cultivation and also to increase cropping intensity. Along with the technical changes, developing countries embraced economic reforms in the 1980s and 1990s. Governments in the developed countries have followed a policy of massive protection to their agriculture. The decades of the 1970s, 1980s, and 1990s witnessed conflicts in international trade in agriculture. To resolve them the World Trade Organization (WTO) was created, but the WTO has its own set of rules. In particular, the TRIPS (trade-related aspects of international property rights) Agreement of the WTO has serious implications for access to technology. Concomitantly, economic reforms and structural transformation of the national economies have paved the way for an enhanced role for the private sector in technology generation, development, and transfer. The WTO and structural transformation through economic reforms were the major watersheds of the 1990s, and they have contributed to a paradigm shift in the approach to technology development and access. The technology is no longer as freely available as it was during the early years of the Green Revolution. Thus the discussion on technology is very important in the context of food security.

This chapter analyzes the issues with regard to food security and technology development and transfer. This chapter is divided into five parts. In the first part, an attempt has been made to understand the meaning of food security. In the second part, the challenges for ensuring food security are discussed. The next part presents the role of technology. In part four, impediments to technology transfer for food security are discussed. The final part presents a conclusion to the chapter.

THE MEANING OF FOOD SECURITY

For an appropriate assessment of the implications of technological development for food production, it is important to consider the term *food security*. The concept of food security has undergone many changes during the past five decades. In 1950s it was closely associated with the total availability of food in the economy. Domestic production performance was the major yardstick of total food availability. In the Indian context, the domestic production of food grain was the measure of food security. Because domestic production of food grain was subject to the vagaries of nature, shortfalls in production due to weather aberrations would create serious food crises in

India. The Indian government resorted to massive importation of food grains in the 1950s and 1960s. Consequently, self-sufficiency in food grain became the cornerstone of India's policy for ensuring food security. Thus, during the initial decade after the country's independence food security implied sufficiency in food grain through domestic production.

The Green Revolution enhanced the productive capacity of Indian agriculture. Since the late 1960s food-grain availability has increased. However, a harsh reality in the context of "food security" was realized in the 1970s—a reality that poverty could coexist amid plenty of food. In other words, poverty and food surplus could coexist in an economy if the poor did not have enough purchasing power. Academicians and international organizations analyzed this problem in detail. Equity considerations therefore became an integral part of food-security policy. The concept of food security changed from "self-sufficiency" in food to physical and economic access to food for all in the economy.

The experience of the 1970s and 1980s also revealed that steady increase in production was not consistent with environmental protection. The environmental capital associated with the production of the Green Revolution crops, soil, and water was under threat. The soil quality in many parts of India degraded and the water quality suffered losses. In addition, the aquifers were depleted and the recurrence of water pollution became quite common in many parts of India. It was soon realized that steady production performance through the Green Revolution was not sustainable. Sustainable use of productive resources became a key element of production policy. Thus ecological consideration became an integral part of the food security in the 1980s. A plea was made for a broader approach for food security. The concern for incorporating ecological and equity consideration was made in a categorical terms for food security by the World Commission on Environment and Development (WCED) (1987).

The WCED raised three major concerns for food security: economics, ecology, and equity. In 1987, the WCED, chaired by Mrs. Gro Harlan Brundtland, Prime Minister of Norway, stated the following:

> The challenge of increasing food production to keep pace with demand, while retaining essential ecological integrity of production systems is colossal both in its magnitude and complexity. But we have the knowledge we need to conserve our land and water resources. New technologies provide opportunities for increasing productivity while reducing pressure on resources. A new generation of farmers combines experience with education. With these resources at our command, we can meet the needs of human family. Standing in the way is the narrow focus of agricultural planning and policies. (WCED, 1987, p. 144)

In 1996, the food and Agriculture Organization (FAO) took a major initiative at the international level for articulating the problems of food insecurity through the World Food Summit (WFS). The action plan and commitments of the WFS are popularly known as the Rome Declaration (1996). Some of the points of the action plan of the Rome Declaration deserve to be mentioned in the context of food security:

1. The World Food Summit has reaffirmed the right of everyone to have access to safe and nutritious food consistent with the fundamental right of everyone to be free from hunger.
2. There will be political and national commitment to achieving food security for all. Nations will make an ongoing effort to eradicate poverty with a view to reducing the number of poor people by half by the year 2015.
3. More than 800 million people throughout the world and particularly in developing countries do not have enough food to meet their basic nutritional needs.
4. The problems of hunger and food insecurity have global dimensions and are likely to persist and even increase dramatically in some regions unless measures are taken.
5. Poverty is a major cause of food insecurity, and sustainable progress in poverty eradication is critical to improved access to food.
6. Increased food production, including that of staple foods, must be undertaken.
7. The role of women in ensuring food security should be recognized.
8. Attaining food security is a complex task for which the primary responsibility rests with individual governments.
9. Food should not be used as an instrument for political and economic pressure.
10. There is a need to adopt policies conducive to investment in human resource development, research, and infrastructure for achieving food security.
11. The summit agrees that trade is a key element in achieving food security.
12. The summit recognizes the fundamental role of farmers, fishers, indigenous people, and their communities.
13. The summit is determined to make efforts to mobilize and optimize the allocation and utilization of technical and financial resources including external debt relief for developing countries to reinforce national actions to implement sustainable food security policies.

14. For achieving food security, the summit has made seven commitments for itself, viz.,

 a. Creation of enabling condition for the eradication of poverty with full participation of women and men,

 b. Implementation of policies for ensuring physical and economic access to food by all,

 c. Encouragement for participatory and sustainable food production,

 d. Ensuring that food and agricultural trade policies are conducive to food security through a fair, market-oriented world trading system,

 e. Endeavor to ensure preparedness on the part of the governments to meet the food requirements in the events of natural calamities,

 f. Promote optimal allocation and use of public and private investment to foster human resources, sustainable food, agriculture, fisheries, and forestry systems, and rural development, and

 g. Implementation, monitoring, and follow up of the plan of action in cooperation with the international community.

Thus it would appear that the concept of food security has evolved from self-sufficiency in the 1950s and 1960s to physical and economic access to food and to nutrition security in the 1970s and 1980s. In the 1990s, after the Rome Declaration, the concept of food security became broad based and now connotes physical and economic access to food for all. It is also consistent with ecological and equity protection, including livelihood security of the poor. Three distinct concepts have emerged in regard to the food problem during the past five decades: (1) self-sufficiency in food, (2) food security, and (3) nutrition security.

In the 2002 World Food Summit, the 1996 plan of action was reaffirmed, but certain issues and challenges were highlighted. That the access to safe and nutritious food is a right was emphasized. A challenge raised was the decline of agricultural and rural development in the national budgets of developing countries. As a result of this trend, investments have been inadequate. The summit also stressed the problem of the lack of access to technology, the depletion of natural resources, particularly water, and a need to address the problems that rural women face. These obstacles and others, such as the lack of political will, it was announced, had made the progress toward reducing the number of hungry people to about 400 million by 2015 disappointingly slow. However, on a positive note, the idea of developing regional strategies to address issues such as agricultural trade facilitation, food safety measures, and hunger and poverty alleviation in rural areas was discussed and seen as one that should be pursued. The summit also pointed to the promising role that trade and biotechnology could play with the establishment of farmers' rights, the conservation and fair and sustainable use of plant genetic resources, and a better position for developing countries in the WTO.

CHALLENGES FOR ENSURING FOOD SECURITY

Ensuring food security is a challenging task. In the developing countries where food insecurity is most pronounced, population has been increasing. A strange demographic pattern has emerged in developing countries. Population growth was characterized by high birth rates and mortality rates—particularly in infant mortality. There has been a perceptible decline in the mortality rate, but birth rates remain high. Thus the population in developing countries has been increasing. This increase poses a serious challenge to food security, indicating that more food will have to be produced by the developing countries.

The best land in developing countries is under considerable pressure. A fair bit of industrialization has occurred in these countries, which has created pressure on land for the creation of shelter, factories, and infrastructure, such as roads, parks, and schools. The urban sprawl resulting from industrialization has taken away good arable land from cultivation in most parts of the developing countries.

These developments mean that increasing amounts of food will have to be produced from a shrinking land base. This problem was analyzed by the FAO in the 1980s. Three sources were identified for augmenting food availability: (1) increase in yield of crops, (2) increase in cropping intensity, and (3) cultivation on land reserves. For a scenario of the year 2000, the study revealed that about 63 percent increase in food production must come from yield increase, about 15 percent through increased cropping intensity, and 22 percent by bringing additional area under cultivation from the land resources. Most of the land reserves are located in Africa and Latin America. Thus for enhancing food-grain production, yield increases through varietal improvement are the main plank of the strategy.

But varietal improvement is not the end in itself. The Green Revolution experience reveals that a given level of productivity cannot be maintained for a long time; very soon fatigue in soils is noticed and more fertilizer, water, and chemicals are needed to maintain the same level of productivity. Intensive use of land for crop cultivation causes ecological damage; soil fertility is eroded and water polluted. Thus the strategy for food security should be such that a sustainable increase in production is possible.

Food security in developing countries is closely associated with developments in the agricultural sector of the developed countries. The developed countries, particularly the United States and those in the European Union, have resorted to subsidizing their agriculture to a high extent. Trade conflicts over this have taken place between the United States and the European Union in the past. These conflicts notwithstanding, the practice of subsidizing agriculture in the developed countries has put the developing countries at a great disadvantage. The prospect of food supplies from the developed countries at cheap rates creates an unfavorable political climate against agriculture in the developing countries. For example, in the Indian context, some hold a view that it is cheaper to import wheat from Australia to Chennai rather than transport the same wheat from Punjab to Chennai. Indian scientist M. S. Swaminathan holds an entirely different view on the subject. He says the import of food is closely linked with livelihood security. According to him, the import of one million tonnes of wheat may render one million families unemployed in India. Thus food security is closely linked with employment security. However, if every country adopts this approach, cuts back on food imports, and closes its borders, autarky may arise in most nations of the world and some countries will be pushed farther toward food insecurity (Stiglitz, 2002).

For solving the problems of trade conflicts, agriculture was included in the Uruguay Round (UR) of talks. A separate agreement, the Agreement on Agriculture (AOA), has been signed by the member countries of the World Trade Organization. The agreement, among other elements, envisaged a reduction in subsidies to a level of 5 percent of the value of agricultural output in the developed countries. However, this has not happened and experience reveals that many loopholes exist in the agreement, given the various clauses can be misused for the justification of subsidies. For example, the subsidies in the green box can be provided in an unrestricted manner to the agricultural sector in the developed countries in the name of environmental protection. Thus the external trade environment of agriculture has a significant bearing on the food security of developing countries.

It would thus appear that the food security of a developing country faces a number of challenges. Associated with the steps for ensuring food security are the problems of ecological damage, equity in physical and economic access to food, empowerment of women, employment security of small and marginal farmers and landless laborers, and the threat of cheap food supplies from the developed countries. These challenges must be effectively met.

THE ROLE OF TECHNOLOGY
IN ENSURING FOOD SECURITY

Technology has played a key role in enhancing agricultural production and thereby ensuring food security. Various disciplines of agriculture, agricultural engineering, and agricultural chemistry have played a role in enhancing agricultural production. Agricultural engineering has provided a wide array of agricultural equipment for tilling the land, pumping irrigation water from surface and subsurface regions, shaping the ponds, ridges, and furrows seeding plants, and applying fertilizers and spraying chemicals. Equipment has been designed to facilitate intercultural operations, harvest the crop, and grade the produce. Agricultural engineering has thus enhanced the capability of farmers to bring larger areas under cultivation in an effective manner. In addition, it has ensured timeliness of agricultural operations and enabled double- and multiple-cropping systems. It has, however, met with criticism. The consequential gains in employment through double or multiple cropping notwithstanding, it is alleged that mechanization displaces labor and causes soil erosion. Nevertheless, the fact is well established that mechanization has played a great role in augmenting food production throughout the world.

Agricultural chemistry has played a significant role in the determination of fertility losses in soils and the development of chemicals that can supplement soil fertility for a desired level of crop production. Various techniques and chemicals have also been developed for preventing the cropping environment from ecological damage. Similarly, the sciences dealing with plant pathology and entomology have enhanced the ability of farmers to control the incidence of disease and pests on their crops. Agronomy has facilitated effective management of production for all crops.

In augmenting food supplies, genetic improvement through plant breeding has also played a great role. The record of plant breeding in this context is extraordinarily impressive, especially since the 1960s. The Green Revolution resulted in a threefold increase in grain production from 1965 to 1975 in Punjab, India; a rice yield increase from 1.8 to 4.4 tonnes per hectare during the same period in Colombia; increased wheat production in Turkey from 7 million tonnes to 17 million tonnes between 1961 and 1977; raised maize yields in Mexico by 30 percent from 1968 to 1972; doubled rice yields and farmers' incomes between 1963 and 1970 in East Pakistan; and boosted rice output in the Philippines by 36 percent from 1973 to 1976. It would thus appear that the record of technology in augmenting food-grain production has been very impressive. Technology has demonstrated its capability for enhancing food production in almost all types of economies, be they developed or developing.

In recent years, new trends have emerged in technological developments. Biotechnology has emerged as a frontier area of research. Although we have known the importance of biotechnology for quite some time through its techniques, namely, tissue culture and fermentation technology, the breakthrough achieved through recombinant DNA technology is of considerable significance. Through this technique, it is possible to transmit genetic traits from one species to another. It offers the promise of developing transgenic varieties that are capable of withstanding biotic and abiotic stresses and, in many cases, of higher production performance. The advances achieved in this area during the mid-1970s have now spread all over the world and paved the way for multinational corporations (MNCs) to enter the seed business. Developed countries, particularly the United States, have gone far ahead in the adoption of transgenic varieties of cotton and oil seeds. Brazil and Argentina and even China have engaged in adoption of transgenic varieties of crops. In India, a beginning has been made, and recently the Genetic Engineering Approval Committee (GEAC) of the Ministry of Environment has granted approval for Bt Cotton for commercial production. Biotechnology thus offers considerable promise for enhancing food production.

In the area of livestock production, technology has played a great role in boosting milk as well as egg and meat production. In India a kind of revolution has taken place. The story of the White Revolution, bringing India to the status of number one milk producer in the world, is well-known. Thus, technology and its effective application are the answers for meeting the challenges of food security.

IMPEDIMENTS TO TECHNOLOGY TRANSFER

The contribution of technology to the growth of agriculture notwithstanding, new developments in trading arrangements are matters of concern in technology transfer. For solving the problems of trade conflicts within the developed countries a special Agreement on Agriculture was included in the GATT Agreement. Through this agreement, the WTO was seen as a vehicle for reducing the support to agriculture in developed countries. However, the WTO Agreement is also made up of another agreement, namely, the Agreement on Trade Related Aspects of Intellectual Property Rights and Trade in Counterfeits Goods, popularly known as the TRIPS Agreement. Under this agreement, arrangements for the protection of intellectual property of various types, including patents, copyrights, industrial designs, trade marks, geographical indications, layout designs of integrated circuits, and protection of undisclosed information, have been discussed. The agreement also includes a provision for the protection of intellectual property rights associated with agriculture. Member countries are required to have legal arrangements for the protection of newly developed varieties. With this caveat of TRIPS it is clear that agricultural technologies that were available in developing countries in an unrestricted manner will no longer be available freely. It may be recalled that the technologies of the Green Revolution were freely available to farmers. The farmers had rights to produce, save, and use the varieties of the Green Revolution in an unrestricted manner. It is because of this freedom that the process of adoption was quickened and the phenomenon came to be known as the Green Revolution. Thus new developments since the creation of WTO do pose a problem in technology transfer, and this may prove to be a threat for food security. India has taken a right step by incorporating a special provision for the protection of farmers' rights in its new legislation for the protection of plant varieties. However, the fact remains that technology transfer has now become a challenging task in the post-WTO era.

SUMMING UP

The conceptual framework of food security has undergone a significant change, particularly in developing countries, during the past five decades. It has evolved from self-sufficiency in food in the decades of the 1950s and 1960s, which essentially was a statistical concept. The need for providing economic and physical access was realized in the 1980s and thus the conceptual frame for food security comprises not only adequate availability of food but also the purchasing power of the poor. It was realized that greater purchasing power was necessary for the poor if everyone in the economy was to enjoy food security. The concept was further modified into the concept of nutritional security, which envisages physical and economic access to balanced nutrition and drinking water for all in the economy so as to enable each and every citizen to exploit his or her full potential. The Food and Agricultural Organization in 1996, through its World Food Summit or Rome Declaration, provided a comprehensive framework for the member countries for ensuring food security. The Rome Declaration envisaged ecological conservation, through the sustainable use of resources, and equity protection as important goals to achieve while production is increased. The role of women has also been recognized in the conceptual framework of food security. However, for achieving food security technology will play a major role, given the impressive record of technology in enhancing food-grain production during the past half century. It has been argued that the challenge of food security cannot be easily met unless research in biotechnology is promoted. It has also been put forth that although biotechnology can offer a major contribution the post-WTO scenario, technology will not be available in an unrestricted manner. Several impediments to technology transfer have arisen due to the provisions of the TRIPS agreement. Since the TRIPS Agreement and WTO are a reality with which developing countries have to exist, ways and means will have to be developed to ensure food security in spite of the problems in technology transfer. These issues need to be discussed further and concrete suggestions need to be made for tackling the problem of food security, which is so important for the entire human race.

REFERENCES

Declaration of the World Food Summit: Five years later. 2002. Rome, June 10-13.

Rome Declaration on World Food Security and World Food Summit Plan of Action. 1996. World Food Summit, Rome, November 13-17.

Stiglitz, J. E. L. 2002. *Globalization and Its Discontents.* New York: W. W. Norton and Co.

World Commission on Environment and Development. 1987. *Our Common Future (The Brundtland Report).* Oxford, UK: Oxford University Press.

Chapter 8

Future Challenges
for the Rural South Asian Economy

Peter Hazell

INTRODUCTION

South Asia has made impressive gains in agricultural growth, food security, and rural poverty reduction since the food crises years of the mid-1960s. Food-grain production has more than doubled since that time, and most South Asian countries now have adequate supplies of food to feed themselves. Although many South Asians still do not have an adequate diet, the average per capita availability of food grains has improved to acceptable levels, and the incidence of rural poverty has fallen from about two-thirds of the rural population to one-third today. But these favorable trends are now stalling and new approaches are needed if agriculture is to contribute to future national economic growth, employment creation, and poverty reduction.

This chapter focuses on five interlinked challenges for the future. The first concerns the recent stagnation in public investment in agriculture that threatens future productivity growth and poverty reduction in the rural sector. There is need to increase both the level and efficiency of public spending on agriculture and rural development. Second, there are new and favorable opportunities for market-driven growth in the agricultural sector with trade liberalization and increasing diversification of national diets. These opportunities need to be seized through more aggressive policy reforms. Third, the largest share of the rural poor now live in less-favored regions that have benefited relatively little from the Green Revolution. It will be particularly important to invest in these regions and to help them participate in new growth opportunities if remaining poverty and food security problems are to be eliminated. Fourth, past patterns of agricultural growth have been environmentally destructive, and there is need to redress this problem at national scales to sustain future productivity growth in agriculture. Fifth, the rural nonfarm must play an increasing role in creating productive employment and livelihood options for rural people as rural populations continue to grow in size, and this will also require appropriate policy reforms.

PUBLIC SPENDING ON AGRICULTURE

Government spending on agriculture and rural development has stagnated in most South Asian countries in recent years. In India, for example, the rate of growth in government "development" expenditure on agriculture, irrigation, transportation, power, and rural development grew at an average annual rate of 15.1 percent during the 1970s, by 5.1 percent in the 1980s, and by only 1.3 percent in the early 1990s (Fan, Hazell, and Thorat, 1999). Despite an increase in private investment in the early 1990s, little evidence suggests that it is substituting for public investment, either in its level or composition. Research shows strong links between government investments in the rural sector and agricultural growth and poverty reduction (e.g., Fan, Hazell, and Thorat, 1999). As such, cutting public expenditure today threatens to slow future growth and poverty reduction in rural areas. Compounding this problem is the fact that many South Asian governments waste a good deal of the resources they do spend on agriculture. They not only divert resources into unproductive subsidies but also fail to adequately prioritize their investments to achieve larger poverty and growth impacts.

Wasted Expenditures

Many governments pay far too much to provide basic services to agriculture (power, fertilizers, water, credit, etc.) and then charge farmers far too little, using subsidies to cover the difference. These subsidies have a number of serious consequences:

- They place a large and growing financial burden on the government. Many of these subsidies played a useful role in launching the Green Revolution in the late 1960s and helped ensure that small farmers gained as much as large farmers. But today the subsidies are largely unproductive and simply detract from the public resources that are available for investment in future agricultural growth. In India, for example, subsidies currently consume more than half of total government spending on agriculture (World Bank, 1999), and Gulati and Narayanan (2000) have estimated that the subsidies for power and fertilizer alone now cost the government about $6 billion per year (equivalent to about 2 percent of national GDP). But there are also large subsidies for water, credit, and the food distribution system; if these are added in, it seems likely that at least $10 billion is spent each year on unproductive subsidies. Enormous opportunities are available for doing much more with these resources.

- A highly subsidized agricultural support system fosters inefficiency in the supply of key inputs and services. In India, the fertilizer industry receives the lion's share of the subsidy, and production costs for urea are well in excess of the world price for the majority of firms in the industry (Gulati and Narayanan, 2000). The public supply systems for power and irrigation water are also notoriously inefficient throughout South Asia, a direct result of the perverse incentives that arise when the agencies responsible are almost fully financed by the government rather than by their clients.

- Because farmers pay too little for the inputs and services they receive, they have little incentive to use them carefully, leading to overuse and waste. This is not only economically inefficient but, as discussed later, also contributes to some of the environmental problems associated with agricultural growth (e.g., fertilizer and pesticide runoff into waterways and waterlogging of irrigated lands).

Suboptimal Priority Setting

IFPRI research in India, China, and other Asian countries shows considerable variation in the growth and poverty impacts of government investments in rural areas, both by type of investment and the type of region in which it is made (Fan, Hazell, and Haque, 2000; Fan, Zhang, and Zhang, 2002). These impacts also vary over time, with impacts often diminishing as the accumulated stock of some investments rise (e.g., irrigation). Rural roads, education, and agricultural research are particularly effective in promoting agricultural growth and reducing poverty (they are the dominant "win-win" investments), while investments in large-scale irrigation and rural development have only modest impacts today. Moreover, as discussed in a later section, many investments in less-favored areas also now have larger impacts per dollar spent than equivalent investments in irrigated and high-potential areas, largely because there have already been heavy investments in the better areas and diminishing returns have set in. These types of results provide useful guidance on how government investment priorities could be adjusted to achieve greater growth and poverty impacts per dollar spent.

In sum, there is considerable scope for using the resources that are available for agriculture and rural development more efficiently for achieving future growth and poverty reduction in rural South Asia. Such efficiency gains may in turn encourage policymakers to reinstate higher and more appropriate levels of growth in total investment.

FROM FOOD SECURITY
TO MARKET-DRIVEN GROWTH

South Asia now has plenty of food grains in aggregate, and with saturated world markets for these commodities, future agricultural growth will be constrained if countries do not move beyond their past concerns with national food self-sufficiency to better exploit their comparative advantage. There is need to keep an eye on the overall food-grains balance, but this seems unlikely to be a major problem within the next couple decades at least (Rosegrant et al., 2001; Bhalla, Hazell, and Kerr, 1999). Food security is primarily a distribution problem now, which requires a different solution than simply growing more food grains. It requires a more focused and targeted effort to raise the incomes of the poor, most of whom are rural and live in less-favored regions, often backward areas with limited agricultural potential and/or infrastructure and market access (Fan, Hazell, and Haque, 2000).

New growth opportunities for agriculture are arising from a number of sources:

- Changes in national diets. With the more rapid national economic growth achieved in recent years by several South Asian countries and the rising affluence of their middle classes, domestic demand for livestock products (especially milk and milk products), fruits, vegetables, flowers, and vegetable oils has shot up. This is creating new growth opportunities for farmers to diversify (even specialize) in higher-value products, especially those farmers who have ready access to markets, information, inputs, and so forth.
- The ongoing domestic policy reforms are also slowly opening up export markets for Asian farmers. This should enable more farmers to specialize in those crops in which they have comparative advantage and can best compete in the market. These opportunities should also improve if future rounds of trade negotiations (including those sponsored by the World Trade Organization) succeed in freeing up more agricultural markets around the world.
- Good opportunities also exist for generating greater value added through agroprocessing, particularly if the agroindustrial sector were also liberalized and became more competitive with imports. In India, for example, oilseed processing is highly protected at present, making domestic vegetable oils noncompetitive with imports. Producers can compete as growers of vegetable oils, but they are penalized when

competing in the vegetables oils market because their products have to be processed by a highly inefficient domestic industry (Gulati and Kelley, 2000).

⟨ Agricultural diversification and value-added activities could make important contributions to income increases in rural South Asia. But similar to the Green Revolution, such growth is likely to leave many poorer regions and poor people behind. Farmers will prosper most in those regions that can best compete in the market. Competitiveness requires investments in rural infrastructure and technology (roads, transport, electricity, improved varieties, disease control, etc.) and improvements in marketing and distribution systems for higher value, perishable foods (refrigeration, communications, food processing and storage, food safety regulations, etc).⟩

If poorer farmers and regions are to benefit from these new opportunities, then policymakers will have to assist them rather than leaving everything to market forces alone. Helping small-scale producers capture part of these growing markets will require that agricultural research systems give adequate attention to the problems of smaller farms and not just large. The private sector seems likely to play a greater role in undertaking the research needed for many higher-value products, but private research firms will be more attracted to the needs of larger farms than small and to regions with good infrastructure and market access. Public research institutions will need to play a key role in ensuring that small farmers and more remote regions do not get left out.

Smallholder farmers will also need to be organized more effectively for efficient marketing and input supply. Although smallholders are typically more efficient producers of many labor-intensive livestock and horticultural products, they are at a major disadvantage in the marketplace because of (1) poor information and marketing contacts and (2) their smaller volumes of trade (both inputs and outputs) that lead to less-favorable prices than larger-scale farmers. Contracting arrangements with wholesalers and retailers has proved useful in some contexts, but for the mass of smallholder farmers in Asia, cooperative marketing institutions probably offer the more realistic option. India's Operation Flood is a good example of what can be done. This project uses dairy cooperatives to collect, treat, and market milk collected from millions of small-scale producers, including landless laborers, women, and smallholder farms, many of whom produce only one or two liters per day. In 1996, Operation Flood reached 9.3 million farmers yet still accounted for only for 22 percent of all marketed milk in India (Candler and Kumar, 1998). The government assists the program through technical support (e.g., research and extension, veterinary services, and the regulation of milk quality), but otherwise the program is run by the cooperatives themselves with no direct financial support from the government.

DEVELOPMENT OF LESS-FAVORED REGIONS

More than two-thirds of South Asia's rural poor now live in less-favored regions that depend primarily on rainfed agriculture (almost 80 percent in India according to Fan, Hazell, and Haque, 2000). Spreading the benefits of new growth opportunities to less-favored areas will be key for eliminating remaining poverty and food insecurity, and this will require targeted policies and investments. These will need to include greater investment in research, infrastructure, and human capital to improve the ability of less-favored areas to compete in the marketplace. Policymakers have been reluctant to do this in the past, preferring to rely on the "trickle-down" benefits from investments in high-potential areas (i.e., increased employment and migration opportunities and cheaper food). But this approach has proved insufficient to resolve the problems of many less-favored areas; people are migrating to better areas and urban jobs, but they are also multiplying faster than they are leaving. Population densities are still increasing in many less-favored areas and seem likely to do so for at least a few more decades. Without adequate investments in basic infrastructure, technology, and human development, less-favored areas will lose out even further as agricultural markets become more liberalized and competitive. They will become victims of market liberalization and globalization, not beneficiaries, with worsening poverty and environmental degradation.

Does investing in less-favored areas have to mean less growth per dollar of investment compared to investing that money in high-potential areas? Few would dispute the possibility of achieving bigger direct reductions in poverty by investing in less-favored areas, but are there significant tradeoffs against long-term growth and poverty reduction? Recent IFPRI research on India says no. In fact, many investments in less-favored areas now offer a "win-win" strategy for India, giving more growth and less poverty (Fan, Hazell, and Haque, 2000). This is even true for investments in R&D, though not in the most difficult agroclimatic zones.

AGRICULTURE AND THE ENVIRONMENT

The Green Revolution played a key role in achieving national food security and reducing rural poverty. By raising yields, it also prevented having to make large increases in the total cultivated area, thereby helping to preserve remaining forests and avoiding rapid crop expansion into other environmentally fragile areas (e.g., hillsides and drylands). Even so, the Green Revolution was also environmentally destructive in many of the areas in which it occurred. The problems include salinization of some of the best-irrigated lands, fertilizer and pesticide contamination of waterways, pesticide poisoning, and declining water tables. The problems began in the 1970s and seem to be getting worse. Mounting evidence shows that yield growth in many of the intensively farmed areas has now peaked and in some cases is even declining (Rosegrant and Hazell, 2000). An ever-increasing number of people are arguing that Indian farmers should revert to the low external- input farming technologies of pre-Green Revolution days (e.g., Vandana, 1991). This would be disastrous for yields and food supplies and would destroy the environment on an even larger scale because of the need to rapidly expand the planted area.

There are realistic prospects for making modern technologies more environmentally benign and reversing resource degradation problems on a national scale (Pingali and Rosegrant, 2000), but it will take significant and determined action by the government. Needed actions include the following:

- Development and dissemination of technologies and natural resource management practices that are more environmentally sound than those currently used in many farmers fields. Some of these technologies already exist and include precision farming, crop diversification, ecological approaches to pest management, pest-resistant varieties, and improved water-management practices. The challenge is to get these technologies adopted more widely in farmers' fields. Managed properly, some of these technologies can even increase yields while they reduce environmental damage. Further agricultural research is needed to create additional technology options for farmers, and this should include interdisciplinary work on pest control, soil management, and crop diversification, but also use of modern biology to develop improved crop varieties that are even better suited to the stresses of intensive farming but with reduced dependence on chemicals (e.g., varieties that are more resistant to pest, disease, drought, and saline stresses). Agricultural research and extension systems will need to give much higher priority to sustainability problems than they have in the past.
- Reform of policies that create inappropriate incentives for farmers in the choice of technology and natural resource management practices. As mentioned previously, the large subsidies for water, power, fertil-

izer, and pesticides make these inputs too cheap and encourage excessive and wasteful use with dire environmental consequences. Pricing these inputs at their true cost would save the government much money while also improving their management. This would reduce environmental degradation and, in the case of scarce inputs such as water, lead to important efficiency gains. Improvements in land tenancy rules would also improve the incentives for many smaller farmers to take a longer-term view in their choice of technologies and management practices. Strengthening community rights and control over common property resources such as grazing areas, woodlots, and water resources could also improve incentives for their more careful and sustainable use. In addition, the farm credit system needs to offer more medium- and long-term loans for investment in the conservation and improvement of natural resources, especially for smallholders and female farmers.

- Reform of public institutions that manage water to improve the timing and amounts of water delivered relative to farmers' needs, and the maintenance of irrigation and drainage structures. When farmers have little control over the flow of water through their fields they have reduced capacity to prevent waterlogging or salinization of their land, or to use water more efficiently. Forestry departments also need to work more closely with local communities, devolving responsibilities where possible, to improve incentives for the sustainable management of public forest and grazing areas.

- Assist farmers in diversifying their cropping patterns to relieve the stress of intensive monocultures. Investments in marketing and information infrastructure, trade liberalization, and more flexible irrigation systems can increase opportunities for farmers to diversify. Unfortunately, the kinds of diversification that the market wants are not always consistent with the kinds of on-farm diversification that are needed for sound crop rotations.

- Resolve widespread "externality" problems that arise when all or part of the consequence of environmental degradation is borne by people other than the ones who cause the problem (e.g., pollution of waterways and siltation of dams because of soil erosion in watershed protection areas). Possible solutions include taxes on polluters and degraders, regulation, empowerment of local organizations, or appropriate changes in property rights. Effective enforcement of rules and regulations is much more difficult to achieve than the writing of the new laws that create them.

DEVELOPMENT OF THE RURAL
NONFARM ECONOMY

The rural nonfarm economy is an extremely important part of South Asia. It accounts for about 20 to 40 percent of total rural employment and for about 25 to 50 percent of total rural income (Rosegrant and Hazell, 2000). It is also an important source of income for women, small farmers, and landless workers, as well as for the urban poor in rural towns. These shares will need to become even larger in the future as the number of rural workers continues to grow and as farm sizes continue to shrink. At the same time, the kinds of jobs needed in the rural nonfarm sector must be productive and contribute to rising living standards.

Formal manufacturing accounts for only a small part of rural nonfarm employment in most Asian countries, typically less than 20 percent, despite the attention it has received from policymakers. The lion's share of rural nonfarm employment comes from service, trade, and household manufacturing activities. These are dominated by small, part-time, mostly family businesses that are highly labor intensive. These activities depend to a large extent on local and regional demand and thus rely on the production of regional tradables to generate their demand.

Agriculture has traditionally served as the primary "motor" for driving the rural nonfarm economy, a role that proved particularly strong during the Green Revolution era. Studies of rural areas in Asia during the 1970s and 1980s showed, for example, that each dollar increase in agricultural income led to an additional $0.5 to $1 of income generated in the local nonfarm economy (Rosegrant and Hazell, 2000). Agricultural growth must continue to serve as an important motor in the future, and this will be particularly important for many less-favored regions. In addition, trade liberalization and the growing forces of globalization can offer new opportunities for many rural regions to capitalize on outside sources of demand for rural nonfarm outputs. These include opportunities for manufacturing some types of intermediate products for urban industry (often under contract arrangements), agroprocessing, and tourism. But liberalization is a double-edged sword because it also "deprotects" much traditional rural nonfarm activity. Greater competition from outside forces changes in the structure and composition of rural nonfarm activity, and this can prove inimical to the welfare of the poor. This was particularly evident in India during the 1990s when market liberalization policies led to a significant short-term contraction in rural nonfarm employment (Bhalla, Hazell, and Kerr, 1999).

To promote growth of the rural nonfarm sector, policymakers need to:

- Give high priority to promoting agricultural growth, especially in regions that have limited opportunities for developing other motors of growth for the rural nonfarm economy.
- Develop rural infrastructure. Villages need to be connected to local towns so that agricultural inputs and outputs can flow freely and so that people can go shopping. Rural towns also need good infrastructure, especially roads, electricity, schools, sewers, water, and communications, in order to attract new firms and to grow. Because the transaction costs of a poor infrastructure are borne unequally by the poor, improvements help level the playing field for the most disadvantaged.
- Promote a legal and regulatory environment (e.g., to secure property rights and enforce contracts) that will help to promote trade, commerce, and manufacturing.
- Provide training in relevant technical, entrepreneurial, and management skills. The rural nonfarm economy provides much of its own training through apprenticeship schemes and on-the-job learning, but in an increasingly technical and communications-oriented world, specialized training schemes (e.g., computing, accounting) are needed, including training for women, who dominate many service and trading activities.
- Pursue industrialization policies that foster the development of all kinds of rural nonfarm firms and not just manufacturing. Despite the relatively small importance of manufacturing in rural employment, policymakers have been enamored of this sector; rural industrialization policies have showered manufacturing firms with all kinds of preferential tax, subsidy, licensing, and regulatory benefits, as well as with targeted and subsidized credit and technical assistance programs. Moreover, these policies have typically favored larger capital-intensive manufacturing firms (the lure of the shiny rice mill or shoe factory) and neglected small labor-intensive firms and informal household manufacturing activities. It is necessary to level the playing field, to revamp rural industrialization policies to (1) be more inclusive, perhaps by redefining them as "rural enterprise" policies, and (2) to remove all unnecessary subsidies and protective policies that prevent rural firms from becoming competitive in the marketplace.

To help ensure that growth of the rural nonfarm economy is pro-poor, it is necessary to offer training and financial services to all kinds of nonfarm businesses, including small, informal, part-time, home-based, and woman-headed establishments. Many of these establishments do not have access to credit, especially in their start-up phase, and rural financial markets need to be strengthened to ensure that all creditworthy firms have access to needed loans.

CONCLUSION

South Asia has made impressive gains in agricultural growth, food security, and rural poverty reduction since the crisis years of the 1960s. Agricultural growth continues to be critical for addressing the livelihood needs of large numbers of rural people, including most of the country's poor. But future growth will need to be different from that of the past. It will be driven less by growth in food-grains production and more by new growth opportunities for higher-value livestock, horticultural, and agroforestry products for the domestic market, by increased value-added opportunities in agro-industry, and by export opportunities. Moreover, if future agricultural growth is to benefit the poor, it will have to be more focused in rainfed areas than in the past, including many of the less-favored and backward regions that gained relatively little from the Green Revolution.

Future agricultural growth will also need to be more environmentally sustainable than in the past, with greater attention to the problems of intensive farming areas. This will require policy reforms to change incentives in favor of more sustainable technologies and natural resource management practices, as well as appropriate types of agricultural research.

Because rural populations will continue to grow in size for at least another two decades, it will also be important to promote growth of productive livelihood opportunities in the rural nonfarm economy. Agricultural growth will need to serve as a key motor of demand for the rural nonfarm economy in many regions, but trade liberalization and the growing forces of globalization will also offer new opportunities for many rural regions to capitalize on outside sources of demand for rural nonfarm outputs. Policy reforms are needed to create a more enabling environment for a broad range of rural nonfarm activity, not just manufacturing, and to assist the poor in accessing opportunities.

Meeting these challenges will require serious policy and institutional reforms, including the phasing out of input subsidies, trade liberalization (including removing trade protection for agroindustry), reform of public institutions serving agriculture, and increases in productive investment in agriculture and the rural sector. India has been flirting with some of these changes for over a decade but faces difficult problems with entrenched interests in the farm, agroindustry, banking, and public sectors that are politically very difficult to resolve at the present time. It may take the same kind of vision to surmount these problems and to rejuvenate the agricultural sector as it did to launch the Green Revolution some thirty-five years ago. It will take true political leadership.

REFERENCES

Bhalla, G. S., P. Hazell, and J. Kerr. 1999. Prospects for India's Cereal Supply and Demand to 2020. 2020 Vision Discussion Paper 29. Washington, DC: International Food Policy Research Institute.

Candler, W. and N. Kumar. 1998. *India: The Dairy Revolution.* Washington, DC: Operations Evaluation Department, World Bank.

Fan, S., P. Hazell, and T. Haque. 2000. Targeting Public Investments by Agro-Ecological Zone to Achieve Growth and Poverty Alleviation Goals. *Food Policy,* 25(4): 411-428.

Fan, Shenggen, Peter Hazell, and Sukhadeo Thorat. 1999. *Linkages Between Government Spending, Growth and Poverty in Rural India.* Research report 110. Washington, DC: International Food Policy Research Institute.

Fan, S., L. Zhang, and X. Zhang. 2002. Growth, Inequality, and Poverty in Rural China: The Role of Public Investments. Research Report 125. Washington, DC: International Food Policy Research Institute.

Gulati, Ashok and Tim Kelley. 2000. *Trade Liberalization and Indian Agriculture.* New Delhi: Oxford University Press.

Gulati, Ashok and Sudha Narayanan. 2000. Demystifying Fertilizer and Power Subsidies in India. *Economic and Political Weekly,* 35(10): 784-794.

Pingali, Prabhu L. and Mark W. Rosegrant. 2000. Intensive Food Systems in Asia: Can the Degradation Be Reversed? In *Tradeoffs or Synergies? Agricultural Intensification, Economic Development and the Environment,* edited by D. R. Lee and C. B. Barrett. Wallingford, UK: CABI Publishing, pp. 383-398.

Rosegrant, M. and P. Hazell. 2000. *Transforming the Rural Asian Economy: The Unfinished Revolution.* Hong Kong: Oxford University Press for the Asian Development Bank.

Rosegrant, M., M. Paisner, S. Meijer, and J. Witcover. 2001. 2010 Global Food Outlook: Trends, Alternatives, and Choices. Food Policy Report 30. Washington, DC: International Food Policy Research Institute.

Vandana, Shiva. 1991. The Green Revolution in the Punjab. *The Ecologist,* 21: 51-60.

World Bank. 1999. Toward Rural Development and Poverty Reduction. Paper presented at the NCAER-IEG-World Bank Conference on Reforms in the Agricultural Sector for Growth Efficiency, Equity and Sustainability, India Habitat Centre, New Delhi, April 15-16.

Chapter 9

Intellectual Property Rights in South Asia: Opportunities and Constraints for Technology Transfer

Anitha Ramanna
Sangeeta Udgaonkar

The extension of intellectual property rights (IPRs) to agriculture is one of the crucial areas determining the future of food security in developing countries. Extensive controversy exists regarding the impact of IPRs on agriculture. One school of thought argues that IPR protection, especially in the context of biotechnological innovations, would provide the means for increasing agricultural output, stimulating agricultural research, and producing technologies that could be utilized for improving nutrition. The opposing viewpoint asserts that increasing IPR protection in agriculture would increase the price of seeds and crops, deny local communities access to genetic resources and knowledge relating thereto, and restrict the traditional sovereignty of farmers over their seeds. The opportunities and constraints for transfer of technology arising from the extension of IPRs to agriculture must be closely evaluated to analyze the impact of the new IPR regime in developing countries.

An examination of the prospects for technology transfer must take into account the new type of IPR regulations being evolved in developing countries and must be understood in terms of the policy changes taking place. Countries in South Asia, along with many other developing countries, are in the midst of framing a new IPR model of protection for breeders, farmers, and traditional communities. These countries are undergoing a policy shift from focus on absorption and adaptation of technology to a system that promotes technology transfer through greater IPR protection. The policy shift is resulting in the emergence of new IPR regimes that not only protect inventions of large corporate breeders but also aim to protect farmers' rights and regulate access to genetic resources. India is one of the first countries in the world to have passed legislation granting both farmers and breeders rights. Bangladesh, Pakistan, and Sri Lanka are in various stages of formulating breeders' rights. Bangladesh is evolving legislation on community knowledge, and Nepal has drafted legislation known as "Genetic Resource Access and Benefit Sharing" that is ready to be tabled in Parliament (<www.grain.org> and personal communication with Dr. Druva Joshy). The impact of IPRs on technology transfer in these countries cannot be analyzed by applying the model found in advanced countries but must be based on an analysis of the new type of regime.

This chapter aims to point out the opportunities and constraints for technology transfer under the new IPR model currently emerging in South Asia. It attempts to understand the political economy and legal reform of the new regime that consists of twin protection systems: (1) protection for breeders and (2) protection for farmers and traditional knowledge. It focuses on the reasons for the policy change and attempts to evaluate the new legal regimes being evolved as a basis for determining the factors that would play a role in promoting or hindering technology transfer.

INTELLECTUAL PROPERTY RIGHTS AND TECHNOLOGY TRANSFER

Intellectual property rights are viewed as an important instrument for promoting innovation. At the heart of the Anglo-American IPR system is the belief that limited monopoly rights are necessary to promote creativity and innovation. Economists generally justify the need for intellectual property based on the theory of public goods. Public goods are distinguished from private goods in two ways: nonexclusion (once a good has been supplied, individuals can access it free of charge) and nonrivalry (the consumption of a good by one individual does not reduce the availability of the good to other individuals) (Panagariya, 1996). Such public goods create the "free rider" problem in which individuals can easily copy the information, discouraging industry to invest in such goods and leading to underproduction of innovative commodities. This logic has been extended to suggest that industry would not invest in countries where IPR protection was weak. Access to technology would be restricted for countries that did not strengthen intellectual property rights. Industrialized nations argue that IPRs should be respected so that private investors who take substantial risks in developing and commercializing new technologies can get fair returns from the innovations (Kabiraj, 1994). The implications of granting IPRs in developing nations on technology transfer have largely been evaluated based on this framework. Weak IPRs in the South, it has been pointed out, create a disincentive for Northern production of intellectual property and/or a disincentive to allow intellectual property flows to the South (Burch, Smith, and Wheatley, 2000).

The impact of IPRs on technology transfer in South Asia cannot be analyzed on the basis of the IPR model found in advanced countries. The implications for technology transfer in South Asia must be evaluated on the model that attempts to extend IPRs while protecting genetic resources, indigenous knowledge, and farmers' rights. The following diagram illustrates the logic of IPR protection in advanced countries:

IPR ───────▶ Innovation ───────▶ Technology and Development

Here, IPRs are seen as an important factor promoting innovation that leads to greater advancement of technology and development.

The developing nations are trying to establish a different type of system, as illustrated here:

In this new model, developing nations are trying to enhance IPRs while protecting genetic resources to promote innovation and uphold the rights of farmers/traditional communities. The aim is to gain technology but also ensure that biopiracy (utilization of resources in developing countries by advanced nations to create profitable products without compensation) does not occur.

To analyze the opportunities and constraints for technology transfer based on the second model, this chapter focuses on the political economy and legal regimes emerging in South Asia. The new type of regime must be understood in terms of policy change. Why and how did this model take shape? What are the new legal mechanisms being devised to implement this new model? The reasons for the policy change and the legal issues arising from this model are used as a basis for pointing out that both opportunities and constraints on technology transfer would arise from this regime.

"COMMON HERITAGE"
AND THE TECHNOLOGY ABSORPTION MODEL

For decades, developing countries have based their agricultural policies on the concept of common heritage and have attempted to focus on a framework of adaptation of technology to national needs. This policy structure must be understood before analyzing the new regimes emerging in South Asia. Genetic resources have generally been viewed as "common heritage," or freely available to all and owned by none. There has long been the universal consensus that the major food plants of the world are not owned by any one country and are a part of our heritage from the past (Wilkes, 1988). Intellectual property rights in agriculture were extremely limited in developing countries, including those in South Asia. As the majority of agricultural research was conducted by the public sector and there was a limited role of the private sector in this field, there was not much demand for IPRs in agriculture in these countries.

The model framed to ensure access to technology also focused on free flow of resources, rather than monopoly rights. Public-sector institutions in India and other South Asian countries have focused on building indigenous capabilities and disseminating them cheaply to industry rather than patenting inventions and exercising ownership rights in tune with government policy. This policy is illustrated in the Indian government's technology statement of 1983, which pointed to the need for development of indigenous technology and efficient absorption and adaptation of imported technology appropriate to national priorities and resources (Ganguli, 1998). Imitation and absorption of technology, sometimes through reverse engineering, formed part of the strategy. In agriculture, the model focused on free transfer of resources internationally and dissemination by national public sector institutions to farmers.

POLICY CHANGE

The common heritage framework and the technology absorption model are now being replaced with a new strategy. Developing countries are currently attempting to revise their policies by replacing the common heritage framework with sovereignty over genetic resources and are aiming to utilize IPRs to promote technology transfer. The new strategy adopts some aspects of the IPR framework in advanced countries but also aims to modify and revise this model by ensuring protection for traditional knowledge and farmers. This type of system is novel, and a focus on the reasons for its emergence provides significant insights into the opportunities and constraints arising for technology transfer.

It is important to analyze the factors that promoted developing nations to revise their policy framework regarding transfer of technology and evolve a new model. External and internal factors led to the policy change. The combination of pressures to protect breeders' rights from industry along with demands for protecting genetic resources from NGOs and other lobbies led to the new IPR model.

External Pressure

The demand for extending IPRs to agriculture in developing countries emerged under the Uruguay Round of GATT negotiations that led to the founding of the WTO. TRIPs, or the Trade-Related Intellectual Property Rights Agreement is the treaty on IPRs framed within the WTO. The United States, led by industry groups, promoted the need for some form of IPR protection for microorganisms, plants, and biotechnology products. The Intellectual Property Committee (IPC), one of the major industry lobbies on IPRs in the Uruguay Round, proposed in 1988 that protection should be afforded for both biotechnology processes and products, including micro-organisms, parts of micro-organisms (plasmids and other vectors) and plants (IPC, 1988). The text on intellectual property rights in agriculture in TRIPs, however, could not be easily negotiated, mainly due to differences between the United States and Europe. Several European nations favored a less stringent form of protection for plant varieties known as plant breeders' rights, while the United States promoted patents. In addition to these differences, the protest by developing nations against extending IPRs to agriculture led to enormous controversy. The resulting text of TRIPs [Article 27.3(b)] dealing with agriculture was not precisely defined. It stated,

> Members shall provide for the protection of plant varieties either by patents or by an effective *sui generis* system or by any combination thereof. The provisions of this subparagraph shall be reviewed four years after the entry into force of the WTO Agreement.

The term "sui generis" in the clause left enormous scope for interpretation on the extension of IPRs in agriculture.

To ensure compliance with TRIPs, various external trade pressures were placed on developing countries, including South Asian nations. India and Pakistan have been targeted by the United States for not increasing IPR protection under the U.S. Special 301 trade law. Special 301 is a specific U.S. trade measure established in 1988 to identify countries that provided weak IPR protection and enforce compliance. Under this law, the United States has annually produced a list of countries that violate IPR standards and has proposed specific trade threats on those countries. The 2002 list includes Pakistan and India. (While India is identified on the "priority watch list," Pakistan is placed on the "watch list" <www.ustr.gov>.)

Trade pressure to implement TRIPs has also been exerted on India and Pakistan through the Dispute Settlement Body of the World Trade Organization. In 1996, complaints were lodged with the WTO against India and Pakistan by the United States for not providing adequate patent protection in agricultural chemicals and pharmaceuticals (Burch, Smith and Wheatley, 2000). Bangladesh and Sri Lanka are facing bilateral pressure to raise IPR protection in agriculture. In the case of Bangladesh, it is in the form of a bilateral agreement being negotiated with the EU in 2001 which stipulates that Bangladesh must accede to UPOV by 2006 (GRAIN, 2001). UPOV is an international agreement on plant breeders' rights. The United States is attempting to ensure compliance from Sri Lanka through a treaty that requires Sri Lanka to provide patent protection to plants and animals. (GRAIN, 2001). Table 9.1 lists the deadline for compliance and the various external pressures exerted on South Asian countries.

Internal Demands

A strong NGO movement within developing countries opposed the implementation of TRIPs and the extension of monopoly rights to breeders. Their main arguments were that TRIPs protects only the rights of breeders and ignores the contribution of farmers and traditional communities. They pointed out that a regime granting only breeders' rights could lead to biopiracy. These lobbies placed enormous pressure not to comply with TRIPs as interpreted by advanced nations but to evolve a new mechanism to protect rights of farmers through interpreting the "sui generis" clause effectively. The stance of actors within developing countries were strongly influ-

TABLE 9.1. External pressure and TRIPs compliance in South Asia

Country	WTO member designation	Deadline for compliance with Article 27.3 (b) of TRIPs	External pressure
India	Developing country	January 2000	Special 301; WTO
Pakistan	Developing country	January 2000	Special 301; WTO
Bangladesh	LDC	January 2006	Bilateral Treaty-EU
Sri Lanka	Developing Country	January 2000	Bilateral Treaty-US
Nepal	LDC	January 2006	

Source: Compiled from <www.wto.org>; GRAIN, 2001, 2002, <www.grain.org>; and Burch, Smith, and Wheatley, 2000.

enced by international agreements that sought to protect the interests of developing countries vis-à-vis TRIPs such as the conclusion of the Convention on Biological Diversity and the movement for farmers' rights within the FAO. The CBD was concluded in 1992 and attempted to replace the common heritage on genetic resources with the principle of sovereign rights. Within the FAO, farmers' rights was promoted as a measure for recognizing the contribution of farmers. It was defined within the FAO as

> rights arising from the past, present and future contributions of farmers in conserving, improving and making available plant genetic resources, particularly those in centers of genetic diversity. These rights are vested in the international community as trustees for present and future generations. (FAO Resolution, 1989, quoted in de Sarkar, 1996)

NGOs and farmers' lobbies in many South Asian countries played an important role in shaping the debate. In India various NGOs and lobbies have been significant in outlining farmers' rights both domestically and internationally. In Pakistan over 30 NGOs under the Sustainable Agriculture Action Group are actively involved in ensuring that Pakistan does not design legislation based solely on UPOV (Rizvi, 2000). In Bangladesh, several NGOs have been participating in the move to draft legislation on genetic resources. The draft is the result of a large-scale participatory process (GRAIN, 1998). Table 9.2 illustrates the international agreements dealing with genetic resources and farmers rights to which various South Asian countries are members.

TABLE 9.2. Membership in international agreements on genetic resources and farmers' rights

Country	FAO	CBD	Adhered to FAO Undertaking
India	*	*	*
Pakistan	*	*	
Bangladesh	*	*	*
Nepal	*	*	*
Sri Lanka	*	*	*
Bhutan	*	*	

FAO: Food and Agriculture Organization;
CBD: Convention on Biological Diversity
Sources: Compiled from Web site of CBD; GRAIN and Kalpavriksh, 2002; and the International Treaty on Plant Genetic Resources: A Challenge for Asia.

Industry/NGO Clash

These internal and external pressures created a controversy on IPRs and agriculture in developing countries between industry and NGOs. The initial positions of both industry and NGOs on this issue were diametrically opposed. The NGOs promoted the view that there should be no patents on plants and opposed the implementation of TRIPs. The NGOs also demanded that TRIPs must be revised in relation to the Convention on Biological Diversity and argued for farmers' rights. The seed industry argued for protection for plants and agriculture in the form of plant breeders' rights as laid down in UPOV (ASSINEL, n.d.). They were opposed to the Convention on Biological Diversity initially and stated that it contradicts TRIPs. They also did not accept farmers' rights and the need for a funding mechanism to reward farmers. The positions of industry and NGOs also came to a head on the implementation of Article 27.3 (b) of TRIPs. While industry promoted the view that the effective "sui generis" system should conform to UPOV, NGOs did not feel UPOV should be taken as the standard for compliance.

Industry/NGO Changes

The positions of both actors were revised on the issue. NGOs in developing countries started adopting a more nuanced position than the earlier anti-TRIPs stance. They expanded the definition of farmers' rights to include (1) benefit sharing; (2) recognizing the farmer on par with the breeder and allowing farmers to register their varieties; (3) ensuring farmers rights to save, use, and sell seed; (4) prior informed consent; and (5) protection for traditional knowledge and communities. They also revised the interpretation of "sui generis" from UPOV to unique legislation that would enable protection for both breeders and farmers. NGOs started taking the position that breeders' rights could be granted if farmers' rights were protected and farmers were given due recognition as breeders.

Industry also began revising its stance. Earlier opposition to farmers' rights was revised to accept limited farmers' rights (ASSINEL, n.d.). Some industry groups conceded that farmers could have the right to limited sale of seed (exchange over the fence) and rights to save seeds. They also changed their positions with relation to the Convention on Biological Diversity and accepted that benefit sharing was important and that it could be done as a commercial venture through biodiversity prospecting agreements. Industry bodies tried to assert that TRIPs and CBD need not be contradictory agreements but could be implemented simultaneously (ICC, 1999). They also agreed to explore means for protecting traditional knowledge through the World Intellectual Property Organization (WIPO).

Industry and NGOs in developing countries are moving toward a compromise position. NGOs have accepted that rights of breeders can be granted if farmers' rights are also upheld. The definition of farmers' rights has also been extended to include the ability to register varieties. Industry has agreed for limited rights to sell seed and rights to register varieties for farmers if breeders' rights are granted. The compromise position illustrates the emergence of the new model of IPR protection in South Asia—one that aims to protect both breeders and farmers.

Policy Change and the New IPR Regimes

This compromise has led to the new model that developing countries are today attempting to evolve with relation to IPR and agriculture. The framework is being established to satisfy different interests groups, and therefore tries to grant protection for both breeders and farmers/communities. The regime is the result of pressures to recognize both the interests of industry through plant breeders' rights and farmers/communities through farmers' rights and protection of genetic resources/traditional knowledge. A regime that aims to protect the interests of various stakeholders creates both opportunities and constraints in terms of technology transfer. The new regime does provide incentives for innovation but also raises a number of barriers in terms of sharing of resources. Figure 9.1 illustrates the interests of various actors that led to the emergence of the new IPR model.

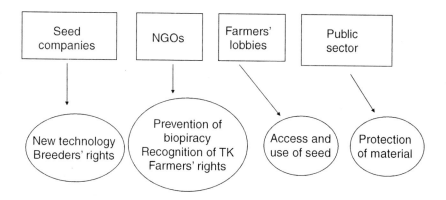

FIGURE 9.1. Stakeholder's interests in breeders' and farmers' rights.

INDIA'S POLICY: A CASE STUDY

A study of India provides an understanding of how the new model evolved as a compromise between various actors and how this has implications for transfer of technology. A focus on the evolution of the 2001 Protection of Plant Varieties and Farmers' Rights Act in India provides a clear understanding of the way that the compromise positions evolved on the issue. India is one of the first countries in the world to have instituted farmers' rights alongside breeders' rights in a legally binding framework. The act is the result of an intricate legal and political tussle between industry, NGOs, and the government in India. The fifth draft of the legislation proposed in the past eight years, the act represents a compromise evolved through a unique process of consultation involving various actors. The first draft of a bill to grant breeders' rights emerged in India in 1994. At this time, both industry and NGOs criticized the bill. The government then initiated a process to revise the bill and drafts were circulated in both 1996 and 1997 (Rangnekar, 2000). The main criticism of NGOs was that the farmers' rights was weak in the legislations. Although the various drafts had a section on farmers' rights, the NGOs were not satisfied because there was no provision for farmers' varieties to be registered. One of their main demands was that farmers' varieties must be treated on par with breeders' varieties and granted similar protection. In 1999 a new version of the bill was introduced in Parliament and sent to a joint committee for examination. This joint committee traveled across the country and gathered views of various NGOs, lobbies, scientists, industry, and institutions on the bill. It incorporated the demands of NGOs for a registration system for farmers' varieties into a revised bill that was submitted in 2000 (Government of India, 2000). In 2001, the bill was passed with the support from prominent NGOs and some seed firms. A crucial compromise occurred with significant agribusiness companies agreeing for farmers' rights and some NGOs accepting breeders' rights. Industry understood that the concept of farmers' rights as seen as an alternative IPR system actually reinforces their position on IPRs and enables them to gain plant breeders' rights in India. NGOs accepted the act as it provided for a mechanism for granting protection for farmers' varieties on par with breeders' varieties.

Recent developments in India point to the shape of this new model in developing countries. Patent applications in India are rising at an enormous rate since 1995. Figure 9.2 denotes patent applications in India from 1970 to 2000. The number of patent applications filed in the Indian patent office

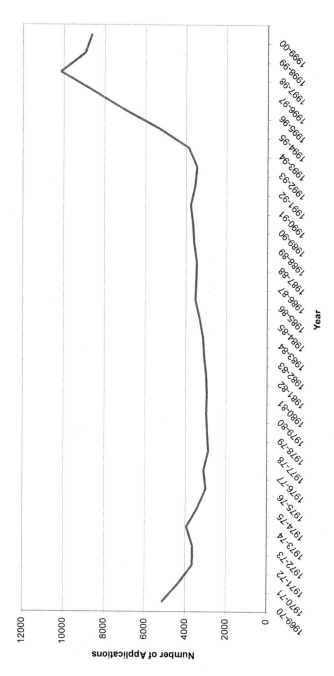

FIGURE 9.2. Trends in patent applications in India, 1970-1999.

has risen approximately 150 percent in 1997-1998 from 1993-1994, crossing the 10,000 mark for the first time in 1997-1998. In the post-1995 period patent applications are more than double than those of previous years. The rising number of patent applications indicates the possibilities for greater investment in India due to increasing IPR protection.

The recent patent applications in India also reveal the potential in agriculture. Table 9.3 illustrates patent applications related to several different agricultural products in India, while Table 9.4 shows patent applications related specifically to neem in India. The figures in Tables 9.3 and 9.4 reveal that there is interest in patent activity in agriculture in India and that the new IPR regime may provide incentives for innovation and technology transfer.

In addition to the rapid rise in patent applications in India, there is also an enormous movement toward documenting traditional knowledge and genetic resources. This is being carried on at the national, state, and local levels and by various actors, including the public sector, NGOs, and government agencies. Table 9.5 depicts some of the documentation activities taking shape in India.

The new IPR regime aims to satisfy the interests of various actors. It attempts to distribute ownership rights to various stakeholders. While one of the reasons for establishing IPRs is to promote transfer of technology through granting limited monopoly rights, the new IPR model attempts to extend the sovereign rights framework to genetic resources. Constraints to transfer of technology could occur with the wide scope of protection, lack of incentive to share resources, and the bureaucratic and legal hurdles arising from the new regime. A complex series of bargaining meetings with several actors may be required for development of new products under the scenario. Overlapping claims are a distinct possibility within this system. Potentially, a specific agricultural resource could be protected under a number of mechanisms, and various actors could claim ownership over a particular aspect relating to the resource. An actor interested in utilizing a resource may have to gain permission from several actors and negotiate on a number of levels which may not be practically feasible. The attempt to regulate access to genetic resources and recognize traditional knowledge involves the creation of new legal mechanisms, as there is no precedent for such legal structures. The following section examines the new legal framework being evolved by various South Asian countries. The concluding section points out the opportunities and constraints for transfer of technology.

TABLE 9.3. Patent applications related to specific agricultural products in India, January 1995 to June 2000

Commodity	Number of Applications
Rice	60
Cotton	51
Neem	47
Wheat	6
Sunflower	2
Tomato	4
Maize	4
Cauliflower	1
Sugarcane	14
Corn	5

Source: Compiled from TIFAC, 1998, updated 2001, Database on patent applications filed in India.

TABLE 9.4. Patent applications related to neem

Applicants	Number
Individual (Indian)	14
Conster Chemicals	8
Chief Controller Ministry of Defense, India	5
Vittal Mallya Scientific Research Foundation	2
CSIR	2
ICAR/IARI	2
Department of Chemistry	2
M/S Synt Drugs Pvt Ltd.	2
Neem Pharmaco	2
E I D Parry India Ltd.	2
Arpita Agro Products P Ltd.	1
M/S JB Chemicals Ltd.	1
Mysore Sandal Products	1
IIT	1
Dalmia Centre for Biotechnology	1
SPIC	1

Source: Compiled from TIFAC, 1998, updated 2001, Database on patent applications filed in India.

TABLE 9.5. Documentation of traditional knowledge

Activity and year launched	Agency	Description
National Biodiversity and Strategy Action Plan, 1999	Ministry of Environment and Forests, UNDP, Kalpraviksh and Biotech Consortium India Limited	Assessment and stock-taking of biodiversity-related information at national, local, and state levels
National Innovation Foundation, 2000	Department of Science and Technology and IIM, Ahmedabad	Register and support grassroots innovations
Biodiversity Plan	Government of Karnataka	State laws regarding biodiversity
Biodiversity Plan	Government of Kerala	State laws regarding biodiversity
Mission Mode Project on Collection, Documentation and Validation of Indigenous Technical Knowledge	Indian Council of Agricultural Research	Documentation and registration of traditional knowledge
Traditional Knowledge Digital Library	Council of Scientific and Industrial Research	International library on traditional knowledge
People's Biodiversity Registers, 1995	Foundation for Revitalization of Local Health Traditions	Records the status, uses, and management of living resources
Honeybee Network, 1996	Sristi	Document innovative practices of farmers/artisans
Database	Swaminathan Foundation	Document contributions of tribal groups for securing benefits
Documentation	Research Foundation, Green Foundation, Gene Campaign	Documenting and collecting traditional knowledge/resources
Village Registry, 1997	Pattuvam Village, Kerala	Produced a registry of genetic resources within their village and declared it their property

Source: Ramanna, 2003.

STRUCTURE OF THE NEW IPR REGIMES IN SOUTH ASIA

The legal structure of the new IPR regimes emerging in South Asia reflects the interests of the various stakeholders. The seed companies have promoted patents or plant breeders' rights as a means of protecting their interests, while farmers' lobbies and NGOs advocate farmers' rights and community rights. One of the worries of the latter stakeholders is that age-old farming practices and traditions might become illegal overnight, making farmers into criminals for doing what they have always done, namely saving, exchanging, and reusing seed. The public-sector groups, on the other hand, are interested in the protection of material of both the genetic diversity and of the new innovations that might come about by the application of technology. All these groups are also concerned about issues of bio-safety and the preservation of the environment, but the emphasis has been different—the seed companies generally emphasize sustainable use; the NGOs generally emphasize conservation. Along with these concerns, which have been reflected in the legal regimes, there has been a growing movement to protect and preserve traditional knowledge. This has found expression in the documentation and recording of traditional knowledge, done differently by different groups from different sectors (primarily the NGO and public sectors), in their areas of interest. Each country in the South Asian region has tried to balance these interests while developing its legislation.

It must be emphasized that almost all the legislations are in draft form and therefore subject to change. Although information is limited, certain draft bills or laws are under contemplation by the countries of the South Asian region. Pakistan has enacted the Plant Breeders' Rights Ordinance, 2000. Sri Lanka has proposed the Protection of New Plant Varieties (Breeders' Rights) (2001). Bangladesh has proposed two closely interrelated legislations: the Plant Varieties Act of Bangladesh (1998) and the Biodiversity and Community Knowledge Protection Act of Bangladesh (1998). India has passed the Protection of Plant Varieties and Farmers' Rights Act, 2001, and the Biological Diversity Bill, 2002. Nepal has drafted a bill on Genetic Resource Access and Benefit Sharing, the details of which are not available as yet. The analysis in this chapter is based on the available bills and acts mentioned here.

These legal regimes may be broadly grouped into two main categories: (1) those intended to deal with the protection of plant varieties and (2) those intended to deal with the protection of biological diversity. A glance at the provisions of the acts and bills in Table 9.6, however, indicates that they do not fit neatly into these two categories. Even within the plant variety protection legislation, considerable attention has been paid to the protection of genetic resources and traditional rights and ways of life. Thus legislation providing for plant breeders' rights may also have provisions relating to the rights of farmers or communities or the protection of the environment. It is therefore necessary to analyze the different legal mechanisms that are found in these legislations in the light of whether they seek to protect new plant varieties or to protect or regulate genetic resources and biological diversity (see Table 9.7). The countries are listed according to the degree of protection that they provide to the breeders of new plant varieties. Sri Lanka provides the maximum degree while Bangladesh provides the minimum degree of protection among these four countries.

As may be seen from Table 9.7, all the countries have some form of plant variety protection, generally termed "breeders' rights," although Bangladesh uses the term "new plant variety certificate." The legal mechanisms used to protect genetic resources, however, show great variation. Even where similar issues are involved, the different countries of South Asia can

TABLE 9.6. Legal regimes of South Asia

Country	Plant varieties	Biodiversity
India	Protection of Plant Varieties and Farmers' Rights Act, 2001	Biological Diversity Bill, 2000
Pakistan	Plant Breeders' Rights Ordinance, 2000	
Bangladesh	Plant Varieties Act, 1998	Biodiversity and Community Knowledge Protection Act, 1998
Sri Lanka	Protection of New Plant Varieties (Breeders' Rights), 2001	
Nepal		Genetic Resource Access and Benefit Sharing

Source: Based on information from <www.grain.org>.

TABLE 9.7. Different protection mechanisms

Country	Plant varieties	Genetic resources
Sri Lanka	Breeders' Rights	Experimental Breeding Exemption
Pakistan	Breeders' Rights	Farmers Exemption, Research Exemption.
India	Breeders' Rights	Farmers' Rights, Community Rights, Researchers' Rights, Biodiversity Authority, Recording Traditional Knowledge.
Bangladesh	New Plant Variety	Community Ownership, Prior Informed Consent, Citation of Award, National Biodiversity Information System.

Source: Based on information from <www.grain.org>.

be seen to have addressed them in slightly different ways, leading to a variation in the legal regimes even within this limited geographical area.

Plant Breeders' Rights

In Pakistan, the applicant for plant breeders' rights may be a national or resident of Pakistan, or a legal person whose registered office is in Pakistan. On the other hand, Sri Lanka's provisions in this regard are more lenient. An application may be filed by a national or resident of Sri Lanka, or by a national or resident of a convention country (not defined), or by a national or resident of a country that provides reciprocal rights to Sri Lanka. In the case of a person who has neither residence nor registered office in Sri Lanka, having an agent resident or with an office in Sri Lanka is sufficient to enable one to assert rights under the Sri Lankan bill.

In India, the Protection of Plant Varieties and Farmers Rights Act provides protection to the breeder upon registration of the variety. The breeder of a registered variety has the exclusive right to produce, sell, market, distribute, import, or export the variety, and in case of extant varieties the government shall be deemed to be the owner unless a breeder establishes his or her right thereto. A farmer also has the right to register his or her variety. It may be a new variety that the farmer has developed, in which case he or she may register it as a breeder. It may be an existing variety that the farmer has developed, in which case he or she may register it as a farmer. Varieties in the public domain may also be registered.

Bangladesh provides that a national of Bangladesh, a juristic person whose headquarters are in Bangladesh, or a person from a country recognizing the Bangladeshi legislation and providing reciprocity may apply for protection. However, anyone who has violated the Biodiversity and Community Knowledge Protection Act of Bangladesh or who belongs to a country that is not a signatory to the Convention on Biological Diversity is not eligible to apply for a new plant variety certificate.

Varieties Protected

In Pakistan, protection is accorded to varieties belonging to a species or genera notified by the government for protection, but the law shall not apply to plant varieties with terminator genes or other similar technology and to microorganisms. In Sri Lanka, notified species are protected, but the government may exclude varieties within a genera or species on the basis of the manner of reproduction or multiplication or by a certain end use. India, in addition to providing for the registration of varieties notified by the government, provides for the registration of extant varieties and farmers' varieties.

Article 7 of the Plant Varieties Act of Bangladesh states that

> breeding by itself . . . shall not be eligible for commercial privilege. To
> be eligible for consideration for such privilege the New Plant Variety
> must meet definite and useful needs of the people of Bangladesh. The
> National Biodiversity Authority shall take necessary measure to re-
> ject New Plant Varieties having no immediate, direct, and substantial
> benefit to the people of Bangladesh.

This is in complete contrast to the other legal regimes proposed for the region, which have no such criteria of usefulness to the people of the country.
The Bangladeshi legislation also has a unique provision titled the "Citation
of Award." This is a form of recognition given to an innovator who is entitled to but does not seek a commercial return from the new variety and is
willing to place it in the public domain. This award gives incentives in the
form of research grants, representation in the National Biodiversity Authority, and a financial award relating to the expense and time involved in research [Article 17(4)].

Research Exemption

Pakistan provides an exemption for scientific research. The exemptions
provided in Sri Lanka are for private, noncommercial use, and for use in experimental breeding of other varieties. India also provides that a researcher
has the right to use a registered variety for conducting experiments or research or as a source for creating new varieties. If the variety needs to be
used as a parental line for commercial production, however, the authoriza-
tion of the breeder is required. Bangladesh has provided an exemption for
personal and noncommercial use.

Farmers' Exemption

There are substantial differences among countries in this exemption to
breeders' rights. Pakistan provides for a farmers' exemption, but Sri Lanka
has no such provision. In fact the bill does not mention farmers at all. The
Indian legislation gives a greater emphasis to farmers' *rights*, as compared
to Pakistan and Sri Lanka, having a separate chapter (Chapter VI) titled
"Farmers Rights." It does not give farmers mere exemption from the plant
breeders' monopoly, but it also recognizes their contribution to the evolution of varieties and gives them positive rights over their varieties, including
the right to registration. Under Section 39(iv), the only restriction is that the

farmer cannot sell branded seed of a protected variety. Branded seed has been explained in the act to mean seed put in a container and labeled in a way that indicates that it is of a protected variety. But if the seed is not branded, the farmer would be entitled to sell it. Bangladesh's legislation makes it clear that the act does not restrict in any way the rights of farmers to access and as innovators. Moreover, it provides a list of positive rights guaranteed to the farmers, including the right to information (Article 22).

Community Rights

The right of the community to benefit sharing is mentioned in the Indian legislation. A community may apply for compensation for its contribution to the evolution of a variety (Section 41). This is implemented through the Protection of Plant Varieties and Farmers Rights Authority. The holder of a plant breeders' right is required to deposit any amount due toward benefit sharing with the National Gene Fund, which will then allot it to the community. Thus the benefit to the community is channeled through the government authority and is not a direct transaction between the breeder and the community. The laws framed by Bangladesh, on the other hand, are based squarely on the notion of the sovereign ownership of the people of Bangladesh over the genetic resources of the country. The community rights are paramount, and the new plant variety certificate is granted only after the breeder obtains prior informed consent from the community concerned.

Administrative Framework

Pakistan has chosen to establish a Plant Breeders' Rights Office as a branch of the existing Directorate General of the Federal Seed Certification and Registration Department for the Ministry of Food, Agriculture, and Livestock. Sri Lanka proposes to use the administrative framework of its National Intellectual Property Office.

India has provided for the establishment of a Protection of Plant Varieties and Farmers Rights Authority and a National Gene Fund under its plant variety legislation. Under its Biological Diversity Bill, 2000, it provides for setting up a national biodiversity authority, state biodiversity boards, and local biodiversity management committees. The permission of the biodiversity authority is required for the use of biodiversity found within the territory of India. It is to determine the method of benefit sharing (Section 21). Where any amount of money is ordered by way of benefit sharing, the authority may direct that it be deposited in the National Biodiversity Fund, or, under certain conditions, it may direct that the amount be paid directly [Section 21(3)]. Under Section 6 of the act, the permission of the Authority is necessary in relation to any application for any intellectual property right by whatever name called for any invention based on any research or information on a biological resource obtained from India. This condition does not apply, however, to any application for any rights under any law relating to the protection of plant varieties enacted by the Indian Parliament. But there is no similar overriding provision in relation to the benefit-sharing requirements, which could lead to bureaucratic hurdles in implementation.

Bangladesh has set up a single National Biodiversity Authority that will implement the Plant Varieties Act as well as the Biodiversity and Community Knowledge Protection Act.

Recording Traditional Knowledge

Both India and Bangladesh have also adopted measures for the recording of traditional knowledge. India has been recording it outside the legal structure, in the form of efforts by different sectors independent of one another. The result has been that these ongoing efforts have varied structures and are located in different parts of the country. Bangladesh, under its Biodiversity and Community Knowledge Protection Act, has made provision for a National Biodiversity Information System (NBIS), which will make a complete inventory of the biological wealth of the country as well as of the knowledge, practices, and cultural expression (Article 12).

Thus an analysis of these legal regimes shows that even within South Asian countries, the rights available with respect to plant varieties vary from country to country. This would have a direct impact on technology transfer and the flow of resources to and between these countries.

OPPORTUNITIES/CONSTRAINTS
FOR TRANSFER OF TECHNOLOGY

The new IPR model creates both opportunities and constraints for technology transfer. The various ways in which greater transfer of technology could take place under the new regime include the following:

1. *Incentive for breeders to introduce new varieties.* The provision of a monopoly right provides incentive for seed companies to introduce new varieties in developing countries, thus enabling transfer of new technology
2. *Incentive for R & D utilizing genetic resources.* The new regime also creates incentives for developing innovative products utilizing genetic resources. The incentive to invest more in R & D of these resources promotes the introduction of new technology.

The constraints for technology transfer could occur in the following ways:

1. *Scope of protection being too wide leading to depletion of resources in the public domain:* The new regime is one that aims to protect the rights of all the actors. As such it attempts to protect both high-technology innovations and traditional knowledge. The danger is that such a system may leave very little in the public domain. The creation of new innovations requires some amount of resources to be freely accessible, particularly in the field of agriculture. There would be a constraint on transfer of technology if the scope of protection becomes too broadly defined.

2. *Lack of incentives to share resources:* The new regime places great emphasis on assigning ownership rights but not enough attention is given to promoting sharing of resources. In the new scenario, there would be high transaction cost of negotiating agreements between various owners who would have different interests. Some actors may also want to keep the knowledge relating to the resource secret.

3. *Bureaucratic hurdles:* The new regime attempts to establish a number of authorities to implement the laws. Some of these authorities have overlapping jurisdictions, which is likely to lead to confusion and delays. Whether the new administrative structures will integrate smoothly with the existing framework remains to be seen.

4. *Costs of implementing the new regime:* It is clear that the new regime would require adequate financial resources. An attempt to document and recognize traditional knowledge would be costly and time consuming. The amount of money, time, and effort involved in the exercise to document traditional knowledge makes it important to make a serious attempt to ensure adequate returns from the knowledge. The information recorded has at least the value of the money spent in gathering it and some emphasis must be placed on recovering the costs (Udgaonkar, 2002). Documentation itself would not promote new technology or give adequate protection to genetic resources. In fact, the costs of the new regime may itself divert resources from other areas that may lead to greater technology transfer and needs to be carefully analyzed.

5. *Benefit sharing:* The difficulties of implementing the benefit sharing provisions found in the legislations could operate as a constraint on the transfer of technology.

6. *Legal constraints:* The legal interpretation of terms such as "breeder" and "farmer" varies within the South Asian region itself, leading to

confusion. South Asia being a closely knit region, varying forms of protection for the same genetic resource could hinder trade and transfer of technology between these countries. This could also act as a constraint on the flow of resources among these countries.

7. *Sharing between countries:* Genetic resources may be similar in several countries, leading to confusion over claims and access, especially in the context of traditional knowledge in South Asia. Competing rather than sharing would be emphasized under the new regime.

POLICY IMPLICATIONS

Various policy implications arise from the opportunities and constraints of the new regime. First, there is a need for a coordinated regulatory regime both within countries and between countries. This is necessary to streamline the process of protecting rights in genetic resources. Second, a public policy intervention is necessary to ensure access and sharing of resources. The new regime has emerged to satisfy the interests of various stakeholders, but it has not given enough attention to overall agricultural growth and development. It is important to focus on initiatives such as the FAO's new international treaty on plant genetic resources for food and agriculture that attempts to establish a multilateral system of exchange of resources (Ramanna, 2001). Third, it is important to ensure that necessary resources are in the public domain to facilitate access and sharing. Fourth, there is a need for coordinated legal regimes to overcome the legal issues that could cause constraints. ASEAN, the OAU, and various other regions are attempting to frame regional agreements on this issue. Finally, it is extremely important to ensure that the flow of resources between countries is facilitated, as agriculture is dependent on this exchange.

The political economy of the IPR regimes and the legal framework emerging in South Asia reveal the need for designing appropriate policy measures to take advantage of the opportunities and overcoming the constraints. Countries of the South Asian region must explore the space for development and growth under the new regime proactively. Technology transfer through IPRs will be dependent on removing the obstacles that may occur under the new IPR regime. Food security and agricultural development in South Asia will be dependent on the ability of these countries to establish an IPR regime that carefully balances the rights of various actors with incentives for exchange of resources between stakeholders.

REFERENCES

Burch, R.K., Smith, P.J.D., and Wheatley, W.P. 2000. Divergent Incentives to Protect Intellectual Property: A Political Economy Analysis of North-South Welfare. *Journal of World Intellectual Property*, 3(2), 169-196.

Commission on Intellectual and Industrial Property, International Chamber of Commerce. 1999. TRIPS and the Biodiversity Convention: What conflict? Policy Statement. Available at <http://www.iccwbo.org/home/statements_rules/statements/1999/trips_and_bio_convention.asp>.

de Sarkar, Dipankar. 1996. Farmers Rights a Reality. *Terra Viva*, 3(June 19).

Ganguli, Prabuddha. 1998. *Gearing Up for Patents: The Indian Scenario*. Hyderabad: Universities Press (India) Ltd.

Government of India. 2000. *Report of the Joint Committee on the Protection of Plant Varieties and Farmers' Rights Bill, 1999*. New Delhi, India: Author.

GRAIN. 1998. Biodiversity and Community Knowledge Protection Act of Bangladesh. Text proposed by the National Committee on Plant Genetic Resources, 29 September. Available at <http://www.grain.org/brl_files/bangladesh-comrights-1998-en.doc>.

GRAIN and Kalpavriksh. 2002. Traditional knowledge and biodiversity in Asia-Pacific: Problems of piracy and protection. Available at <http://www.grain.org/publications/tk-asia-2002-en.cfm>.

Intellectual Property Committee, Keidanren, UNICE. 1988. *Basic Framework of GATT Provisions on Intellectual Property: Statement of Views of European, Japanese and United States Business Communities*, June.

International Association of Plant Breeders for the Protection of Plant Varieties (ASSINEL). n.d. Are Farmers' Rights and Plant Breeders' Rights Compatible?, available at <www.worldseed.org>.

International Association of Plant Breeders for the Protection of Plant Varieties (ASSINEL). 1999. Fostering Plant Innovation: Assinel Brief on Review of TRIPs 27.3b, available at <www.worldseed.org>.

Kabiraj, Tarun. 1994. Intellectual Property Rights, TRIPs and Technology Transfer. Centre for Studies in Social Sciences, Calcutta, October.

Kuyek, Devlin, Biothai, GRAIN, Kilusang Magbubukid ng Pilipinas, Farmer Scientist Partnership for Development (MASIPAG), Pesticide Action Network Indonesia, Philippine Greens, Unnayan Bikalper Nitinirdharoni Gobeshona, Romeo Quijano, and Oscar B. Zamora. 2001. Intellectual property rights: Ultimate control of agricultural R&D in Asia. Briefing. Barcelona: GRAIN.

Panagariya, Arvind. 1996. Some Economic Aspects of TRIPs (Revised December 1995). Prepared for presentation at the Conference on Law and Economics Interface, January 11-13, New Delhi.

Ramanna, Anitha. 2001. India's Policy on IPRs and Agriculture: Relevance of FAO's New International Treaty. *Economic and Political Weekly*, Commentary, 36(51)(December 22), 4689-4692.

Ramanna, Anitha. 2003. *India's plant variety and farmers' rights legislation: Potential impact on stakeholder access to genetic resources*. Environment and Production Technology Division Discussion Paper 96. Washington, DC: International Food Policy Research Institute.

Rangnekar, Dwijen. 2000. A Comment on the Proposed Protection of Plant Varieties and Farmers' Rights Bill, 1999. Submitted to the Ministry of Agriculture, India.

Rizvi, Muddassir, Pakistan: Monsanto Fiddles with Plant Protection Act, available at <www.twnside.org.sg>.

TIFAC. 1998, updated 2001. Database on Patent Applications filed in India.

Udgaonkar, Sangeeta. 2002. The Recording of Traditional Knowledge: Will it prevent "bio-piracy"? *Current Science,* 82, 413-419.

Wilkes, Garrison H. 1988. Plant Genetic Resources Over Ten Thousand Years: From a Handful of Seed to the Crop-Specific Mega-Gene Banks (pp. 67-89). In Jack Kloppenburg, Jr., ed., *Seeds and Sovereignty: The Use and Control of Plant Genetic Resources.* Durham, NC: Duke University Press.

PART IV:
THE CHALLENGE OF WATER
FOR FOOD SECURITY IN SOUTH ASIA

Chapter 10

Emerging Water Issues in South Asia

Ruth S. Meinzen-Dick
Mark W. Rosegrant

INTRODUCTION

For thousands of years, irrigation has played a central role in South Asia. Rajas, colonial empires, and national governments have built massive infrastructures to store and convey water to fields to increase production where rainfall was insufficient or stabilize it against the fluctuations of monsoons. No less significant, farmers' groups have developed and maintained elaborate systems, ranging from the tanks of Sri Lanka and India to the hill systems of Nepal and Pakistan. In the past two decades, the availability of small mechanized pumps has led to massive farmer investment in groundwater irrigation from the arid zones of Pakistan, across the Indo-Gangetic plain and Nepali Terai, to the deltas of Bangladesh.

These investments in irrigation, together with modern crop varieties, have led to very substantial increases in agricultural production. Although millions of poor and food-insecure people remain in South Asia, this is not because of lack of food production at the regional or even national levels. Agricultural prices and concerns over food self-sufficiency have declined over the past two decades. Although the achievements of irrigation in ensuring food security and improving rural welfare have been impressive, past experiences also indicate problems and failures of irrigated agriculture, often related to environmental issues, including groundwater overdraft, water quality reduction, waterlogging, and salinization.

South Asian economies are shifting from agrarian and rural bases to industrial and urban concentrations. Consequently, demands for municipal and industrial (M&I) water uses are growing many times faster than demands for irrigation. At the same time, developed water resources are al-

We are grateful to Ximing Cai for his contribution to the water impact model results and summary.

most fully utilized in many places, and the financial, environmental, and political costs of developing new water-control systems are rising. The combination of rising demand and limited supplies is creating scarcity and competition among water uses, as well as users. These developments raise the question of whether water scarcity will constrain food production growth and what effects increasing municipal and industrial water demand will have on rural livelihoods. This chapter reviews trends in water use in South Asia as well as assesses the impact of water supply on future food production growth and the consequences for livelihoods. The analysis of food supply projections uses an integrated global water and food model and employs alternative scenario assessments to examine the degree to which water will constrain future food production and the implications of competition for water. The discussion of livelihood consequences draws upon research on the multiple uses of water in irrigation systems and a new study of intersectoral water transfers in South Asia.

TRENDS IN WATER USE IN SOUTH ASIA

South Asia has some of the oldest and the most extensive water-control structures in the world. The edict of the twelfth-century Sri Lankan King Parakramabahu I to "Let not even a small quantity of water that comes from the rain flow into the ocean without being made useful to man"[1] has resonated throughout the region through the beginning of the twenty-first century. Responses range from single wells to small-scale systems built by groups of farmers to the Indus Valley irrigation system, the largest contiguous irrigation system in the world. Most have been built to supply water for crops, but many have also met a wide range of water uses, including domestic use, livestock, and other enterprises (and, in the case of large dams, hydropower generation).

Although irrigation development has been occurring for thousands of years, the pace of development increased dramatically in the latter half of the twentieth century, as governments placed high priority on feeding their growing populations. Irrigation investments for large dams and canals were among the top five types of government expenditure in India, Pakistan, and Sri Lanka. Financial assistance through multilateral and bilateral aid and loans helped to fuel this trend through the 1980s. The increases in irrigation, together with modern varieties of rice and wheat, led to marked increases in food availability throughout the region. However, the rate of public investment in irrigation fell during the 1980s. Part of the reason was the rising cost of surface irrigation development, as the most favorable sites are used up, costs of land acquisition rise, and lags in project development add to the costs (Gulati, Svendsen, and Choudhary, 1995). Disillusionment with the poor performance of many irrigation systems, along with concerns over waterlogging and salinity, further eroded support for public irrigation investment, which increasingly has to compete for financial resources with education, health, roads, and other types of investment. As the output price of food grains fell in real terms (due, in large part, to the increases in production that irrigation made possible), it became increasingly difficult to justify irrigation investments in economic terms. Outright opposition to many large-scale irrigation systems in India has been voiced on the grounds that the reservoirs will cause environmental damage and dislocation of thousands of people from their land.

Government construction and management of irrigation systems eclipsed, and in many cases supplanted, collective farmer-managed irrigation systems through much of this period, but private investment was far from lacking. As mechanized pumps became available, farmers throughout the subcontinent made massive investments in private wells for groundwater irrigation, often subsidized by government loans and free or reduced-price rural electrification programs. For a time, increases in groundwater irrigation made up for declining rates of surface irrigation development, but without further recharge (much of which comes from surface irrigation systems), groundwater irrigation has already surpassed its sustainable limits in many parts of the Indo-Gangetic plain and peninsular India.

By the close of the twentith century, irrigation was having difficulty competing with other sectors for financial resources. Although irrigated area and water withdrawals continue to grow, the pace of irrigation development has slowed. The annual growth in irrigated area for South Asia as a whole dropped from 2.8 percent per year in 1975-1980, to 1.8 percent for 1980-1985, and 0.1 percent in 1985-1988 (Rosegrant and Svendsen, 1993).

As we enter the twenty-first century, irrigation will increasingly have difficulty competing with other sectors for the water resource itself. Irrigation currently accounts for over 90 percent of water withdrawals in India, Nepal, Pakistan, and Sri Lanka, and 86 percent in Bangladesh (Table 10.1). However, with the rapid urbanization and industrialization taking place, municipal and industrial demands for water are increasing at a much higher pace than demands for irrigation. With the limitations of available water (especially around most major cities) and the high economic, environmental, and political cost of developing new systems, these new demands compete with irrigation for the water presently used in agriculture. Infrastructure development and tapping of nonconventional water sources such as desalination and wastewater reuse have roles to play in the future in some regions, but it will be a reduced role compared to past trends, when dam building and expansion of irrigated area drove a rapid increase in irrigated area and crop yields.

TABLE 10.1. South Asian water use, mid-1990s

Country	Renewable freshwater m³/capita	Renewable freshwater km³	Total withdrawals (km³)	Sectoral withdrawals (% of total withdrawals) Domestic	Sectoral withdrawals (% of total withdrawals) Industry	Sectoral withdrawals (% of total withdrawals) Agriculture
Bangladesh	813	105.0	14.64	12	2	86
Bhutan	44,728	95.0	0.002	36	10	54
India	1,244	1,260.6	500.00	5	3	92
Nepal	8,282	198.2	29.00	1	0	99
Pakistan	541	84.7	155.60	2	2	97
Sri Lanka	2,656	50.0	9.77	2	2	96

Source: Adapted from Gleick, 2002.

In India, the current domestic water use of about 25 billion cubic meters (BCM) is expected to more than double by 2025 to 52 BCM, while current water demand by Indian industry and energy generation of about 67 BCM is projected to increase to 228 BCM by 2025. Thus Indian domestic and industrial water withdrawals will more than double over the next twenty-five years, accounting for 27 percent of total withdrawals for the country by 2025, compared to 17 percent in the mid-1990s (World Bank, 1998). Although at the aggregate level these changes may not seem very significant, given the large amount of water that still goes for irrigation, rural areas around cities and industrial growth points are already experiencing shortages as water is transferred from irrigation to other sectors.

In the past, most increases in food production to meet growing populations in South Asia have come from irrigated production. What effect will the limitations on irrigation have for food production and prices? In the following section we examine results from the IMPACT-WATER model on these questions.

WATER AND FOOD SECURITY: THE IMPACT-WATER MODEL

Methodology

To explore the relationships among water, environment, and food production, a global modeling framework, IMPACT-WATER, has been developed that combines an extension of the International Model for Policy Analysis of Agricultural Commodities and Trade (IMPACT) with a newly developed Water Simulation Model (WSM).[2] IMPACT is a partial equilibrium agricultural sector model representing a competitive agricultural market for crops and livestock. Demand is a function of prices, income, and population growth. Growth in crop production in each country is determined by crop and input prices and the rate of productivity growth. World agricultural commodity prices are determined annually at levels that clear international markets. IMPACT generates projections for crop area, yield, production, demand for food, feed and other uses, prices, and trade, and for livestock numbers, yield, production, demand, prices, and trade (Rosegrant et al., 2001).

In this chapter, the IMPACT model is integrated with a basin-scale model of water resource use, the Water Simulation Model, to create a linked model, IMPACT-WATER. The linkage is made through (1) incorporation of water in the crop area and yield functions and (2) simultaneous determination of water availability at the river basin scale, water demand by irrigation and other sectors, and crop production. IMPACT-WATER divides the world into 69 spatial units, including macro river basins in China, India, and the United States, and aggregated basins in other countries and regions. Domestic and industrial water demands are estimated as a function of population and income. Water demand in agriculture is projected based on irrigation and livestock growth, climate, and basin-level irrigation water use efficiency (WUE). Then water demand is incorporated as a variable in the crop yield and area functions for each of eight major food crops, including wheat, rice, maize, other coarse grains, soybean, potato, yam and sweet potato, and cassava and other roots and tubers. Water requirements for all other crops are estimated as a single aggregate demand.

Water availability is treated as a stochastic variable with observable probability distributions. WSM simulates water availability for crops at a river-basin scale, taking into account precipitation and runoff, WUE, flow regulation through reservoir and groundwater storage, nonagricultural water demand, water supply infrastructure and withdrawal capacity, and environmental requirements at the river basin, country, and regional levels. Environmental impacts can be explored through scenario analysis of committed instream and environmental flows, salt leaching requirements for soil salinity control, and alternative rates of groundwater pumping. For a detailed description of the integrated model, see Cai and Rosegrant (2002).

IMPACT-WATER thus provides a modeling framework that allows a wide range of analysis of relationships among water, environment, and food at the basin, country or region, and global levels. In this chapter, IMPACT-WATER is used to examine the likely scenarios for agricultural production and food security under existing trends, as well as the impact that eliminating groundwater overdraft throughout the world would have.

Baseline Assumptions and Projections

The starting point for the analysis is a baseline scenario that incorporates our best estimates of the policy, investment, technological, and behavioral parameters driving the food and water sectors. In the water component, the model utilizes hydrologic data (precipitation, evapotranspiration, and runoff) that re-creates the hydrologic regime of 1961-1991 (Alcamo, 2000). Nonirrigation water uses, including domestic, industrial, and livestock water uses, are projected to grow rapidly. Total nonirrigation water consumption in the world is projected to increase from 370 cubic kilometers (km^3) in 1995 to 620 km^3 in 2025, an increase of 68 percent. The largest increase, of about 85 percent, is projected for developing countries. Moreover, instream and environmental water demand is accounted as committed flow that is unavailable for other uses, and ranges from 10 percent to 50 percent of the runoff depending on runoff availability and relative demands of the instream uses in different basins. The model fully allocates minumum stream flow for navigation as a committed flow prior to any withdrawals for consumptive purposes.

Moderate increases are projected for water withdrawal capacity, reservoir storage, and water management efficiency, based on estimates of current investment plans and the pace of water management reform. Water demand can be defined and measured in terms of withdrawals and actual consumptive uses. The potential demand or consumptive use for irrigation water is defined as the irrigation water requirement to meet the full evapotranspirative demand of all crops included in the model over the full potential irrigated area. Potential demand is thus the demand for irrigation water in the absence of any water supply constraints. The global potential irrigation water demand is 1,758 km^3 in 1995 and 1,992 km^3 in 2021-2025, increasing by 13.3 percent. The developing world is projected to have much higher growth in potential irrigation water demand than the developed world between 1995 and 2021-2025, with potential consumptive demand in the developing world rising from 1,445 km^3 in 1995 to 1,673 km^3 in 2021-2025, or 15.8 percent.

Actual irrigation consumptive use is the realized water demand, given the limitations of water supply for irrigation. Total global water withdrawals are projected to increase by 23 percent between 1995 and 2025, from 3,906 km^3 (groundwater pumping 817 km^3) in 1995 to 4,794 km^3 (groundwater pumping 922 km^3) in 2025. Reservoir storage for irrigation increases by 621 km^3 (18 percent) over the next 25 years. The worldwide average basin efficiency increases from 0.56 in 1995 to 0.61 in 2025.

Global consumptive use of water will increase by 16 percent, from 1,800 km^3 in 1995 to 2,085 km^3 in 2025. Assuming nonirrigation water demand will be satisfied with the first priority, water available for irrigation water consumption will increase by only 3.9 percent, from 1,430 km³ to 1,485 km³, which is considerably lower than the 13.3 percent increase in potential irrigation demand. Therefore, of critical importance, irrigation water demand will be increasingly supply constrained, with a declining fraction of potential demand met over time. The situation is especially serious in developing countries, where potential demand increases by 15.8 percent, and the increase of supply, and therefore the increase in actual consumptive use of irrigation water, is only 4.4 percent.

This tightening constraint is shown by the irrigation water supply reliability index (IWSR), which is defined as the ratio of water supply available for irrigation over potential demand for irrigation water. For developing countries, Table 10.2 shows that the IWSR declines from 0.79 to 0.71 in 2025. Within South Asia, the largest drop is in India, where IWSR falls from 0.80 to 0.69. Relatively dry basins that face rapid growth in domestic and industrial experience, slow improvement in river basin efficiency, or rapid expansion in potential irrigated area show even greater declines in water supply reliability. For example, the Ganges basin declines from 0.83 to 0.65. The developed countries as a whole show a sharp contrast to the developing world. Irrigation water supply in the developed world is projected to increase by 7.0 km³, while the corresponding demand decreases by 5.0 km³. As a result, after initially declining from 0.86 to 0.84 in 2010, the IWSR improves to 0.89 in 2025 as domestic and industrial demand growth slows in later years and efficiency in agricultural water use improves. The divergence between trends in the developing and developed countries indicates that agricultural water shortages will become worse in the former even as

TABLE 10.2. Irrigation water supply reliability index 1995, 2010, and 2025

Basin/Countries/Regions	Water available/water requirement		
	1995	2010	2025
India	0.80	0.72	0.69
Indus	0.83	0.78	0.75
Ganges	0.83	0.70	0.65
Pakistan	0.78	0.75	0.72
Bangladesh	0.79	0.76	0.73
Other South Asia	0.88	0.86	0.85
Developed countries	0.86	0.84	0.89
Developing countries	0.80	0.73	0.71

Source: IMPACT-WATER simulations.

they improve in the latter, providing a major impetus for the expansion in virtual water transfers through agricultural trade.

The global yield growth rate for all cereals is projected to decline from 1.5 percent per year during 1982-1995 to 1.0 percent per year during 1995-2025; in developing countries, average crop yield growth will decline from 1.9 percent per year to 1.2 percent. Increasing water scarcity for agriculture is a significant cause of the slowdown in cereal yield growth in developing countries, as shown by the projected relative crop yields for irrigated cereals. Relative crop yield is the ratio of the actual projected crop yield with respect to the economically attainable yields at given crop and input prices under conditions of zero water stress. Table 10.3 shows the relative crop yield for cereals in irrigated areas for selected basins, countries, and aggregated regions. The relative crop yield for cereals in irrigated areas in developing countries is projected to decline from 0.86 in 1995 to 0.74 in 2021–2025. Within South Asia, the sharpest decline is in Pakistan (from 0.84 to 0.68). The fall in the relative crop yield index is a significant impediment to future yield growth. For developing countries as a group, the drop from 0.86 to 0.74 represents an annual loss in crop yields due to increased water stress, compared with the base year of 0.72 metric tons (mt) per hectare, or an annual loss of cereal production by 2025 of 139 million mt, more than the total rice production in China in 1995.

Elimination of Groundwater Overdraft

Many regions in the world, including northern India, northern China, some countries in West Asia and North Africa (WANA), and the western

TABLE 10.3. Relative cereal yield index 1995, 2010, and 2025

Basin/Countries/Regions	Realized yield/potential yield		
	1995	2010	2025
India	0.84	0.76	0.72
Indus	0.84	0.75	0.73
Ganges	0.83	0.70	0.69
Pakistan	0.84	0.74	0.68
Bangladesh	0.88	0.85	0.82
Other South Asia	0.85	0.81	0.79
Developed countries	0.89	0.86	0.86
Developing countries	0.86	0.80	0.74

Source: IMPACT-WATER simulations.

United States have experienced significant groundwater depletion due to pumping in excess of groundwater discharge. In any given aquifer, groundwater overdraft occurs when the ratio of pumping to recharge is greater than 1.0. However, given the large macrobasins utilized in the IMPACT-WATER model and the unequal distribution of groundwater resources in these basins, in some areas within these basins available groundwater resources are subject to overdraft, even if the whole-basin ratio shows pumping to be less than recharge. Postel (1999), drawing upon several sources, estimates that the total annual global groundwater overdraft is 163 km^3, of which 104 km^3 is from India alone. Utilizing the Postel estimate, the threshold point at the whole-basin level at which localized groundwater overdraft occurs is set at 0.55. Using this benchmark, a number of basins and countries experience groundwater overdraft in the base year, including the Rio Grande River Basin and the Colorado River Basin in the western United States, where the ratio of annual groundwater pumping to recharge is greater than 0.6; the Hailuan River Basin in northern China, where the ratio is 0.85; several river basins in northern and western India with ratios in excess of 0.8; Egypt, with a ratio of 2.5; and other WANA, with a ratio of 0.8.

It is possible for regions and countries that are unsustainably pumping their groundwater to return to sustainable use in the future. The low groundwater pumping (LGW) scenario assumes that groundwater overdraft in those countries/regions unsustainably using their water will be phased out over the next 25 years through a reduction in the ratio of annual groundwater pumping to recharge at the basin or country level to below 0.55.

Compared with levels in 1995, under LGW, groundwater pumping in these countries/regions will decline by 163 km^3, including a reduction by 11 km^3 in the United States, 30 km^3 in China, 69 km^3 in India, 29 km^3 in WANA, and 24 km^3 in other countries. The projected increase in pumping for areas with more plentiful groundwater resources remains almost the same as under the baseline scenario, and total global groundwater pumping in 2025 falls to 753 km^3, representing a decline from the value in 1995 of 817 km^3 and from the baseline 2025 value of 922 km^3.

Phasing out groundwater mining would reduce total cereal production by 17 million mt from baseline projections in 2025 (Table 10.4). Irrigated cereal production will decline by 35 million mt, but rainfed production will increase by 18 million mt, because the shortfall in irrigated production results in price increases that stimulate increased rainfed production. International wheat prices are projected to be 11 percent higher in 2025 than baseline projections, rice prices 7 percent higher, and maize prices 6 percent higher (Table 10.5). Total developing-country cereal production declines by 27 million mt in the LGW scenario compared to the baseline, with irrigated production declining by 34 million mt and rainfed production increasing by

TABLE 10.4. Elimination of global groundwater overdraft: Change in cereal production from baseline, 2021/2025

	Change in million mt (and percentage)					
	Irrigated		Rainfed		Total	
World	−35.1	(−3.0)	17.9	(1.3)	−17.2	(−0.7)
Developed countries	−0.5	(−0.2)	10.7	(1.4)	10.2	(1.0)
United States	−1.2	(−1.4)	3.9	(1.1)	2.7	(0.6)
Developing countries	−34.6	(−3.9)	7.2	(1.1)	−27.4	(−1.8)
China	−17.4	(−4.4)	1.2	(0.9)	−16.2	(−3.1)
Haihe	−8.4	(−24.7)	0.1	(0.5)	−8.3	(−15.5)
Yellow	−6.0	(−23.9)	0.1	(0.8)	−5.9	(−15.4)
India	−14.4	(−8.1)	1.6	(2.0)	−12.8	(−4.9)
Ganges	−8.3	(−10.8)	0.3	(1.0)	−8.0	(−7.5)
Indus	−4.4	(−10.9)	0.1	(1.6)	−4.3	(9.2)
West Asia/North Africa	−2.9	(−5.7)	1.2	(1.6)	−1.7	(−1.4)

Source: IMPACT-WATER simulations.

TABLE 10.5. Elimination of global groundwater overdraft: World prices, 2021/2025

Crop	Baseline (US$/mt)	Elimination of overdraft (US$/mt)
Wheat	123	136
Maize	106	111
Rice	236	251
Other coarse grains	83	87
Soybean	252	256
Potato	179	187

Source: IMPACT-WATER simulations.

7 million mt. Cereal production actually increases by 10 million mt in developed countries because the increase in world prices generates production increases that more than compensate for the direct reductions due to the fall in groundwater pumping. The elimination of groundwater overdraft would also cause a fall in global soybean production of 1.9 million mt (0.8 percent), in potato production of 4.3 million mt (1.1 percent), and in sweet potatoes of 1.5 million mt (0.7 percent). Production of cassava and other sweet

potatoes, which is virtually all rainfed, will actually increase by 1.8 million mt (0.7 percent), as increases in prices of other staples shifts some demand to cassava, pushing up prices and inducing slightly higher production.

Although substantial, the estimated global production cost of elimination of groundwater overdraft is much less than that calculated by some analysts. For example, using an estimate that it takes 1000 m³ of water to produce 1000 kg of cereal, Postel estimates that the approximately 160 km³ water deficit is equal to 160 million tons of grain. But a significant amount of overdraft water is for noncereal crops; the amount of reduction of cereal production due to 1000 m³ less overdraft of water is considerably less than 1000 kg (560 kg/1000 m³ for rice and 390 kg/1000 m³ for nonrice cereals in the developing world according to our estimates); and the cereal price effects of the cutback in pumping induces partially offsetting increases in area and yield.

Instead, the biggest impacts under LGW are concentrated in the basins that currently experience large overdrafts. Cereal production falls by 14 million mt in India, with a few basins particularly hard hit, including the Ganges basin, which has a decline in cereal production of 8 million mt, and the Indus basin, where cereal production falls by 4 million mt. Cereal demand also falls by 3.5 million mt in India, with the difference made up by increased imports. Compared with the baseline scenario, LGW results in an increase of cereal imports of 7.0 million mt in India, 6.5 million mt in China, 1.7 million mt in WANA, and 0.8 million mt in other developing countries in 2021-2025.

Simulations of improvements in basin irrigation efficiency targeted within the overdrafting basins show that water sector policies to compensate for the loss in production and consumption from reduced groundwater pumping would require a focus beyond the basins most affected. Although improvements in irrigation efficiency in the specific overdrafting basins could in theory compensate for these declines in groundwater use, the required efficiency improvements would be huge and likely unattainable. In the Indus basin an improvement in BE in 2025 from 0.59 to 0.76 would be required to generate enough cereal production to compensate for the reduction in groundwater overdraft (Table 10.6). In the Yellow River in China, the required improvement in BE in 2025 would be from 0.62 to 0.82.

Alternatively, could increased rainfed cereal production within the overdrafting countries compensate for the irrigated production decline due to reduced groundwater pumping? This question is addressed by a scenario combining elimination of groundwater overdraft with higher rainfed agriculture development. The reductions of irrigation production due to reduced groundwater pumping can be offset by an increase in rainfed area and yield within the same regions, but the required increase in rainfed cereal

TABLE 10.6. Basin-specific efficiency gains required to compensate for LGW production losses, 2021-2025

	Baseline efficiency	Compensating efficiency
China		
Haihe	0.75	0.86
Yellow	0.62	0.82
India		
Ganges	0.62	0.85
Indus	0.59	0.85

Source: IMPACT-WATER simulations.

yields would be very large. In 2025, average rainfed cereal yields would need to be 20 percent or 0.3 metric tons per hectare higher in India than baseline projections. In addition, rainfed cereal area would need to increase by 0.8 million hectares in India. But the greater expansion in rainfed area is itself environmentally damaging, requiring encroachment on fragile lands. Moreover, the yield increase would require substantial additional investments in agricultural research and management for rainfed areas, water harvesting, and market access, and it is not clear that these yield increases are achievable in rainfed areas, even with increased investments.

The global food production impact of the elimination of groundwater overdraft is relatively small, but the impacts for specific countries and basins are quite large. In India and China, significant reductions in cereal production and consumption are accompanied by increased cereal imports. Although the seriousness of these country-level shortfalls in demand and increases in imports should not be minimized, they may be a worthwhile tradeoff for restoring sustainability of groundwater supplies; and it may be necessary for these countries to rely more on imports to meet the decline in irrigated production compared to the baseline. Agricultural research investments should be increased, and, particularly in the hardest-hit river basins (such as the Ganges), investments and policy reforms (including elimination of power subsidies for pumping) should be implemented to increase basin efficiency and encourage diversification out of irrigated cereals into crops that give more value per unit of water.

Improvement of basin water use efficiency depends on both technological improvements in irrigation systems, domestic and industrial water use and recycling systems, and institutional settings related to water allocation, water rights, and water quality (Cai, Ringler, and Rosegrant, 2001). In the irrigation sector, technical improvements can include advanced irrigation systems such as drip and sprinklers, conjunctive use of surface and groundwater, and precision agriculture including computer monitoring of crop water demand; managerial improvements include adoption of demand-based irrigation scheduling systems and improved maintenance of equipment; and institutional improvements include establishment of effective water user associations, establishment of water rights, introduction of water pricing, and improvement in the legal environment for water allocation. In the industrial sector in developing countries, the amount of water used to produce a given amount of output is far higher than in developed countries. A major technology backlog exists that can be tapped by developing country industry to save water. The domestic water sector has the potential to improve water use efficiency through leak detection and repair in municipal systems to installation of low-flow showerheads and toilets and ecological sanitation, and to adjust lifestyles to consume less water.

CONSEQUENCES FOR RURAL LIVELIHOODS

It is not only crop production, in the aggregate, that is affected by water scarcity for irrigation. Water is a vital asset and the basis for livelihoods, especially in rural areas. Although most irrigation development policy has considered only the production of field crops in designated command areas as the output of irrigation systems, anyone who walks through an irrigation system in South Asia can observe a wide range of water uses (Meinzen-Dick and van der Hoek, 2001). Rural people—who may or may not have irrigated holdings—water their livestock and grow or catch fish in the canals and storage areas. Others may be using the water for brick making, clay pots, or other enterprises. Homestead gardens draw their water either from seepage of diversion from surface systems or percolation into wells. Within some systems, especially in semiarid areas, irrigation water may be the only source of domestic supplies available to the households. Even if piped drinking water is available in rural areas, it is usually only enough for drinking and cooking, leaving people to bathe and wash clothes with irrigation water.

One reason that these multiple water uses have been overlooked is that the volume of water used is often quite small in comparison to the volume of water used in irrigation of field crops (Bakker et al., 1999). However, the total value of these other uses can be quite high. Renwick (2001) estimated that the value of fish production in Kirindi Oya Irrigation System in Sri Lanka provides an additional 18 percent beyond the value of water for paddy irrigation alone. Homestead gardens seem individually small, but they produce high-value horticultural products that provide important micronutrients for household consumption as well as items for sale.

Another reason that these other water uses are overlooked is that it is often women, landless pastoralists, or other marginal groups that are involved in the activities. They may also have no water rights recognized by the state but claim access to water based on a range of customary water rights and livelihood needs (Bakker et al., 1999). Thus the *uses* are overlooked because the *users* are overlooked. Yet the production is all the more important for poverty reduction because it is under the control of these groups, who often have few other assets or sources of income. A recent study in Bangladesh found that improved vegetable cultivation on homestead gardens had a significant effect on empowerment of women (Hallman et al., 2002).

How are these other water uses affected by water scarcity? Because such water uses rarely have state-recognized water rights, they are likely to lose out when the government makes decisions about water allocation between sites or between sectors.[3] Even farmers with land in recognized irrigated areas rarely receive compensation when they lose access to water through increasing competition, but they may have greater leverage with politicians and government agencies to block some reallocations. Fishers, livestock keepers, and others generally have a harder time in protecting their claims to water.

Irrigation also supports rural livelihoods through forward and backward linkages in agriculture. As irrigated production increases, so do jobs for traders and suppliers of goods and services to farmers. But industrial production can also create jobs. In fact, with diversification of livelihoods, many farm households may have some members employed in factories or other jobs. As a result, interests in water may be split within a family. Thus water transfer from irrigation to other uses is likely to create some winners and some losers. What is important to note is who loses and who gains— whether it is the rich or poor, men or women. To do this requires recognizing the full range of water users as stakeholders.

Externally imposed water transfers are increasingly evoking protests from those who are asked to give up water. More inclusive negotiation processes could enhance the equity and acceptability of transfers. The outcome is more likely to be seen as equitable because it includes all stakeholders, and when people take part in the process they are more likely to accept the outcomes (Meinzen-Dick and Pradhan, 2002). Negotiated transfers can also identify creative means for benefit sharing. If water is to be transferred because nonagricultural uses are of higher value, then rural water users should be able to share in those gains from trade. On the other hand, if the apparent increase in value of water is due to an underestimate of the value of water use in rural areas, then recognizing a broader spectrum of water users will give a more accurate picture of the benefits of water for rural livelihoods.

CONCLUSION

Rapid growth in nonagricultural demand, a slow down of growth in water supply investments, and moderate progress in water use efficiency derived from water policy and management reforms will lead to growing water shortages for agriculture in much of South Asia. The resulting water supply constraints will reduce food crop yield growth. With irrigation deficits becoming more severe in many basins and countries, a further decrease in water available for agriculture, whether due to increased environmental reservation, reduced groundwater pumping, or growth in other nonagricultural demands, will further reduce agricultural production growth.

However, local and river basin effects are more severe than global aggregates, and concerted policy efforts could significantly mitigate the negative effects. Although local impacts on agricultural employment and related sectors can be severe under a scenario of rapidly increasing competition for scarce water resources, the most important impacts occur in specific basins where production shortfalls are concentrated. It would be here that interventions are needed to compensate farmers and other rural water users who are negatively affected. An additional possibility would be to introduce demand management through conservation education campaigns and policy reforms, such as more aggressive water pricing, to constrain municipal and industrial uses, which are assumed to be the first claimant on water. Increased investments in agricultural research could induce increases in agricultural productivity that would further compensate for the diversion of water from agriculture. Technology, pricing, and institutional reforms all needed to grow enough food, create livelihoods so that all can obtain food, and protect the environment for the next generation.

NOTES

1. Quotation inscribed in the entrance foyer of the building constructed for the Sri Lankan Irrigation Department, now headquarters of the International Water Management Institute.

2. This section draws upon Rosegrant and Cai (2002).

3. At the local level, religious or customary rights to water (e.g., for drinking water) may be just as strong as rights deriving from state law, but when the government becomes involved, local law is generally less influential.

REFERENCES

Alcamo, J.P. 2000. *Personal Communications on Global Change and Global Scenarios of Water Use and Availability: An Application of Water GAP.* University of Kassel, Germany: Center for Environmental System Research (CESR).

Bakker, Margaretha, Randolph Barker, Ruth S. Meinzen-Dick, and Flemming Konransen (Eds.). 1999. Multiple Uses of Water in Irrigated Areas: A Case Study from Sri Lanka. SWIM Report 8. Colombo, Sri Lanka: International Water Management Institute. Available at <http://www.cgiar.org/iwmi/pubs/SWIM/Swim08.pdf>.

Cai, X., C. Ringler, and M.W. Rosegrant. 2001. Does Efficient Water Management Matter? Physical and Economic Efficiency of Water Use in the River Basin. Environment and Production Technology Division discussion Paper 72. Washington, DC: International Food Policy Research Institute.

Cai, X. and M.W. Rosegrant. (2002) Global Water Demand and Supply Projection: Part 1. A Modeling Approach. *Water International* 27(2): 159-169.

Gleick, P.H. 2002. *The World's Water: The Biennial Report on Freshwater Resources, 2002-2003.* Washington, DC: Island Press.

Gulati, A., M. Svendsen, and N.R. Choudhary. 1995. Capital Costs of Major and Medium Irrigation Schemes in India. In *Strategic Change in Indian Irrigation,* eds. M. Svendsen and A. Gulati. New Delhi: Macmillan India Ltd., pp. 37-72.

Hallman, K., D. Lewis, S. Begum, and A.R. Quisumbing. 2002. Impact of Improved Vegetable and Fishpond Technologies on Poverty in Bangladesh. Paper presented at the International Conference on Impacts of Agricultural Research and Development: Why Has Impact Assessment Research Not Made More of a Difference? San José, Costa Rica, February 4-7.

Meinzen-Dick, Ruth S. and Rajendra Pradhan. 2002. Recognizing Multiple Water Uses in Intersectoral Water Transfers. Paper presented at Workshop on Asian Irrigation in Transition—Responding to the Challenges Ahead, Bangkok, Thailand, April 22-23.

Meinzen-Dick, Ruth S. and W. van der Hoek. 2001. Multiple Uses of Water in Irrigated Areas. *Irrigation and Drainage Systems* 15(2): 93-98.

Organization for Economic Cooperation and Development (OECD). 1998. *The Athens Workshop, Sustainable Management of Water in Agriculture: Issues and Policies.* Paris: Author.

Postel, S. 1999. *Pillar of Sand: Can the Irrigation Miracle Last?* New York: W.W. Norton.

Renwick, M.E. 2001. Valuing Water in a Multiple-Use System: Irrigated Agriculture and Reservoir Fisheries. *Irrigation and Drainage Systems* 15(2): 149-171.

Rosegrant, M.W. and X. Cai. 2002. Water Constraints and Environmental Impacts of Agricultural Growth. *American Journal of Agricultural Economics* 84(3): 832-828.

Rosegrant, M.W., M.S. Paisner, S. Meijer, and J. Witcover. 2001. *Global Food Projections to 2020: Emerging Trends and Alternative Futures.* Washington, DC: International Food Policy Research Institute.

Rosegrant, M.W. and M. Svendsen. 1993. Asian Food Production in the 1990s: Irrigation Investment and Management Policy. *Food Policy* 18(1): 13-32.

World Bank. 1998. *India-Water Resources Management Sector Review: Report on Intersectoral Water Allocation, Planning and Management, Volume 1: Main Report.* Washington, DC: Author.

Chapter 11

Pricing, Subsidies, and Institutional Reforms in Indian Irrigation

K.V. Raju
Ashok Gulati

INTRODUCTION

The irrigation sector in India suffers from at least two interrelated problems. The first is a severe pressure on the resources for normal operation and maintenance as the cost recovery from canal irrigation is extremely low and the state budgets are unable to allocate more funds because of the overall fiscal crunch. The second is the reduction in construction funding for new or ongoing canal networks, leading to undue delays in the completion of projects, and in turn, the rise in costs and reduction of benefits. These issues are also linked to the low price of water. Unless urgent steps are taken to reverse these trends, such as innovative institutional reforms, the Indian irrigation sector could be headed for near collapse or at least a situation in which it would remain below its potential for food production. This chapter examines two issues: (1) irrigation subsidies and their implications, and (2) emerging institutional reforms in the irrigation sector.

The issue of irrigation subsidies is far more complex than ordinarily imagined. Less discussion takes place on irrigation—perhaps because the apparent magnitude of irrigation subsidies does not compare with that for either fertilizer or power. However, because water is a critical component in Indian agriculture,[1] subsidies on irrigation deserve just as much attention as the other two resources. Irrigation has undoubtedly been instrumental in achieving self-sufficiency in food production, but it has come with a cost. The neglect of rational pricing of canal waters has resulted in rising subsidies. Public irrigation subsidies may be less of a mystery than power and fertilizer in terms of who really benefits from them. Of greater concern are the ramifications of these subsidies in different spheres, such as financial status of input-supplying agencies, environmental consequences, efficiency in resource use, and others. Trends suggest that these may have grave impli-

231

cations for the agricultural sector if left unattended. Hence the need for reform.

Before opening the Pandora's box of issues, it is essential to first conceptualize, define, and then quantify the notion of an irrigation subsidy. How does one do this? In the past there have been varying views on this issue and a whole range of estimates have been put forth. Several of these may significantly understate the magnitude of irrigation subsidies and hence tend to undermine the gravity of the problem—in particular the financial burden of these subsidies on the input-supplying agency and the state. A major part of this chapter is therefore devoted to this exercise. Accordingly, we will discuss the concept and different approaches to quantifying irrigation subsidies. We also present our estimates of irrigation subsidies and compare them with some possible alternatives.

The next step is to link the issue of irrigation subsidies with the larger scenario of the irrigation sector, specifically to examine the diverse consequences of these subsidies and review the problems that have occurred. We will examine in detail the impact of irrigation subsidies on different spheres and sum up the state of Indian irrigation today. Following this, we outline an agenda for reforming the regime of irrigation subsidies with a hope that it will lead to a sustainable, financially sound, and efficient use of the resource—canal water. The issues addressed are: What kind of pricing policy should be followed to keep the growth of subsidies under check and ensure that irrigation has self-generated funds? What other reforms are necessary and feasible in the area of irrigation subsidies? The concluding section wraps up the discussion with a map of possible future reforms.

IRRIGATION SUBSIDIES

The subsidy on public irrigation is not stated explicitly by the government. It has to be determined from government data sources and is often done so by more than one method. The method commonly employed for this purpose is based on the concept that the irrigation subsidy can be approximated as the losses that the input-supplying agency bears on account of supplying irrigation water at concessional rates. These estimates implicitly define the irrigation subsidy as the difference between cost of supplying water for irrigation and the revenue received as payment from the users of irrigation water. However, this definition of the subsidy is incomplete and the estimates are inaccurate for a number of reasons. First, the cost of providing the service is not a true reflection of the cost of delivering irrigation water. Apart from the operations and maintenance costs, and other current expenses, huge capital expenditures are incurred on the provision of irrigation. Second, most state governments are beneficiaries of loans forwarded to them by the central government. Although a part of them are grants (normally one-third), a larger part is in the form of a loan (usually, two-thirds). This then entails an interest burden as well (at 8-10 percent). Third, some scholars stress that depreciation also needs to be taken into account, which would then reflect the consumption of fixed irrigation assets on a replacement cost basis (Dhawan, 1999). Only its inclusion will yield the true extent of the subsidy.

Approaches to Quantifying the Irrigation Subsidy

The irrigation subsidy concept adopted in this chapter is based on the difference in the cost of supplying irrigation water and what the farmers pay for irrigation water as its direct price. It is based on the perspective of the supplying agency. Accordingly, our focus is on estimating the cost of providing public irrigation water through major, medium, as well as minor irrigation schemes, and the payments made by farmers for this water. The cost of irrigation in major and medium schemes consists of three components: (1) the capital cost, (2) the working cost of operation and maintenance (O&M), and (3) depreciation.

The capital cost is incurred over a number of years. In a representative major irrigation project, it would be quite common to find that capital costs are spread over roughly 20 years, out of which there might be some years, say the first seven years, when no potential is created and only capital expenditure is incurred. Thereafter, some irrigation potential may be forthcoming even as capital expenditure continues to be incurred. Gestation lags—the time gap between expenditure incurred and irrigation potential created—vary from project to project, but are generally longer for major projects and shorter for medium projects. Thus, in estimating the true capital cost of each hectare of potential created, one has to adjust for this lag factor by incorporating an appropriate social rate of discount. Depending upon value of this discount factor, the capital cost of irrigation will differ significantly. (Gulati et al., 1994). The O&M costs are added to capital cost, and then payments made by farmers for irrigation water are deducted to obtain the public subsidy for major and medium irrigation. Therefore, the minimum subsidy on every hectare of irrigation utilized through major and medium irrigation schemes would be above rupees (Rs.)10,000 at 1997-1998 prices. This is at the national level, an average for irrigation potential created during 1963-1995, and is obviously dependent on the way one calculates the capital cost of irrigation.

So far the focus has been specifically on government major and medium irrigation works covering the area of canals. However, it is important to consider other sources of irrigation as well and examine the extent of subsidization through them. Minor irrigation[2] has already become "major" in that it irrigates a larger area than the so-called major and medium irrigation schemes. The notion of a subsidy on public minor irrigation is difficult to conceptualize. Among the various categories of "minor irrigation" the largest share is well irrigation, which has a large number of private owners. Here the subsidy is primarily through cheap power supplies. The government may also provide subsidies that serve to defray part of their capital costs for wells, pump sets, etc.

In defining and quantifying public irrigation subsidies for major and medium irrigation, we use the Vaidyanathan Committee report (GOI, 1992a) on pricing of canal irrigation water. The Committee suggested that pricing of canal irrigation water must cover the entire O&M expenses and 1 percent of cumulative capital expenditures incurred in the past at historical prices.[3] This can be viewed as the cost of canal irrigation that needs to be recovered from the farmers. By deducting the payments made by farmers from this cost of irrigation water, one would obtain the relevant public irrigation subsidy on canal waters. The same methodology used for major and medium irrigation is applied to public sector minor irrigation as well. Thus, the capital expenditures and the O&M expenses of minor irrigation are grouped with major and medium irrigation subsidies to arrive at an aggregate public irrigation subsidy for the country.

Quantification of Irrigation Subsidies

The exercise of quantifying subsidies on major, medium, and minor irrigation schemes was thus carried out along the approach previously described. Irrigation subsidy estimates are presented in Figures 11.1 and 11.2(a) and (b). The share of the southern region (Andhra Pradesh, Karnataka, Kerala, and Tamil Nadu) increased marginally from 23 percent in the triennium ending 1982-1983 to 24 percent in 1999-2000 (projected) as did the northern region to 32 percent from 27 percent during the same period. Although the share of the eastern region declined from 17 percent to 12 percent, the western region (comprising Gujarat, Maharashtra, Rajasthan, and Madhya Pradesh) has a share of around 26 percent. The western region has maintained such a high share possibly because the costs of canal irrigation are much higher in this region due to undulating terrain. It must be emphasized that the coverage for the computation of subsidies is not complete since information is not available for some states in some years.

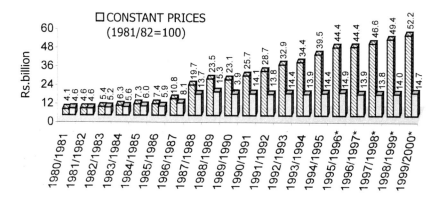

FIGURE 11.1. Subsidies on major, medium, and minor irrigation (1980-1981 to 1999-2000) (*Notes:* Irrigation subsidy is computed as O&M + 1 percent cumulative capital costs at historical price receipts of major, medium, and minor irrigation).
*indicates estimates for these years are based on projected values. Coverage is staggered.

It would be interesting to see how these estimates compare with those obtained from other approaches, particularly those constructed from government documents. To do this, three different series are computed. The first is the government estimate drawn from the National Accounts Statistics. The second is the approach used in this study that follows the Vaidyanathan Committee recommendation that O&M expenses and 1 percent of cumulative capital cost must be charged as the price of irrigation water. Yet another estimate approach uses only O&M expenses minus gross receipts. These three approaches in juxtaposition (Figure 11.3) reveal that government estimates are significantly higher than our estimates of irrigation subsidy. This is particularly true of the later years. However, our estimates are somewhat higher than the O&M method estimates, which consider only O&M costs.

The approach used in this chapter has its drawbacks; these estimates take into account neither depreciation nor interest rates. However, this approach has the advantage of simplicity. In the past, the finance commissions had been pleading for recovering at least 2.5 percent of the cumulative capital costs, which was reduced to 1 percent. The unanimous decision of the Committee on the Pricing of Irrigation Water to charge only 1 percent of the historically accumulated capital expenditures, with full O&M expenses of the

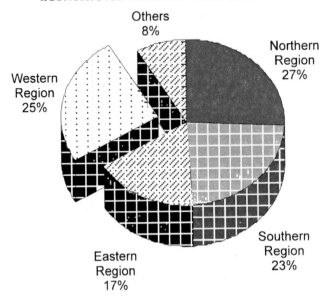

FIGURE 11.2(a). Regional shares in irrigation subsidies-TE 1982-1983.

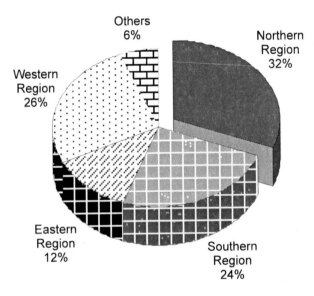

FIGURE 11.2(b). Regional shares in irrigation subsidies-TE 1999-2000.

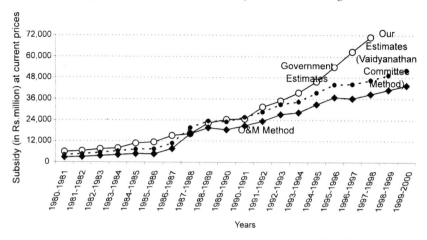

FIGURE 11.3. Estimates of subsidies on major, medium, and minor irrigation: A comparison.

current year, perhaps reflects that the remaining capital expenses have to be borne by the rest of the society, maybe as a price for food security, or to keep food within the reach of the masses. This is broadly in line with the practice that currently prevails in most countries.

Conseqences of Irrigation Subsidies

The role of irrigation in enhancing agricultural production is well established and a significant increase in production, especially food grains, over the years can be attributed to increasing irrigation. It promotes faster adoption of HYV (high-yielding variety) seeds, fertilizer consumption, and other inputs associated with intensive agriculture. As a result of this package use of inputs, the yield on irrigated plots in 1992-1993 tended to be 2.3 times greater than the yield on nonirrigated plots (Dhawan, 1999). Econometrically, it is not an easy exercise to segregate the impact of each of these highly complementary inputs such as water, HYV seeds, fertilizers, insecticides, etc., on agricultural production. Although it is the combined impact of all the inputs that is reflected here, irrigation is the facilitating factor for all the other inputs. It has thus led to the growth of self-sufficiency in food production. However, the cost at which this has been achieved has not been studied in detail. Slowly but surely, bulging subsidies have taken a toll on the irrigation sector itself; if neglected any further its impact on finances, environment, and investments and agriculture as a whole could be disastrous.

Financial and Economic Consequences

The price of canal water is pathetically low and totally unrelated to the productivity or scarcity value of water, or the cost of delivering it. It accounts for just 8 percent of cropping expenses and is equal to barely 5 percent of the average incremental production of irrigated areas over rainfed lands (World Bank, 1997). Moreover, water charges are fixed in nominal terms that remain unchanged for years, so that in real terms they have been falling. In most states, the agency levying the water charges and those responsible for its collection (usually the responsibility of the revenue department) are different. As a result, collection has tended to remain low. The overall loss amounts to around 7 percent of the total plan expenditure on all irrigation schemes. The inability to recover costs has led to growing state revenue deficits so that currently, irrigation alone is responsible for a third of the states' revenue deficit (World Bank, 1997). With inadequate cost recovery and inability to generate funds, the irrigation departments have to increasingly rely on the state government to meet even their O&M activities. Since irrigation subsidies have to be absorbed at the state level and the budgetary situation of most of the states is under severe strain, it has resulted in increasing cuts in further state expenditures on irrigation. A subsidy reduction of 20 percent could have helped raise expenditure by at least 20 percent. Alternatively, a subsidy reduction of even 5 percent in 1986-1987 would have doubled expenditures on O&M of Rs.4.93 billion incurred that year. This would have entailed an over threefold rise in the collection of gross revenues from farmers in that year, i.e., Rs.167 billion (Vaidyanathan, 1993).

Physical Constraints

The result of curtailed expenditures on irrigation is poor maintenance of projects and neglect of existing irrigation systems (leading to poor quality of service), and the inability to complete ongoing projects because of paucity of funds. These factors have led to rapid deterioration of physical infrastructure, particularly of surface irrigation facilities and drainage infrastructure. Although this is partly due to poor design and construction, it is to a larger extent due to the lack of adequate maintenance. Broken-down distribution systems and siltation of canals and drains tend to reduce irrigation efficiency and lead to irregular supply.

Wastage and Inefficiencies in Water Use

The contribution of water is generally about a quarter to two-fifths of the additional agricultural production. When a farmer uses water as an input until its marginal value product is zero, he overexploits the resource. This leads to large-scale emergence of water-intensive crops such as paddy or sugarcane in irrigated tracts. In almost all of the canal commands, the actual area under paddy or sugarcane turns out to be much more than was initially planned. Surface water use efficiency under the existing projects in India is estimated to be as low as 40 percent (Navalawala, 1994). Farmers tend to waste 27 percent through excessive irrigation, and only 29 percent is actually used by crops (Veeraiah and Madamkumar, 1994). In contrast, in the advanced systems of the West, as much as 60 to 70 percent of the water diverted in large surface systems is available for plant use (Repetto, 1986). This enormous wastage of water during conveyance and on the field is as much due to poor design of structures as on account of lack of incentives to conserve water. In the absence of financial accountability and operation autonomy, project authorities do not have any motivation to take water-conserving measures, such as the lining of canals for supplying water on a volumetric basis. In the absence of large-scale investments in drainage, vast areas either become unfit for cultivation or returns on them are much below their potential.

Inequity

A disquieting consequence of underpricing of surface water is the intensive watering of fields by farmers at the head, leaving those at the tail end with sparse supplies. Apart from lowering the productivity per unit of water used, this leads to inequity. The intensive water use particularly at the head is responsible to a significant extent for shrinking effective command areas of the system compared to what is originally envisaged. This is one of the major reasons why the irrigation potential utilized in canal commands is usually 15 to 20 percent less than the potential created. Moreover, only 26 percent of the villages were reported as having a government canal in 1997-1998. This reflects considerable inequity in the distribution of irrigation service and therefore of subsidies.

Decline in Public Investment

Rapidly increasing subsidies on canal irrigation results in a strong negative impact on public sector investments in agriculture. The mid-1960s saw a considerable increase in public investment in response to the need to achieve food self-sufficiency. The 1970s were characterized by the contributions of private investment in agriculture with the introduction of the profitable HYV technology. The decline in the 1980s was at the rate of 1.73 percent per annum (Rao, 1994). The outcome of this was that resources were spread thinly over a number of projects in the pipeline. As many as 500 major and medium irrigation projects at various stages of completion at the end of the seventh plan entailed a spillover cost of Rs.390 billion (GOI, 1992a). These time overruns contribute to the higher real cost per hectare of irrigation potential created. The current trend suggests that it would take another 50 years to fully exploit the remaining irrigation potential of 25 million hectares through major and medium schemes. Even if no new projects were undertaken, the potential of ongoing projects would take two decades to realize.

"THE VICIOUS CIRCLE" AND THE NEED FOR REFORM

The consequences of irrigation subsidies delineated form elements of a vicious circle (see Figure 11.4; Oblitas et al., 1999). The water pricing policy in India is such that it does not even cover the cost of O&M of the irrigation systems, let alone the full capital cost including O&M. This leads to severe financial pressure on the state since it has to absorb the subsidies. The fiscal constraints of the irrigation service agency and the state leads to inadequate budgetary allocation toward O&M of these systems resulting in physical deterioration of the irrigation system, which affects water delivery and supply. The poor irrigation service is also caused by institutional constraints such as the lack of incentive and accountability on the part of the monopolistic government agency to assure quality supply. There is no link between irrigation quality provided, revenues generated, and staff incentives. Furthermore, there is lack of coordination among departments dealing with agriculture and those dealing with irrigation, within the irrigation department itself, and between agencies dealing with different types of irrigation such as lift irrigation, canal projects, groundwater schemes, etc.

Irrigation departments are highly centralized and function with a top-down approach resulting in a failure to establish any linkages with the farmer/users. Lack of farmer involvement results in inappropriate design of irrigation systems, which also leads to poor irrigation service. Farmers are as a result dissatisfied. Unreliable supply with iniquitous distribution of water leaves many disgruntled and unwilling to pay (higher) water rates. Indirectly, the poor irrigation service also affects the farmers' ability to pay, since inadequate irrigation (combined with inefficient water use technolo-

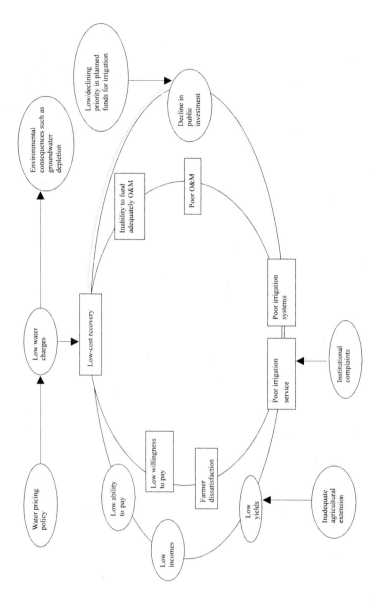

FIGURE 11.4. The vicious circle in Indian irrigation. (*Source:* Based on Oblitas et al., 1999.)
Note: ———— indicates weak link.

gies) results in low yields and incomes. The water pricing policy too may impact incomes in another way. This happens when severe inefficiency and wastage in water use leads to environmental problems in the long run, which have an adverse impact on yields and incomes. The unwillingness of farmers to pay more for irrigation services coupled with the possible inability to do so precludes any change in the water policy in terms of raising the water rates charged by irrigators. No policymaker would want to risk such an undertaking. Nor is it fair to increase water rates without concomitant improvement in the quality of service. Improvements in quality in turn are constrained by funds and the inefficiency of the input irrigation agency. Thus the vicious circle perpetuates.

REFORMING THE VICIOUS CIRCLE

As far as the irrigation sector goes, we find ourselves in a catch-22 situation. Given that the input-supplying agencies are in the financial doldrums as are the states that ultimately bear the burden of the subsidy, it is imperative that these agencies recover costs so that they become financially viable. This would entail a manifold increase in water rates. To reiterate, farmers would be unwilling to accept such a step unless they derive some benefit in terms of better delivery of the input. For this to happen, the physical conditions of the irrigation systems need improvement. However, this is itself predicated on the availability of funds and an institutional overhaul.

If the vicious circle in Indian irrigation is to be turned into a virtuous circle a multifaceted reform program must be realized. One of the elements of the vicious circle that should be targeted is price reform with the aim of ensuring that the irrigation agency is made financially self-sustaining with the capability of providing efficient irrigation service. Since price reform without improvement in quality of service is inconceivable, it must be accompanied by fundamental changes in the institutional framework. Simultaneously, good agricultural practices and efficient water use technologies must be promoted so that excessive use and wastage of water are prevented. How can these be accomplished?

Price Reform

Pricing of canal waters is a state subject (like power) and hence tends to differ widely across states. In addition, prices also vary across crops within the same state and across the seasons for the same crop. Pricing can also differ across different regions or projects within the same state. Although there are some states, such as Punjab, which give free irrigation water from canals, in most states, pricing is based on crop area and season basis. Since technically it is difficult and expensive to measure water supplies on a volumetric basis to millions of small cultivators, volumetric pricing seems to be unfeasible at least in the foreseeable future.

The debate on the principles underlying the pricing of irrigation water has held the view that the farmers must pay at least the short-run marginal cost of providing water comprising O&M charges and a small part of the interest on capital invested (Rao et al., 1999). The Fifth, Sixth, and the Seventh Finance Commissions in agreement with the Jakhade Committee had recommended that pricing of canal irrigation water should recover 2.5 percent of the capital invested besides the working expenses. However, given the poor financial performance of canal irrigation the subsequent Finance Commissions, the Eighth and the Ninth, recommended only the recovery of O&M expenses. However, the Tenth Finance Commission did reiterate the need to recover at least 1 percent of the capital cost besides the working expenses. The Vaidyanathan Committee (GOI, 1992b) was the last major report on this subject.

In practice, the pricing of canal waters did not cover more than 20 percent of the O&M expenses in the mid-1990s. Over the years the capacity of the farmers to pay for higher irrigation charges has increased due to spread of HYV seeds, commercial and high-value crops coming in the commands of canals, and higher productivity through better cropping operation.[4] Besides, despite overstaffing, the actual expenditure on O&M per hectare of irrigated area is considerably below the accepted norms. Against the generally accepted principle of appropriating as water charges between 25 to 40 percent of the additional net income generated per hectare on account of irrigation, only about 2 to 5 percent of such income is being collected as water rates. It appears therefore that political resistance notwithstanding, at least a fivefold increase in the existing water rates may be necessary, both theoretically speaking and from a practical viewpoint of managing the projects.

The next obstacle is in the collection of water charges. Collections remain low and mechanisms to collect (done by the revenue department) are not effectively coordinated with the irrigation agencies. Cost of collection is more than the sum collected in some states (e.g., Bihar). That speaks volumes in terms of the urgent need to usher in institutional reforms in these irrigation agencies/departments that are saddled with enormous staff and rampant corruption. Price reforms in Indian irrigation, if they were so simple, would have been carried out already. The fact that the subsidy situation has worsened over time should compel us to recognize the issue is far from simple. It is intertwined with the state and nature of politics and a lack of understanding of who is being subsidized and by how much. As a result, the suggested approach of reforming the regime of subsidies is often divorced from the ground reality, and therefore, remains a non-starter.

Institutional Reform

Why Institutional Reform?

In the context of irrigation subsidy reform, the preceding section brings to the fore two points about canal irrigation. First is the implicit willingness to pay with respect to irrigation charges as imposed by the state. However, in practice, due to the monopolistic position of the state with regard to water, actual payments made by consumers (inclusive of "scarcity rent" paid in one form or the other) may be higher. Estimates of these payments do exist for some parts of the country (Wade 1987). A large part of willingness to pay may already be captured by these payments. Second, higher willingness to pay is related to access to the services of the input in amounts and at the time the farmer requires. As the consumer pays more, he or she also expects qualitative improvements in supply. It is exactly for this reason that price reforms in Indian irrigation must be accompanied by institutional reforms. Price reform is essential but not sufficient for a well-functioning irrigation system.

Recent Efforts at Institutional Reform

A number of expert committees in India starting with the Taxation Enquiry Commission in 1953 to the Committee on Pricing of Irrigation Water (GOI, 1992b), have expressed the desirability of improving the management of irrigation systems so as to make them more responsive to the needs of user cultivators. However, experience shows that this shall not succeed unless these systems are distanced from political interference and also de-bureaucratized. The recent debate on these issues among the experts all over the world has resulted in a remarkable consensus on the need to (a) make the project authorities financially accountable by according them operation autonomy; (b) associate the user farmers with the decision-making process in the projects at various levels; (c) entrust the water users' associations with the tasks of managing the systems in their area of operation as well as collecting the water charges on the basis of some workable formula linking the rates with the quantity of water consumed; and (d) allow the private sector participation in irrigation systems renovation and modernization and collection of fees. In the following sections we focus on two major reforms: (1) setting up financially autonomous irrigation agencies as a part of the financial reforms, and (2) irrigation management transfer to water users' associations.

Financially autonomous irrigation agencies. Canal irrigation financing in India suffers from two distinct problems: (1) the funding for construction of ongoing or new canal networks has been shrinking, leading to undue delay in completion of projects, which in turn raises costs and reduces benefits; and (2) the resources for normal operation and maintenance are also under severe pressure as the cost recovery from canal irrigation is extremely low, and the state budgets are not able to allocate more funds because of the overall fiscal crunch. Much less attention has been given to the potential of domestic financial markets to provide such funding. Since the 1980s, the Indian capital markets have emerged as an important source of funds for corporate units in the private and public sectors. Primary capital mobilization by private sector companies in the form of equity and debt rose from less than Rs. 2 billion in 1980 to over Rs. 43 billion in 1990-1991 and then recorded a quantum jump to over Rs 260 billion by the end of 1994-1995 (GOI, 1996). During this period, several state governments tapped into this domestic financial market to finance irrigation development.

This is not the first time that such institutional reform has been proposed. Indeed, the working group on major and medium irrigation projects for the Eighth Plan considered the issue of inadequate funding for projects in the Seventh Plan. In contrast to the spillover liability of Rs.260 billion for major and medium projects that remained uncompleted from previous plans, the outlay was only Rs.115 billion. To enable the central government to assume a more positive role, in 1988 the Ministry of Water Resources formulated a proposal for establishment of an irrigation finance corporation to provide financial assistance to projects of national importance in the irrigation sector (GOI, 1995). Though this proposal was supported by a large number of states, the planning commission did not approve it. Over the years, the states that had important ongoing projects established autonomous irrigation finance corporations. In southern India, Karnataka's Krishna Bhagya Jal Nigam Limited (KBJNL)[5] is one of them (Gulati et al., 2002).

In a normal course, the state budget could have supported the entire Upper Krishna Project (UKP) execution, but then the project completion could have been anywhere from 15 to 20 years, since the state budgetary allocation of around Rs. 10 billion is meant for all major and medium projects in the state. In 1995, the government contemplated an outlay of Rs. 57 billion (then revised to 82 billion in 2001) for the completion of UKP. It included Rs. 30.5 billion from market borrowing, Rs. 24.5 billion from the government of Karnataka, and another Rs. 2.45 billion from internal generation.

KBJNL focused primarily on mobilizing funds and completing the physical work before 2003. Encouraging support from the government's top functionaries (such as chief secretary and finance secretary) and having the right persons in the key positions (such as executive director, finance, with good experience) helped it to move faster in the desired direction. This has produced three main outcomes: (1) successful mobilization of funds; (2) less immediate financial burden on state; and (3) project implementation more or less on schedule. Although KBJNL has made considerable progress in mobilizing capital for construction, it has not made structural reforms within the organization, nor has it paid attention to repayment (Gulati et al., 2002). The organization depends on the government's budgetary support even for interest and principle payment to bond subscribers and shareholders. Though the KBJNL was originally designed to be a financially autonomous body, its function is mainly along the lines of a government agency. In spite of some financial success, the main objectives of the financially autonomous irrigation agency (FAIA) remains unfulfilled. Clearly there is a lack of vision among the management staff about what a financially autonomous irrigation agency can do. It also indicates inadequate conceptualization of KBJNL as an autonomous body. However, one more corporation known as Karnataka Neeravari Nigam Limted (KNNL) has been formed to raise funds and manage eight irrigation projects in the Krishna basin of Karnataka. Four more corporations are being planned also.

This is not the first attempt in India in this direction. Andhra Pradesh State Irrigation Development Corporation was registered in 1974 to function as a corporation and access private and institutional finances. But cost recovery never even approached actual expenses; the corporation accumulated heavy losses and could not service its bank loans. It no longer acquires bank financing due to its arrears. The Gujarat Water Resources Development Corporation, wholly owned by the Government of Gujarat and registered under the Companies Act, engaged in groundwater exploration, construction, and management of the public tube wells, but faced worsening financial and operational conditions ever since its inception in 1975.[6] The 1994 finance committee suggested that the corporation should be ended (Kolavalli and Raju, 1995; Shah et al., 1995).

Four Indian states (Gujarat, Maharashtra, Karnataka, and Andhra Pradesh) have now set up corporations, or Nigams, that focus on mobilizing funds for surface irrigation. All four states started their corporations mainly to overcome the reduced budgetary allocations for their irrigation sectors. These corporations were broadly established along the lines of public sector companies to mobilize funds. Emphasis was on mobilizing funds from institutions, particularly those that are directly or indirectly regulated and/or are linked to governments rather than individuals.

Irrigation management transfer to water users associations: The Vaidyanathan Committee (GOI, 1992b) recommended that on the institutional front, user groups be involved in the management of the irrigation systems and that their role be gradually increased from management of minors to distributaries and then to main canal systems. The preconditions of carrying out this recommendation are (1) that there exist water user associations who can take delivery of water from irrigation authorities at the wholesale level, and (2) that there are measurement devices installed to measure volume of water at the distributory level. At present, both these preconditions are not satisfied in most of the irrigation projects. As a result, the recommendations of the committee can only be implemented in the long run with the gradual development of water users association (WUAs) and fixing measurement devices.

The government has been quite slow in moving in that direction. Nevertheless, the Ninth Five Year Plan set up a working group on participatory irrigation management (PIM), which recommended that farmers' involvement in the management of canal irrigation works should be given high priority. To start, 2,000 pilot projects should cover at least 2 to 3 percent of canal area irrigated in the country. Gradually, this should be increased to cover 50 percent of irrigated area under PIM. Although it requires quite a bit of spade work in terms of defining water rights, the role and jurisdiction of WUAs vis-à-vis that of the state irrigation departments, and how the disputes would be settled, and so on, a beginning has been made in some states. The announcement of a one-time management subsidy to states for the forming of WUAs in the central government budget of 1999-2000 is another positive step in inducing institutional reform.

Efforts at forming WUAs in India have so far been isolated attempts on a small scale. There were 4,420 WUAs functioning in the early 1990s (before the "big bang" in institutional reforms in Andhra Pradesh). The area under their operation was however only 0.33 percent of total irrigated area in the country (Rao et al., 1999). In many of these institutions in India, the main focus of these associations was the management of the irrigation systems through the involvement of farmers. Cost recovery and other financial aspects were not the motivating factors for such organizations. Gulati, Meinzen-Dick, and Raju (2002) point out that this aspect may be of much greater importance to the future of irrigation systems since state funds are shrinking and central government support is limited. This is true especially because few of these WUAs have really emerged as robust institutions and most die out once external support is withdrawn. In this context, it is noteworthy that so far, the impetus for irrigation management transfer in different states in India has come from external agencies, Indian government policy, and donor pressures (Brewer and Raju, 1995). This may influence the type of WUA and the legal framework within which these institutions operate. The need is, however, a statewide policy wherein the institutions are designed to suit the physical, technical, and sociopolitical framework of the individual state. One way to sustain these institutions is by making farmers co-owners of the systems through, for instance, equity shares, in a way that would allow them to participate in the management, design, and construction of the irrigation systems. This must be backed by a strong legal framework.

Only very recently has there been a large-scale effort at institutional reform initiated by the states themselves. The most progressive state in this regard is Andhra Pradesh (Raju, 2001). It has taken a lead in passing an act to transfer the management of irrigation systems to farmers' organizations. By 1999-2000, Andhra Pradesh alone had more than 10,000 WUAs; nearly 80 percent of them are minor irrigation tanks. So far, the WUAs have been set up on minor and distributory levels. Beyond the distributory level, the WUAs manage distribution of water and collection of dues thereof. Early indications show that the institutional reforms undertaken in Andhra Pradesh are reasonably successful in some respects (irrigated area expansion, water efficiency, small reduction in agency expenditure, users' participation in operation and maintenance) although it is too early to tell. However, macro-level issues are yet to be touched:

1. Project-level users' organization (federation of three tier organization) has not been formed.
2. It is feared that each project-level federation may encompass more than one administrative district and their "power" to regulate and operate the water distribution may go beyond the elected leaders of the state assembly. Some federations' budgets may be larger than a budget of a minister at the state level. This is a direct threat to the elected leaders of the local area.
3. Reservoir operations have to be restructured to meet the demands of the WUAs and their federations. Agencies have to provide the guarantee to supply at the agreed levels. They may be unprepared and find it difficult due to lower efficiency in the main system and lower than designed storage levels in the reservoirs.
4. Most important, financial sustainability of WUAs has to be obtained. Even after four years of WUA formation and functioning, they are not empowered to collect water fees and retain the agreed portion; it is still untested.

Other states would do well to watch Andhra and draw lessons from its experience. Many other states are indeed inching ahead on this line of action, most notably Rajasthan, Karnataka, Haryana, besides Gujarat and

Maharashtra, which already have an informal system of PIM. The donor agencies, such as the World Bank, are also insisting on forming farmers' groups and upward revision of canal irrigation water rates under their water resources consolidation (WRC) projects in states such as Tamil Nadu, Orissa, and Haryana.

Direction of Change in User Involvement

Overall, it appears that the change is in the right direction. But the speed of change is slow, and time will tell the exact nature and design of this experiment. The degree of success will depend upon the degree to which the user groups get interested in managing the canal networks, how much autonomy is granted to project authorities, and how much transparency is introduced in the management of funds. Learning from the experiments tried out so far, conditions for the success of WUAs can be outlined (Kolavalli, 1997). The results of a number of studies of Indian WUAs suggests that the major factors influencing the viability of WUAs are: wide-ranging and comprehensive changes in the legal framework and policies, autonomy of the WUAs, a new accountability of the irrigation department to the WUAs, and attitudinal changes in bureaucracy (Navalawala, 1994; GOI, 1997). With the Constitution (Seventy Third) Amendment Act passed by the Indian parliament, and the strengthening of grassroots institutions such as the panchayati raj, it is possible to think of transferring management to local-level institutions. The case of West Bengal points to the efficacy of decentralized management of infrastructure (Sen, 1993). Overall, performance of WUAs in canal, lift, and tank irrigation in the states of Gujarat, Maharashtra, and Tamilnadu have also been studied. The major results obtained from this and other studies include the following:[7]

1. When faced with a legal fiat that water shall only be sold to groups of users, farmers are quick to come together.
2. Some flexibility in determining water charges helps as allocation rules in successful WUAs often differ from region to region and depend on crops grown and corresponding irrigation schedules.
3. External support for WUAs may be necessary to create capability. A nongovernmental organization (NGO) or the irrigation department may play this role for a limited period of time
4. The creation of WUAs changes the strategic position that the irrigation department and its line agencies have had for a long period of time. Likely reactions of this set of vested interests must be taken into

account. In addition, their experience needs to be tapped within the new institutional framework.

Although, certain factors have been identified as crucial to the success of WUAs, ultimately the key factor is designing institutions that are appropriate for a given socioeconomic, legal, and political context.

Full agency control is often reported as a form of management, although in practice, it does usually involve some user representation, informal though it may be. Similarly, full WUA control is also rare in practice. "Agency O&M, user input" is the more common form of irrigation management. Under shared management, agencies are responsible for O&M, but not completely. WUAs share some O&M responsibilities while chiefly representing users. Many irrigation management transfers today are characterized by the WUAs subsuming the responsibilities of O&M while the state agencies continue to own and regulate the system. Another interaction system is when the WUA not only manages the system (that is, has O&M responsibilities), but also owns the system. The state agencies have only a regulatory function here. What model is followed depends largely on the system level. At the river basin level for instance, the state usually plays a dominant role and the users, very little. In the main system level, again the state retains ownership and O&M responsibilites, although user representation may enable them to participate in decision making. Shared management and O&M by WUA is usually found at the system or distributory level. WUA ownership is often seen as the culmination of management transfer programs at the distributory level. In addition to the formation of WUAs, the ownership of canal networks starting from the distributories through the issue of water bonds must be given priority. Another policy is the establishment of tradeable water rights (Meinzen-Dick and Bruns, 2000; Rosegrant and Binswanger, 1994). This would require investment in irrigation technology for conveyance, metering, diversion, and institutional improvement and would result in more efficient water use.

It has also been suggested that individual irrigation systems be made financially autonomous, like corporations, so that their income depends chiefly on the revenues that they collect for the irrigation service they provide. This would provide incentive for stricter collection of revenue from users apart from superior service that would facilitate better recovery. Although in terms of efficiency these corporations would be better performers than government departments, they are likely to be natural monopolies. It would therefore be essential to ensure transparency in the transactions and capital expenditures of these agencies. The need to keep their expenses transparent and under control would be important if private sector participation were introduced. In this context, each state should have an independent regulatory commission such as an independent regulatory commission for canal irrigation (IRCCI) like those for electricity supply, with decentralized agencies at the user group level.

Institutional reforms must necessarily accompany price reforms and must focus on the input-supplying agencies, the modes of their operation, and also extend to the formation of new institutional mechanisms where considered necessary (Gulati and Moench, 1997). Institutional reforms would provide the right environment for undertaking price reforms by depoliticizing or disengaging the state from the management of irrigation systems. It would make individual irrigation systems financially autonomous so that their incomes are dependent on the revenue they collect. In addition, it would enable the linking of the payments for irrigation service with the quality of service offered by the agency in charge of the system, which has largely been absent so far. Unless this functional link between the revenue and service and performance (Gulati, Svendsen, and Choudhury, 1995) is accomplished, the chances of successful reforms in this critical sector will remain very low.

CONCLUDING REMARKS

Several facts come to the fore in the context of irrigation subsidies in India:

1. Pricing of water is way below the level that any theory would suggest, be it demand-side pricing based on the marginal valve product (MVP) or supply-side pricing based on long run marginal cost (LRMC).
2. The collection of charges is poor, making the actual receipts per unit of water even lower than their price.
3. The quality of service provided by irrigation agencies is not satisfactory, so that the farmers often have to resort to hiring/buying water from fellow farmers. This alternative costs the farmer more than what he pays to the irrigation authorities.
4. In spite of this, the farmer is unwilling to pay higher charges since he does not anticipate a related improvement in the quality of the service.
5. Raising canal water charges under the given institutional structure and quality of service is a point of contention between the bureaucracy and policymakers on the one side and farmers and their representatives on the other. Input-supplying agencies are inefficient, and are the project authorities in case of canal irrigation.

This offers an opportunity for reform and is surely a win-win situation for the farmers as well as policymakers. This can be achieved by ensuring that the quality of irrigation service is linked to the price being charged and that the costing of this service is transparent and there is an effort to keep these costs down through innovative methods. Canal irrigation subsidies can be reduced/rationalized without adversely affecting agricultural output. Farmers have the capacity to pay for higher irrigation charges, and many are also willing to pay, but they need to be assured of better irrigation service and plugging of leakages in irrigation funds. For this to occur greater autonomy must be given to irrigation authorities, farmers must be involved in management and decision making, an independent regulatory commission must be established, and the system must be more transparent than what it is today. With these institutional reforms, one hopes that canal irrigation in India will be able to overcome not only the issue of subsidies, recovering O&M expenses and 1 percent of cumulative capital expenditures at historical prices, but also be on a sustainable path of higher efficiency—both physical and financial.

NOTES

1. This is reflected in plan outlays on irrigation. During the last 45 years, starting with the first Five Year Plan (FYP) in 1951-1952 to 1996-1997, the nation spent almost Rs.920 billion at historical prices on irrigation. This includes the expenditure by the government and from institutional sources, but excludes the expenditure financed from farmers' own resources, on major and medium irrigation, minor irrigation, command area development (CAD) and flood control. At 1996-1997 prices, this figure stands at a staggering level of Rs.2313.87 billion (GOI, 1997).

2. Minor irrigation, officially defined as projects with less than 2,000 ha command area, includes surface irrigation through tanks, watersheds, lift irrigation, and even small canal networks, but largely well irrigation consisting of dugwells, and shallow and deep tubewells.

3. The choice of 1 percent of cumulative capital cost is itself highly debatable and arbitrary. But this was decided on unanimously by the Vaidyanathan Committee based on the "Delphi Principle."

4. In the case of sugarcane in Maharashtra, for example, irrigation cost per hectare as a ratio of gross revenue from sugarcane farmers declined from 11.2 percent in 1968 to only 5.9 percent in 1995. Its share in net revenue decreased from 19.3 percent in 1968 to 9.7 percent in 1995. In the case of paddy in Punjab, the ratio of irrigation cost to net revenue per hectare has fallen from 38 to 13-14 percent.

5. *Krishna Bhagya Jal Nigam Limited:* At the root of the KBJNL formation lies the sharing of the Krishna river water. The river flows through Maharashtra, Karnataka, and Andhra Pradesh states. In 1971, the Krishna Waters Dispute Tribunal (KWDT) was set up to allocate utilization levels of Krishna waters. The tribunal

reported its findings by 1973, and the states provided the answers for the queries raised by the tribunal. In 1976, the tribunal said that the award (popularly known as the Bachawat Award) may come under review by May 2000. (However, due to lack of initiative from these states, the award has not been reviewed and the old status is continuing into 2002.) Thus a deadline was set to utilize the given water allocations by three states. Under this award, Karnataka is to utilize 734 TMC (20.7 million ha m) of water from Krishna river.

The Upper Krishna Project (UKP) was developed to take advantage of the award. The state government sought World Bank assistance for UKP during 1980. The World Bank gave two credits: one expired by 1986 and another by June 1997, for a total loan of Rs. 5.48 billion. Meanwhile, in 1988, the state felt the need for an authority to look into required land acquisition, which was posing a major problem in project implementation.

6. The corporation has accumulated a loss of over Rs. 700 million and depends on the government for large subsidies to continue its operations. It faces constraints on what it can charge for its services and cost escalation add to the deficit every year. Nearly 20 percent of the deep tubewells that were not being adequately utilized have been closed down; the corporation began leasing out the tubewells to users in 1987 to reduce costs. It had a staggering wage bill of Rs. 220 million for a staff of 6,400, while its annual gross income was only Rs. 60 million.

7. The requisites of robust WUAs have been studied in some detail based on the empirical evidence from several countries (Subramaniam et al., 1995).

REFERENCES

Brewer, J., S. Kolavalli, A.H. Kalro, G. Naik, S. Ramnarayan, K.V. Raju, R. Sakthivadivel. 1997. *Irrigation Management Transfer in India: Policies, Processes and Performance,* report, Colombo: Indian Institute of Management Ahmedabad and International Irrigation Management Institute.

Brewer, J.D. and K.V. Raju. 1995. *Irrigation Management Transfer Policies and Law.* Paper presented at the Workshop on Status of Irrigation Management Transfer in India. March 13, New Delhi.

Dhawan, B.D. 1999. *Studies in Indian Irrigation.* New Delhi: Commonwealth Publishers.

Government of India. 1992a. *Eighth Five-Year Plan, 1992.* New Delhi: Author.

Government of India. 1992b. *Report of the Committee on Pricing on Irrigation Water, Planning Commission.* New Delhi: Author.

Government of India. 1995. *Private Sector Participation in Irrigation and Multipurpose Projects: Report of the High Level Committee* (Chaired by P.V. Rangayya Naidu). Volumes I and II, Main Report and Annexures. New Delhi: Ministry of Water Resources.

Government of India. 1996. *The India Infrastructure Report: Policy Imperatives for Growth and Welfare.* Report submitted by the Expert Group on the Commercialisation of Infrastructure Projects (Chaired by Rakesh Mohan). 3 volumes. New Delhi: National Council for Applied Economic Research.

Government of India. 1997. *Economic Survey, 1996-97.* Author.

Gulati, A., R. Meinzen-Dick, and K.V. Raju. 2002. *Institutional Reforms in Indian Irrigation.* New Delhi: SAGE Publications.

Gulati, A. and M. Moench. 1997. *Towards Reforms in Groundwater Irrigation.* The World Bank.

Gulati, A., M. Svendsen, and N.R. Choudhury. 1994. Major and Medium Irrigation Schemes: Toward Better Financial Performance. *EPW,* 29(26).

Gulati, A., M. Svendsen, and N.R. Choudhury. 1995. *Capital Costs of Major and Medium Irrigation Schemes in India.* In M. Svendsen and A. Gulati (Eds.), *Strategic Change in Indian Irrigation,* New Delhi: MacMillan India Limited.

Kolavalli, S. 1997. *Assessing Water User's Associations.* Paper presented at the Seminar on Agriculture and Environment, Delhi School of Economics, Delhi, December.

Kolavalli, S. and K.V. Raju. 1995. Turnover of Public Tubewells by Gujarat Water Resources Development Corporation. In *Irrigation Management Transfer: Selected Papers from the International Conference on Irrigation Management Transfer, Wuhan, China, September 20-24, 1994,* Water Reports, no. 5, edited by S.H. Johnson, D.L.Vermillion, and J.A. Sagardoy. Rome: FAO and IIMI.

Meinzen-Dick, R. and B.R. Bruns (Eds.). 2000. *Negotiating Water Rights.* London: Intermediate Technology Publications; New Delhi: Vistaar Publications.

Navalawala, B.N. 1994. *Planning Perspective of Farmer's Participation in Irrigation Management in India.* A paper distributed at the International Conference on Irrigation Management Transfer, Wuhan, China, September 20-24.

Oblitas, K.P., J.R. Pringle, G. Qaddumi, M. Halla, and P. Jayanatha. 1999. Transferring Irrigation Management to Farmers in Andhra Pradesh, India. World Bank technical paper No. 449. Washington, DC: The World Bank.

Raju, K.V. 2001. Participatory Irrigation Management in Andhra Pradesh: Promise, Practice and a Way Forward. Working Paper No. 65, Institute for Social and Economic Change, Bangalore.

Rao Hanumantha, C.H. 1994. *Agricultural Growth, Rural Poverty and Environmental Degradation in India.* New Delhi: Oxford University Press.

Rao Hanumantha, C.H, B.D. Dhawan, and A. Gulati. 1999. Toward Reforms in Indian Irrigation: Price and Institutional Policies. Paper prepared for the NCAER-IEG-World Bank Workshop on Reforms in Indian Agriculture for Growth, Efficiency, Equity and Sustainability held on April 15-16, 1999 at Delhi.

Repetto, R. 1986. *Skimming the Water: Rent Seeking and the Performance of Public Irrigation Systems.* Research Report No. 4, Washington, DC: World Resource Institute.

Rosegrant, M.W. and H.P. Binswanger. 1994. Markets in Tradable Water Rights: Potential for Efficiency Gains in Developing Country Water Resource Allocation. *World Development,* Vol. 22, No. 11.

Sen, A. 1993. *Agriculture in Structural Adjustment.* Paper presented at the Seminar on Agricultural Reforms in India, Indira Gandhi Institute of Development Research, India.

Shah, T., V. Ballabh, K. Dobrial, and J. Talati. 1995. Turnover of State Tubewells to Farmer Cooperatives: Assessment of Gujarat's Experience. In *Irrigation Man-*

agement Transfer: Selected Papers from the International Conference on Irrigation Management Transfer, Wuhan, China, 20-24 September, 1994, Water Reports, no. 5., edited by S.H. Johnson, D.L. Vermillion, and J.A. Sagardoy. Rome: FAO and IIMI.

Subramaniam, A., N.V. Jagananthan, and R. Meinzen-Dick (Eds.). 1995. User Organisations for Sustainable Water Services, World Bank Water Resources Seminar, Virginia, December 11-13.

Vaidyanathan, G. 1993. Consumption, Liquidity Constraints and Economic Development. *Journal of Microeconomics,* 15: 591-610.

Veeraiah, C. and N. Madamkumar. 1994. *Waste of Irrigation Water Under Major Irrigation Projects in Andhra Pradesh and Its Prevention.* Paper presented at the Seminar on Wastage of irrigation Water Under Major Irrigation Projects, Sri Venkateswara University, Tirupati, September.

Wade, R. 1987. The Management of Common Property Resources: Finding a Cooperative Solution, *Research Observer,* 2(2): 219-234.

World Bank. 1997. India: Water Resources Management Sector Review. Final Draft, May 16, 1997.

PART V:
MARKET REFORMS,
DIVERSIFICATION,
AND FOOD SECURITY

Chapter 12

Agriculture Diversification in South Asia: Patterns, Determinants, and Policy Implications

P.K. Joshi
Ashok Gulati
Pratap S. Birthal
Laxmi Tewari

INTRODUCTION

Most of the South Asian economies have been undergoing a process of economic reform since the late 1980s. They are gradually adopting trade liberalization as a policy plank. The unfolding globalization of agriculture, however, has thrown new challenges and opportunities to the agrarian sectors in these countries. Although there are apprehensions that the influx of subsidized cheap imports from the developed countries will adversely affect their agriculture, on the other hand, there is evidence that these countries are able to increase their agricultural exports, especially of high-value and labor-intensive commodities. This opens up a window of opportunity as South Asian agriculture experiences shrinking of its holdings, decelerating technological advances in staple crops, declining investment in agriculture, and increasing degradation of natural resources.

Diversification of agriculture in favor of more competitive and high-value commodities requires an important strategy to overcome many of these emerging challenges. If carried out appropriately, diversification can be used as a tool to augment farm income, generate employment, alleviate poverty, and conserve precious soil and water resources. Several micro-level studies support this proposition (von Braun 1995; Pingali and Rosegrant 1995; Ramesh Chand 1996; Ryan and Spencer 2001).

A sound understanding about the patterns of agricultural diversification and the constraints it faces would help in crafting appropriate policies regarding institutional arrangements and creation of adequate infrastructure, which could benefit a large mass of small and marginal holders. This study is an attempt in this direction. Specifically, the study intends to (1) examine the extent, nature, and speed of agricultural diversification in South Asian countries, (2) identify determinants of agricultural diversification, and (3) assess implications of agricultural diversification on food security, employment, and sustainable use of natural resources.

The study is confined at two levels: (1) macrolevel, and (2) mesolevel. At the macrolevel Bangladesh, Bhutan, India, Maldives, Nepal, Pakistan, and Sri Lanka have been studied in terms of diversification of agriculture. At the mesolevel, more disaggregated analysis across different regions in India has been attempted.

MAPPING PATTERNS OF DIVERSIFICATION

The concept of diversification conveys different meaning to different people at different levels. For example, at the national level, it generally conveys a movement of resources, especially labor, out of agriculture to industry and services, a sort of structural transformation. Within agriculture, however, diversification is considered a shift of resources from one crop (or livestock) to a larger mix of crops and livestock, keeping in view the varying nature of risks and expected returns from each crop/livestock activity, and adjusting it in such a way that it leads to optimum portfolio of income.

This definition of diversification needs to be distinguished from movement of resources from low-value commodity mix (crops and livestock) to a high-value commodity mix (crops and livestock), as it may often be reflected in an increasing degree of specialization (reducing diversity) to high-value activities, especially at the farm level. It is precisely this movement to high-value agriculture that is of great interest to us in this chapter because it indicates yet another way to augment income, besides the traditional ways of increasing yield, area, or cropping intensity.

Thus, based on these various definitions, the nature of diversification can be broadly described as (1) a shift of resources from farm to nonfarm activities, (2) use of resources in a larger mix of diverse and complementary activities within agriculture, and (3) a movement of resources from low-value agriculture (crops and livestock) to high-value agriculture (Hayami and Otsuka 1992; Vyas 1996; Delgado and Siamwalla 1999).

Measuring Diversification

Quite a few methods explain either concentration (i.e., specialization) or diversification of commodities or activities in a given time and space by a single indicator. Important ones include:

1. Index of maximum proportion,
2. Herfindal index,
3. Simpson index,
4. Ogive index,
5. Entropy index,
6. Modified Entropy Index, and
7. Composite entropy index (Kelley, Ryan, and Patel 1995; Pandey and Sharma 1996; Ramesh Chand 1996).

Each method has some limitation and/or superiority over the other. Considering our objective of assessing the extent of diversity in crop, livestock, and fisheries activities, we used the Simpson Index. The index provides a clear dispersion of commodities in a geographical region and ranges between 0 and 1. If there exists complete specialization, the index moves toward 0. The index is easy to compute and interpret, as follows:

$$SID = 1 - \sum_{i-1}^{n} P_i^2 \qquad (12.1)$$

Where, SID is the Simpson Index of Diversity, and P_i is the proportionate area (or value) of ith crop/livestock/fishery activity in the gross cropped area (or total value) of output.

The nature and patterns of diversification were examined by looking into temporal changes in area, production, and value of different crops, and quantity (and/or value) of livestock and fisheries activities. To estimate the speed of diversification in favor of high-value commodities, annual compound growth rates of area, production, and yield of different crop/livestock activities were computed.

For South Asian countries, the SID was computed for crop sector only, while the livestock sector was assessed separately to examine its performance in different countries. The diversity was probed within the crop sector, dividing it into broad subsectors such as cereals, pulses, oilseeds, fruits, vegetables, spices, and other crops. For India, the SID was computed for entire agricultural sector-comprising crop, livestocks, and fishery subsectors, as well as within each subsector.

Determinants of Diversification

Several forces influence the nature and speed of agricultural diversification from staple food to high-value commodities. Earlier evidence suggests that the process of diversification out of staple food production is triggered by rapid technological change in agricultural production, improved rural infrastructure, and diversification in food demand patterns (Pingali and Rosegrant 1995). These are broadly classified as demand- and supply-side forces. The demand-side forces that have been hypothesized to influence the diversification include per capita income and urbanization. On supply-side forces, the diversification is largely influenced by infrastructure (markets and roads), technology (relative profitability and risk in different commodities), resource endowments (water and labor), and socioeconomic variables (pressure on land and literacy rate).

The Generalized Least Square (GLS) technique with a fixed-effect model was applied to examine how different forces have influenced crop and livestock diversification in India. The analysis is based on pooling of cross-section and time-series information from major states (19 out of 28) in India for the period 1980-1981 to 1998-1999.[1] The GLS eliminates the effect of heteroscedasticity arising due to cross-section data, and autocorrelation as a result of time-series data. The following model was used to examine the determinants of diversification:

$$D_c \text{ or } D_l = f \text{ (TECH, INFR, PROF, KNOW, DEMA, RAIN)} \quad (12.2)$$

The variables were defined as follows: The dependent variable, D_c or D_l was defined in two ways: (1) Simpson Index of Diversity in crop sector (SID_c) and livestock sector (SID_l), and (2) index of output values of horticultural commodities and livestock commodities at constant prices with base 1980-1981. Results for the latter were found statistically superior, and therefore, used for discussion.

Independent variables were broadly grouped into

1. technology (TECH) related,
2. infrastructure (INFR) related,
3. profitability (PROF) related,
4. resources and information (KNOW) related,
5. demand (DEMA) side, and
6. climate (RAIN) related.

To capture their effect, few proxy variables were used in the model. For technology (TECH), these included: proportionate area under high-yielding varieties of food grain crops (percent), fertilizer use (kg per ha), proportion of gross irrigated area to gross cultivated area (percent), and mechanization (number of tractors per 1,000 ha area). For infrastructure (INFR) the proxy variables were market density (number of markets per 1,000 ha of gross cropped area), and roads length (square km per 1,000 ha of gross cropped area). Relative profitability of high-value enterprises with cereals and other crops was the proxy for profitability (PROF) related variables. Average size of land holding (ha) and proportion of small landholder in total holdings were proxys for available resources, and rural literacy (percent) for information- (KNOW) related variables. On demand-side (DEMA) variables, urbanization (percent urban population) and per capita income (rupees per person) were used in the model. Annual rainfall (mm) was used to define the climate- (RAIN) related variable in the model. The specification of variables and their expected signs are given in Table 12.1. Different combinations of independent variables were tried to arrive at the best-fit equations. Both linear and double log equations were estimated and the best ones were selected.

Data

The study covers a period of two decades from 1980-1981 to 1999-2000. These two decades were divided into two periods: (1) 1980-1981 to 1989-1990, and (2) 1990-1991 to 1999-2000. There were two obvious reasons for studying the past two decades. First, the historical evidence showed that the impact of the "green revolution" in South Asian countries was gradually fading during the 1980s. Second, the process of economic reforms started in most of the South Asian countries in the late 1980s or early 1990s, and also most of them bound themselves to the commitments under the World Trade Organization (WTO), which is likely to have serious implications for their respective agricultural sectors. Hypothetically, the slowing down of the green revolution and gradual opening up of economies will lead to greater diversification of agriculture in favor of high-value commodities.

The data for the study were collected from various published sources. For South Asian countries, the most important data source for crops and livestock was the Food and Agricultural Organization (FAO) statistical database (FAOSTAT). This was complemented by the country-specific statistical yearbooks. For India, the study covered more disaggregated analyses by including the states, therefore, the study relied heavily on the national statistical bulletins (CMIE 2001).

TABLE 12.1. Specification of variables and their expected signs for diversification

Drivers	Indicator	Unit	Expected sign
Technology	Area under HYV of food grains	Food grain HYV area to total food grain area (%)	−
	Fertilizer use	Kg/ha	−
	Irrigated area	Proportionate irrigated area to the gross cropped area (%)	−
	Mechanization	No. of tractors 1,000/ha	−
Infrastructure	Market density	No. of markets/1,000 ha of gross cropped area	+
	Road length	Square km/1,000 ha of gross cropped area	+
Profit	Relative profitability	Profit from fruits and vegetables in relation to cereals, pulses, oilseeds, sugarcane	+
Resources and information	Holding size	Proportion of small holders in total holdings (%)	+/−
	Literacy	Percent literate population in rural areas	+
Demand side	Urbanization	Urban population in percent	+
	Per capita income	Rs./person	+
Climate	Rainfall	`millimeter	−
Period	Dummy	1981-1990 = 0; 1991-1999 = 1	+

Note: In case of livestock, proportion of crossbred cattle of total cattle (percent) was used as a proxy for technological advancement, with an expected negative sign for diversification.

PATTERNS OF AGRICULTURAL DIVERSIFICATION IN SOUTH ASIA

Agriculture is the mainstay of economic growth in South Asia. A large proportion of the population depends on agriculture for income, employment, and food security. Agricultural performance in South Asia is improving. The annual compound growth rate of agriculture was 3.7 percent during the 1990s compared to 3.2 percent in 1980s. Besides the continuing role of high-yielding rice and wheat varieties in South Asian countries, the agricultural growth is attributed to diversification in favor of high-value commodities. South Asia is diverse in climate, soils, and other agroecological features. Diversity permits South Asian farmers to cultivate a variety of crops, rear different species of livestock, and catch a wide range of fish species from various sources. The Simpson Index of Diversity (SID) for South Asia was 0.64 in the triennium ending (TE) 1999-2000, up from 0.59 in TE 1981-1982. This shows that South Asia is gradually diversifying its crop sector in favor of high-value commodities, especially fruits and vegetables (Tables 12.2 and 12.3). Bangladesh, Bhutan, and Nepal show less diversity as compared to other countries. Bangladesh has specialized in rice. More than three-fourths of the area in the country is under rice; the remaining one-fourth is highly diversified, which is a result of some policy initiatives in different plan periods. Nepal and Bhutan are aiming to have a higher degree of self-sufficiency in basic food grain, and therefore, concentrating more toward cereals, particularly rice, wheat, and maize.

Two sources of crop diversification are used: area augmentation and crop substitution. Area augmentation comes through utilization of fallow lands and rehabilitation of degraded lands, or increasing cropping intensity.[2] Ta-

TABLE 12.2. Extent of diversification and its sources in South Asian countries

Country	Simpson Index of Diversity in triennium ending			Sources of diversification 1991-1992 to 1999-2001(%)*	
	1981-1982	1991-1992	1999-2000	Cropping intensity	Crop substitution
Bangladesh	0.39	0.36	0.35	64.67	35.33
Bhutan	0.37	0.48	0.44	97.82	2.18
India	0.61	0.65	0.66	36.63	63.37
Maldives	0.77	0.77	0.77	83.22	16.78
Nepal	0.39	0.40	0.41	84.79	15.21
Pakistan	0.54	0.56	0.57	76.56	23.44
Sri Lanka	0.76	0.77	0.75	78.90	21.10
South Asia	0.59	0.63	0.64	42.98	57.02

Source: Computed by authors from the data derived from FAOSTAT.
*The columns were computed as follows: Gains in Cropped Area (A) = Change in Gross Cropped Area (B) + Crop Substitution (C). Since (A) and (B) are known, (C) is the residual.

TABLE 12.3. Annual compound growth rates (%) of area, production, and yield of major commodity groups in South Asian countries

Commodity group	1980-1990			1991-2000		
	Area	Production	Yield	Area	Production	Yield
Cereals	−0.01	3.08	3.09	0.34	2.45	2.11
Pulses	0.04	2.37	2.33	−0.02	0.72	0.74
Oilseeds	1.72	5.46	3.68	0.95	2.05	1.09
Vegetables	1.41	3.33	1.89	2.44	2.59	0.14
Fruits	1.71	2.61	0.89	2.40	5.61	3.14
Dry fruits	1.98	3.56	1.55	3.62	4.30	0.66
Spices	1.46	4.27	2.77	0.68	2.47	1.78

ble 12.2 shows that in most of the countries, crop diversification is coming from area expansion, with some exception of crop substitution in India and Sri Lanka. Incidentally, in Nepal, Pakistan, and Sri Lanka, area expansion is also coming from deforestation, which is a cause of concern from the environmental point of view.

To examine the nature and speed of agricultural diversification, production performance and area expansion of different commodities was assessed. Annual compound growth rates in area, production, and yield of major commodity groups in South Asia during the decades of the 1980s and 1990s are given in Table 12.3. Production performance of nonfood commodities was superior to the food commodities. Among food grain groups, cereals performed better than pulses. The cereal sector was specializing in favor of rice and wheat because of overriding concern for food self-sufficiency in all the South Asian countries. Availability of improved and high-yielding rice and wheat varieties induced specialization in favor of these crops. These replaced sorghum, millets, and barley. Performance of pulses was disappointing during 1990s. These were relegated to marginal environments. With the availability of irrigation and improved varieties of rice and wheat, a large share of the pulses area was shifting in favor of rice and wheat. There are some exceptions as well. For example, lentil and pigeon pea are increasing in Nepal. Black gram and green gram and to some extent chickpea are emerging in Indian rainfed regions. In Pakistan, chickpea is gaining importance.

Different countries grow a large number of vegetable and fruit crops. Fruits (both fresh and dry) and vegetables showed good performance during the 1980s and 1990s. Fruits and vegetables are highly diversified in all the countries. Livestock and fisheries sectors also flourished during the past two decades (Table 12.4).

TABLE 12.4. Growth performance of livestock activities and fisheries in South Asia

Commodity group	Annual compound growth rates (% per annum)					
	1981-1990			1991-2000		
	Number	Production	Yield	Number	Production	Yield
Milk						
Cow	2.33	4.86	2.53	2.10	5.50	3.40
Buffalo	4.11	4.84	0.73	2.53	5.10	2.57
Total	—	4.93	—	—	5.17	—
Poultry						
Chickens	9.26	10.51	1.25	5.72	5.66	−0.06
Eggs	4.43	7.19	2.76	4.76	4.49	−0.27
Total Meat	—	4.30	—	—	2.12	—
Fish	—	5.20	—	—	3.50	—

Agriculture is gradually diversifying in the subcontinent with some intercountry variation. Diversification was observed in favor of high-value commodities. Since their share in area and production was too low in comparison to food grain crops, the extent of diversification was unnoticed. It came despite a lack of policy initiatives and poor infrastructure in the subcontinent, therefore, its pace slowed down. It is reflected from the performance of different commodity groups, which was better during the 1980s than 1990s. During the 1980s, growth in production was mainly attributed to yield increase, while area expansion was the major source during the 1990s. Slowing down in yield levels is ascribed to (1) technological slack, (2) weak input delivery system, and (3) poor infrastructure. To accelerate the pace of diversification and harness its potential benefits, there is a need to introduce appropriate technologies and create suitable institutions and infrastructure. Domestic market reform to support agricultural diversification is necessary. This calls for correcting several outdated market acts, which impede the pace of agricultural diversification in favor of high-value enterprises. For example in India, the existing Agricultural Produce Market Committee (APMC) empowers the state governments to set up markets for agricultural development of efficient and transparent agrimarketing, use modern pre- and postharvest techniques, and set up quality standards and enforcement. It discourages private sector participation in developing markets and has led to inefficiencies in marketing. Similarly, many processing units/products are still reserved for small-scale and cottage industry.

PATTERNS OF AGRICULTURAL DIVERSIFICATION IN INDIA

Agricultural diversification in India is gradually picking up momentum in favor of high-value crops/livestock/fishery activities to augment incomes rather than a coping strategy to manage risk and uncertainty. However, the nature of diversification differs across regions due to wide heterogeneity in agroclimatic and socioeconomic conditions. Therefore, it would be interesting to delineate the key regions and subsectors of agriculture where diversification is catching up fast. This section is an attempt to unfold these features and diagnose the regional patterns of agricultural diversification in India.

Crops, livestock, fisheries, and forestry constitute the core subsectors of agriculture. The crop subsector is the principal source of generating income in agriculture followed by the livestock sector (Table 12.5). Strong synergy exists in crop and livestock subsectors, both being complementary to each other. The fisheries subsector is prominent in the coastal areas, and forestry in the hilly regions.

The share of crop sector in the agricultural gross domestic product marginally declined during 1980s (from about 76.25 percent in TE 1981-1982 to 73.65 percent in TE 1990-1991) and then recovered slowly during 1990s (rising to 74.91 percent in TE 1997-1998). There are two obvious reasons for this: (1) normal monsoons during most of the years in the 1990s, and (2) greater emphasis on horticultural crops, which led to higher production. On the other hand, there was a quantum jump in the share of the livestock subsector during 1980s, which escalated from about 18 percent in TE 1981-1982 to 23 percent in TE 1990-1991. Although the value of livestock (at constant prices) during the 1990s nearly doubled, its share in agriculture remained stagnant at 23 percent. This was because the value of the bigger crop subsector increased more than that of the livestock subsector. The

TABLE 12.5. Share of individual sectors (%) in gross value of agricultural output in India at 1980-1981 prices

Region	Crop			Livestock			Forestry			Fishing		
	1981-1982	1990-1991	1997-1998	1981-1982	1990-1991	1997-1998	1981-1982	1990-1991	1997-1998	1981-1982	1990-1991	1997-1998
Eastern	73.90	70.65	75.84	16.83	24.44	21.36	6.65	2.35	0.89	2.62	2.56	1.92
Northeast	77.48	75.95	78.96	13.74	18.44	18.26	5.68	3.08	1.19	3.10	2.54	1.60
Northern	75.73	73.98	72.87	21.62	24.94	26.45	2.44	0.75	0.41	0.21	0.33	0.27
Southern	80.06	78.46	77.38	15.64	19.10	21.19	2.29	1.28	0.56	2.01	1.16	0.88
Western	75.71	71.65	73.01	18.95	23.79	24.53	4.08	3.40	1.60	1.25	1.16	0.86
All India	76.25	73.65	74.91	18.27	23.09	23.24	3.95	1.91	0.85	1.53	1.35	1.00

same was true for the fisheries subsector. The value of the fisheries subsector swelled by about 50 percent during the 1990s, but its share in agricultural gross domestic product marginally reduced to about 1 percent in TE 1997-1998 from 1.35 percent in TE 1990-1991. This is despite the fact that fisheries production during the decade of 1990s increased at an annual rate of 5.35 percent.

Regionally, the patterns, by and large, reveal shifts from the crop to the livestock subsector during 1980s and 1990s. The exceptions were eastern and northeastern regions, where the shares of both crop and livestock subsectors in total value of agricultural output were rising at the cost of fisheries and forestry. In the southern region also, the share of fisheries and forestry in total output during the 1980s and 1990s was diminishing. The livestock subsector across different regions increased as a result of growing demand for livestock products, such as milk, meat, eggs, etc. The "cooperative model" linking growers, processors, and retail distribution seems to have attained a reasonable degree of success contributing to increased livestock production.

Diversification Within the Crop Subsector

The crop sector is steadily diversifying in India. The SID slowly moved-up from 0.63 in TE 1981-1982 to 0.66 in TE 1998-1999 (Table 12.6). Nonfood-grain crops have gradually replaced food grain crops, with the latter

TABLE 12.6. Share of food and nonfood crops in cropping pattern and value of output in India at constant prices

Region	Simpson Index of crop diversity TE 1981-1982	Simpson Index of crop diversity TE 1998-1999	TE 1981-1982 Food crops Area	TE 1981-1982 Food crops Value	TE 1981-1982 Nonfood crops Area	TE 1981-1982 Nonfood crops Value	TE 1998-1999 Food crops Area	TE 1998-1999 Food crops Value	TE 1998-1999 Nonfood crops Area	TE 1998-1999 Nonfood crops Value
Eastern	0.50	0.53	81.63	51.73	18.37	48.27	73.83	43.04	26.17	56.96
Northeast	0.43	0.46	70.11	44.43	29.89	55.77	65.06	35.80	34.94	64.2
Northern	0.53	0.51	77.42	54.92	22.58	45.08	76.86	53.74	23.14	46.26
Southern	0.68	0.75	62.86	41.82	37.14	58.18	53.08	28.20	46.92	71.80
Western	0.66	0.72	71.92	44.44	28.08	55.56	61.85	36.10	38.15	63.90
All India	0.63	0.66	70.34	48.05	29.66	51.95	65.44	39.85	34.56	60.15

going up from about 30 percent of area in TE 1981-1982 to 35 percent in TE 1998-1999, but in value terms it went up significantly from about 52 percent to 60 percent in respective periods. Non-food grain crops, such as oilseeds, fruits, vegetables, spices, and sugarcane have taken over coarse cereals.

Regional patterns in diversification of the crop sector were quite stark (Table 12.6). The southern region was highly diversified followed by the western region. The process of diversification was modest in these regions during the 1980s and 1990s. These were the only regions that accomplished higher agricultural growth during the 1990s over the preceding decade. These regions swiftly moved toward more non-cereal crops, which perhaps contributed to accelerating agricultural growth. These regions are relatively less developed in irrigation and largely rely on rainfall. Pulses and oilseeds are low-water-requirement crops, and therefore found a niche in these regions. Like pulses and oilseeds, these regions also witnessed substantial increases in area under fruits and vegetables. Government-supported programs promoted the cultivation of fruits and vegetables. Among others, a watershed program facilitated conservation of rainwater and gave higher priority for cultivation of fruits and vegetables. Among cereals, maize is picking up fast in the southern region and to some extent, in the western region. Maize is gaining importance in meeting the growing demand for poultry feed in these regions. Among all the crops gaining in these regions, oilseeds are under serious threat in the wake of import liberalization of edible oils as the cost of imported oils, especially palmolein, is much lower than the prevailing domestic price. To sustain oilseed production, technical efficiencies in production and processing will have to be improved through better management and technology adoption.

The northern region favors rice and wheat crops. Favorable government pricing policies, assured procurement, high-yielding technologies, and irrigation development have encouraged farmers to allocate more area to these crops. Rice and wheat have replaced coarse cereals and pulses from the region. Over time, the region has concentrated more on cereals and only marginally diversified in noncereal commodities. With the availability of short duration black gram, green gram, and pigeon pea, pulses are slowly regaining in this region. Other crops that are gaining importance in this region are sugarcane, vegetables, and fruits. Ironically, reports show that extensive cultivation of rice and sugarcane are causing negative externalities related to soil and water resources. The soil fertility with respect to macro- and micronutrients is declining with continuous cultivation of these crops. The water resource of the region is also depleting. The negative externalities have adversely affected the total factor productivity of rice-wheat-based cropping systems in this region (Kumar et al. 1998). To sustain the food security and further augment export of rice (both basmati and nonbasmati), there is a need to improve the water use efficiency. The region has potential for cultivating a variety of fruits and vegetables, but the future of these crops relies on developing appropriate infrastructure suitabe for linking production and consumption.

The eastern region is the most backward region with respect to per capita income, growth in agriculture, and development of infrastructure. The yield levels are comparatively low because of uncertain production environment and poor adoption of improved varieties and technologies. Overall, the region is food based and the extent of diversification is relatively low as compared to other regions. This region is largely concentrating on rice. The humidity and high rainfall makes its cultivation more favorable. However, there is high diversity in nonrice-area allocation. This region is an important vegetable growing area in the country with a share of about 44 percent in the total vegetable area in TE 1998-1999 (Government of India, 2001). Cultivation of fruits is also gradually increasing. This region has also emerged as an oilseed-producing region with crops such as rapeseed-mustard, groundnut, sesame, and soybean.

During the post-green-revolution period, oilseeds, fruits, and vegetables performed impressively in all the regions. Although the success of oilseeds was largely inluenced by high tariff barriers on imports of edible oils, the National Horticultural Board in 1984 encouraged horticulture production by coordinating production and processing of fruits and vegetables. Exports of different fruits and vegetables also grew during the past decade as a result of emerging infrastructure facilities (cold storage and cargo handling) at international airports.

Diversification Within the Livestock Subsector

The livestock sector is growing at a fast rate and therefore its share in total value of agricultural output is progressively rising in India (Birthal and Parthasarthy, 2002; Birthal et al., 2002). However, the SID within the livestock subsector is modest (0.508), and also slowly decreasing. The modest SID is mainly due to a large share of milk in total value of livestock products (around 68 percent during the past two decades; Table 12.7). Remaining shares of livestock products (32 percent) are distributed to meat, poultry, wool, etc. Milk production more than doubled, from 33 to 71 million tons over the period TE 1981-1982 to TE 1998-1999, with an annual compound growth rate of about 4.62 percent during the past two decades. The growth of milk production was much higher (5.23 percent) during 1980s than 1990s (3.46 percent).

TABLE 12.7. Share of individual commodities (%) in gross value of livestock subsector in India at 1980-1981 prices

Region	Milk			Meat			Poultry			Miscellaneous		
	TE 1982-1983	TE 1991-1992	TE 1998-1999	TE 1982-1983	TE 1991-1992	TE 1998-1999	TE 1982-1983	TE 1991-1992	TE 1998-1999	TE 1982-1983	TE 1991-1992	TE 1998-1999
Eastern	54.63	53.65	47.14	11.96	20.24	22.08	11.91	11.38	11.19	21.50	14.74	19.59
Northeast	53.81	55.84	56.83	15.63	16.61	17.58	18.92	18.21	17.05	11.65	9.33	8.54
Northern	76.32	77.87	79.87	2.67	4.54	5.06	2.93	5.13	5.46	18.07	12.46	9.62
Southern	63.75	64.64	64.72	9.91	7.47	7.20	14.94	16.26	16.58	11.40	11.64	11.50
Western	69.16	73.81	74.13	5.61	3.30	3.85	6.27	6.20	7.31	18.96	16.68	14.71
All India	68.09	69.22	68.96	6.57	7.92	8.39	7.85	8.97	9.58	17.48	13.88	13.07

Note: Meat includes cattle meat, buffalo meat, sheep meat, goat meat, and pig meat. Poultry includes poultry meat and eggs.

Meat and poultry subsectors have also registered good performances, increasing from a low of 0.80 million tons in TE 1982-1983 to 2.73 million tons in TE 1991-1992 and finally to 4.41 million tons in TE 1998-1999, giving an annual compound growth rate of about 5.81 percent during the 1980s vis-à-vis just 3.90 percent during the 1990s. The high increase in meat production during the 1980s was partly due to the severe drought in 1987 (often claimed to be the severest of the century) in most parts of the country. Acute shortages of green and dry fodder forced people to dispose of less productive animals for slaughtering at a large scale. Poultry also flourished during the 1980s contributing to higher growth of the livestock sector during the 1980s. The share of poultry and goat meat in total value of meat production went up from 66 percent in TE 1982-1983 to 77 percent TE 1998-1999. Similarly, egg production also increased by 8.46 percent annually during the 1980s compared to 4.60 percent annually during the 1990s. Unlike dairy, the poultry sector grew at the instance of the private organized sector, which controls roughly 80 percent of total poultry production in the country.

Regional patterns are dissimilar due to agroclimatic variability, food habits, and status of economic development (Table 12.7). Diversification of livestock activities was least in the northern region and highest in eastern and northeastern regions of the country. In the northern region, there is more specialization of livestock subsectors primarily in dairy, with some emergence of poultry. The share of milk in total value of livestock was as high as 80 percent in TE 1998-1999 in the northern region. The western region is also concentrating in favor of milk production. The state of Gujarat, located in the western region, witnessed the evolution of dairy cooperatives and led a revolution often called the "white revolution."

The southern region is showing relatively higher diversity in livestock sectors as compared to northern and western regions. Milk and poultry together contributed about 81 percent of the total value of livestock in the southern region in TE 1998-1999. After milk, poultry has emerged as an important activity in the southern region.

Eastern and northeastern regions showed highly diversified livestock sectors distributed between milk, meat, and poultry. With few exceptions, the share of milk production is decreasing and that of meat and poultry increasing.

Livestock production brought out revolutionary changes in the country during the 1980s and these continued during the 1990s, though at a slower pace. In the dairy sector, the breakthrough is ascribed to the implementation of "Operation Flood Program." The National Dairy Development Board (NDDB) developed a cooperative model for procuring and marketing milk and milk products. The program established about 170 cooperative milk unions operated in over 285 districts and covered nearly 96,000 village-level societies in different states. Nearly 10.7 million farmers were members until 1999-2000 (NDDB, 2002). Realizing the success of the program, an Integrated Dairy Development Program was launched in the hilly and backward areas in 1992-1993 to enhance production, procurement, and marketing of milk, and to generate employment opportunities in those areas.

The future of the livestock sector is quite promising in the country, as there still exists huge potential to augment production, consumption, and export of different livestock commodities. Meat production is mostly confined to the unorganized sector, and will require modern slaughter facilities and development of refrigeration facilities.

Strengthening the livestock sector would benefit the small farmers in rural areas. This sector can significantly contribute to farm income, offer employment opportunities in rural areas, and meet the food and nutritional needs of small farmers.

Diversification Within the Fisheries Subsector

The fisheries subsector has also diversified over the years. The SID of the fisheries sector has shown marginal improvement to 0.49 in TE 1999-2000 from 0.47 in TE 1981-1982 (Table 12.8). It is mainly due to a gradual shift from marine to inland fisheries. Traditionally, the marine fisheries dominated the fish production in the country, which was more than 75 percent in 1960-1961. Recognizing the importance and potential of the fish sector in

TABLE 12.8. Temporal changes in fish production (000 tons) and diversity in India

Period	Marine fish*	Inland fish*	Total	Simpson index of diversity
TE 1981-1982	15 (62.50)	9 (37.50)	24	0.47
TE 1991-1992	21 (60.00)	14 (40.00)	35	0.48
TE 1999-2000	29 (53.70)	25 (46.30)	54	0.49

Source: Kumar et al. (2001). Reprinted with permission from the WorldFish Center.
*Figures in parentheses are the percentage share in total fish production.

the inland areas, a greater impetus was accorded to the inland fisheries. The share of marine fish in the total production has fallen to about 54 percent in TE 1999-2000, while that of inland fisheries has risen to about 46 percent in TE 1999-2000 from less than 25 percent in 1960-1961. The annual compound growth rate of inland fisheries was higher (6.54 percent) during the 1990s than during the 1980s (5.27 percent). Marine fish production, which performed poorly during the 1980s (0.12 percent), improved during the 1990s (2.53 percent). The inland fish potential is still higher due to many rivers, canals, and reservoirs.

The higher growth in inland fisheries was mainly attributed to overwhelming progress in aquaculture, both in freshwater and brackish water. The share of culture fisheries in inland sectors increased about 43 percent in 1984-1985 to a high level of about 84 percent in 1994-1995 (Kumar et al., 2001). A bulk of growth in culture fisheries has come from freshwater aquaculture (Krishnan and Birtha, 2000). Production of culture and other products in the brackish-water areas could be expanded. Only 10 percent of the available brackish-water area (12 million ha) in the country was exploited until 1995-1996 (IASRI, 2001). The expansion of inland fisheries has also led to some negative externalities related to degradation of arable lands due to salinity.

The remarkable progress in the fisheries sector was the outcome of a well-knit strategy to accomplish multiple goals of augmenting production, enhancing exports, and overcoming poverty of fishermen. The outlay in the fisheries sector was raised from around 2 to 3 percent of total agricultural outlay during the 1970s to over 5.5 percent during the 1980s and 1990s. Several production- and development-oriented programs were launched in the potential areas and included Development of Freshwater Aquaculture, Integrated Coastal Aquaculture, and Development of Coastal Marine Fisheries. Under these programs, fish farmers' development agencies (FFDAs) were established in freshwater areas, and brackish water fish farmers' development agencies (BFDAs) in brackish water areas. To encourage aquaculture, the programs were initiated to upgrade technology, and to encourage involvement of the private sector in activities such as quality seed, feed, and other inputs, and creation of suitable infrastructure for storage, transport, marketing, and credit.

Seed production is very important to fisheries production. To meet this objective, more than 50 seed hatcheries at the national level were established. The results were quite rewarding: the seed production rose from only 409 million fry in 1973-1974 to about 20,000 million fry in 1999-2000. To develop better infrastructure facilities, "fisheries industrial estates" were developed by grouping clusters of fishing villages. The major accomplishment until 1998-1999 was construction of 30 minor fishing harbors and 130 fish landing centers in the major fishing harbors at Cochin, Chennai, Paradeep, Roychowk, and Visakhapatnam (Kumar et al., 2001).

The future of the fisheries sector is bright with the opening of the economy. A good export market exists for both marine and inland fish and aqua products. In this context, sanitary and phytosanitary (SPS) issues are more important in exploiting export potential. The need is to focus more on quality control, modernize the crafts used in marine areas, and utilize the full potential of the inland fisheries.

DETERMINANTS OF DIVERSIFICATION

Agricultural diversification is influenced by a number of forces both from the supply side and the demand side. This section examines the determinants of diversification in crop and livestock sectors separately, and results are given in Tables 12.9 and 12.10.

Determinants of Diversification in the Crop Subsector

To examine the forces that influence diversification in favor of high-value commodities a number of explanatory variables were studied. The variables were related to infrastructure development, technology adoption, relative profitability, resource endowments, and demand-side factors including urbanization and income level. The estimated double-log equations of Generalized Least Square are given in Table 12.9.

To capture the effects of infrastructure development, two important variables, namely markets and roads, were included in the model. Both the variables yielded positive and significant influence on diversification of the crop sector. Obviously, better markets and road networks induced diversifi-

TABLE 12.9. Determinants of diversification in favor of horticultural commodities: Double-log estimates of Generalized Least Square

Explanatory variables	Dependent variable: Index of gross value of horticultural commodities at 1980-1981 prices		
	Equation 1	Equation 2	Equation 3
Irrigation	−0.4575*** (0.0614)	−0.4697*** (0.0607)	−0.5073*** (0.0564)
Relative profit-ability	0.3549*** (0.04450)	0.3329*** (0.0411)	0.3152*** (0.0441)
Roads	0.2873*** (0.0664)	0.2843*** (0.0665)	—
Markets	0.1261* (0.0710)	0.1870*** (0.0528)	—
Rural literacy	−0.7976*** (0.1458)	−0.8415*** (0.1419)	−0.5497*** (0.1389)
Small landholders	1.1964*** (0.2283)	1.2016*** (0.2285)	1.6043*** (0.2002)
Urbanization	0.1840 (0.1438	—	0.3050*** (0.1094)
Income	0.4892*** (0.0668)	0.5082*** (0.0652)	0.4671*** (0.0686)
Rainfall	−0.0583 (0.0422)	−0.0712* (0.0411)	−0.0949** (0.425)
Time dummy: 1981-1990 = 0; 1991-1999 = 1	0.8944*** (0.0700)	0.8839*** (0.0696)	0.8960*** (0.0722)
R-square	0.7735	0.7722	0.7572
Adjusted R-square	0.7642	0.7637	0.7490
F-statistic	82.82***	90.00***	91.40***

Note: Figures in parentheses are standard errors of the respective coefficients; ***, **, * significant at 1%, 5%, and 10%, respectively.

cation in favor of horticultural commodities. Better market and road networks mean low marketing costs and easy and quick disposal of commodities. It also reduces the risk of postharvest losses in case of perishable commodities.

The technology was defined by area under high-yielding varieties of cereals, irrigated area, and extent of mechanization. However, irrigated areas turned out to be significant and represented the technological advancement in the region. The regression coefficient of this variable was showing a neg-

TABLE 12.10. Determinants of diversification in favor of livestock subsector: Double-log estimates of Generalized Least Square

Explanatory variable	Dependent variable: Index of gross value of livestock at 1980-81 prices		
	Equation 1	Equation 2	Equation 3
Irrigation	−0.1993*** (0.0214)	−.0294 (0.0332)	—
Relative profit-ability	0.2009*** (0.0222)	—	0.1736*** (0.0288)
Roads	—	0.0534* (0.0292)	—
Markets	0.0368 (0.0260)	0.0906*** (0.0312)	—
Rural literacy	−0.2049*** (0.0479)	—	−0.0672 (0.0545)
Small landholders	0.6790*** (0.0705)	—	0.5689*** (0.0760)
Urbanization	0.1114** (0.0482)	—	0.0569 (0.0455)
Income	0.1521*** (0.0239)	—	0.2033*** (0.0216)
Rainfall	−0.0029 (0.0148)	0.0524** (0.0221)	—
Time dummy : 1981-1990 = 0; 1991-1999 = 1	0.2479*** (0.0255)	—	—
R-square	0.7472	0.0637	0.5299
Adjusted R-square	0.7376	0.0448	0.5188
F-statistic	78.45***	3.37***	47.67***

Note: Figures in parentheses are standard errors of the respective coefficients; ***, **, * significant at 1%, 5%, and 10%, respectively.

ative relationship with diversification, which meant that the crop diversification in favor of horticultural commodities was declining with increasing irrigated area. This suggests that crop diversification is more pronounced in rainfed areas, which are deprived of technological advancement in terms of irrigation. These areas are characterized as rainfed, low-resource endowed, with abundant labor force, and bypassed during the green-revolution period.

Relative profitability of horticultural commodities with other crops is also an important determinant for diversification in their favor. The regression coefficient is significant and positive. Obviously, the higher profit of these crops would induce farmers to diversify in their favor. Fruits and vegetables are highly profitable in comparison to cereals and other crops. Relative profitability of fruits was more than eight times higher than cereals. The corresponding figure for vegetables was 4.8. Although high profits of horticultural crops encourage their cultivation, uncertain prices and high yield instability limit their widespread cultivation. The price instability is more in fruits and vegetables compared to cereals (Subramanian et al., 2000). The high price variability of fruits and vegetables is due to poor vertical linkage between production, marketing, and processing. Appropriate institutional arrangements must be developed for minimizing the price uncertainty. Some scattered success stories are available for strengthening farm–firm linkages. These are contract farming by firms such as Pepsi and Hindustan Lever for potato and tomato, and cooperative societies under the banner of "Safal" for fruits and vegetables. Contract farming is becoming popular in many developing countries, but it is still based on informal arrangements. This issue needs legal changes to encourage processing industries to give further impetus to agricultural diversification. Evidence suggests that well-managed contract farming has proven effective in linking the small farm sector to sources of extension advice, mechanization, seeds, fertilizer, and credit, and to guaranteed and profitable markets for produce (FAO, 2001).

A positive relationship exists between growth of horticultural commodities and the proportion of small farmers. This indicates that diversification in favor of horticultural commodities is confined to the small farmers. Such a move by small farmers in favor of high-value commodities is expected to enhance their income. Cultivation of horticultural crops suits the small farmers. The advantage is that these are labor intensive and generate regular flow of income. The caution is that absence of appropriate markets and a rise in supply may adversely affect the prices and opportunities for higher income (Tewari et al., 2001).

Rainfall is another variable, which was included in the model to assess the effect of climate on crop diversification. The variable was highly significant, indicating that crop diversification is limited in higher rainfall areas. Obviously, high rainfall areas specialize toward rice, while farmers go for diversification in medium- and low-rainfall areas to increase income and minimize risk. Demand-side factors such as urbanization and per-capita income showed positive and significant impact on crop diversification.

This discussion suggests that assured markets and good road networks could stimulate agricultural diversification in favor of high-value crops as they help maximize profits and minimize uncertainty in the output prices.

Inadequate markets may discourage farmers to exploit potential benefits of cultivating high-value crops. Encouraging appropriate institutional arrangements for better markets through cooperatives or contract farming would go a long way in strengthening farm–firm linkages. In addition, the role of technology cannot be ignored. The high-yielding and more stable genotypes in fruits and vegetables need to be propagated through development of a strong seed sector.

Determinants of Diversification in the Livestock Subsector

The results of Generalized Least Square for the livestock subsector are given in Table 12.10. Technological progress in the crop subsector has strong influence on the diversification of the livestock subsector. Irrigated area, one of the proxy variables for technological progress, was significant with negative effect on expansion of the livestock subsector.

The farmers' resource endowment (particularly the size of land holding) was captured by the proportion of small landholders in the total holdings in the region. The regression coefficient was highly significant and positive, indicating that the prospects of livestock activities are higher on small farmers. Livestock activities are often well integrated with crop activities and generate regular income and quick returns to the small farmers.

Rural literacy, which is a proxy for level of knowledge, is significant with a negative sign. Obviously, higher literacy indicates a shift from agriculture (including livestock) to other job opportunities.

Relative profitability of livestock in comparison to crops was also found significant with positive relationship for the growth of livestock activities.

Infrastructure development also plays a crucial role in influencing the prospects of the livestock subsector. Road networks and markets were included in model to represent infrastructure development. These variables were significant and have positive bearing on the growth of livestock.

The two demand-side variables, per capita income and urbanization, showed positive and significant influence on the growth of livestock activities. Rising per capita income and growing urbanization have increased the demand for livestock products leading to diversification in the livestock sector (Kumar and Mathur, 1996).

Annual rainfall was significant with a positive sign indicating that higher rainfall areas have more inclination toward livestock activities. This may be due to availability of green fodder from fallow areas. This is unlike the crop sector, where diversification was higher in low rainfall areas. This has important implications for designing appropriate strategies for promoting the livestock subsector.

IMPLICATIONS OF DIVERSIFICATION

Several benefits of agricultural diversification are reported in the literature:

1. shifting consumption pattern,
2. improving food security,
3. increasing income,
4. stabilizing income over seasons,
5. generating employment opportunities,
6. alleviating poverty,
7. improving productivity of scare resources (e.g., water),
8. promoting export, and
9. improving environmentally sustainable farming systems through conservation and enhancement of natural resources (Jha, 1996; Ramesh Chand, 1996; Vyas, 1996; Delgado and Siamwalla, 1999; Ryan and Spencer, 2001).

These short-run benefits have implications for the prospects of long-run growth in agriculture, regional equity, and sustainable farming systems. The benefits are more clearly captured at the microlevel. In the present study, we have assessed implications of diversification at the macrolevel on (1) improving food security, (2) generating employment opportunities, and (3) promoting export.

Food Security

Food security at the national and household level is an important issue in the context of agricultural diversification. Producing additional food is a major challenge when populations and incomes are rising, and natural resources degrading. Some believe that a shift in the crop portfolio from food to nonfood crops may lead to food insecurity. Incidentally, the diversification in a majority of the states in India came as a result of expansion in cropping intensity. Crop substitution was also taking place, which was diverting to areas in favor of high-yielding cereals from low-yielding inferior cereals. Rice, wheat, and maize gained while sorghum and millets lost in area. The high-yielding nature of food-grain crops has improved their availability. The production trends reveal that the per capita daily availability of food grains has increased from 448.56 grams in TE 1981-1982 to 475.4 grams in TE 1999-2000. Similarly, the per capita daily availability of milk has substantially risen from 128 grams in 1980-1981 to 214 grams in 1999-2000 (Government of India, 2002).

Interestingly, the consumption basket is changing over time. Food consumption is shifting from cereals to noncereals in both rural and urban areas (Table 12.11). The per capita cereal consumption in rural and urban areas has declined, while that of milk, milk products, vegetables, and fruits has increased significantly (Kumar, personal communication). The most remarkable increment in consumption was witnessed in fruits.

Evidence clearly reveals that diversification of crop and livestock sectors has not only increased production of noncereal commodities, but also raised their consumption pattern. A more favorable environment for diversifica-

TABLE 12.11. Per capita consumption pattern of food items (kgs/person/annum)

Item	1977	1987	1993	1999
Rural				
Rice	86.5	88.1	85.4	81.0
Wheat	49.4	61.6	53.5	53.9
Coarse cereals	56.7	29.8	24.1	17.7
Total cereals	192.6	179.5	163.0	152.6
Pulses	8.7	11.5	9.2	10.1
Milk and milk products	24.6	58.0	51.4	50.5
Edible oils	2.7	4.3	4.6	6.0
Vegetables	24.7	50.8	53.2	66.0
Fruits	2.6	10.3	9.8	17.0
Meat, eggs, fish	2.7	3.3	4.1	5.0
Sugar and gur	13.5	11.0	9.2	10.1
Urban				
Rice	67.6	68.1	64.2	62.5
Wheat	64.6	60.4	57.4	55.4
Coarse cereals	14.8	10.6	7.7	7.1
Total cereals	147.0	139.1	129.3	125.0
Pulses	11.7	12.2	10.5	12.0
Milk and milk products	39.7	64.9	68.3	72.4
Edible oils	4.8	6.8	6.3	8.6
Vegetables	39.7	66.4	63.1	70.0
Fruits	5.9	18.8	20.1	19.0
Meat, eggs, fish	4.8	4.9	6.3	6.8
Sugar and gur	17.1	12.3	11.8	12.0

Source: Kumar, personal communication (2002).

tion toward high-value commodities will not only ease the pressure of storing huge surpluses of rice and wheat but also accelerate growth of the agricultural sector through high-value commodities.

Employment

Generating employment avenues in rural areas is critical. Some information was collated from labor use in production of different crop activities (Table 12.12). Note that labor use for cultivation of noncereals is substantially higher than cereals (except rice).

Area shifts from cereals to vegetables would generate substantial employment opportunities in rural areas. Rough estimates suggest that a 1 ha shift in area from wheat to potato would generate 145 additional labor days. Similarly, a 1 ha area shift from coarse cereals (sorghum and pearl millet) to onion would generate seventy days more employment in rural areas. Substitution from coarse cereals to other vegetables (for example cabbage, cauliflower, eggplant, tomato, lady's finger), would generate 70 days/ha additional employment. A marginal shift in area from wheat and coarse cereals in favor of high-value crops can thus generate enhanced employment opportunities.

Export

Indian exports during the 1990s grew at an annual rate of 10.1 percent, as against 7.4 percent during the 1980s (Government of India, 2001). The exports of agricultural commodities during the 1990s, however, grew at an an-

TABLE 12.12. Average labor use in vegetables, cereal, and noncereal crops (human days per ha) in India

Vegetables		Cereals and noncereals	
Crop	Labor use (human days/ha)	Crop	Labor use (human days/ha)
Potato	200	Rice	105
Onion	125	Wheat	55
Cabbage	110	Sorghum	55
Cauliflower	120	Pearl millet	50
Eggplant	70	Cotton	100
Tomato	195	Sugarcane	190

Source: Derived from several sources, including Subramanian et al. (2000) and Government of India (2000).

nual rate of 8.1 percent, as against only 3.3 percent during the 1980s. However, share of agriculture in total exports has declined from 24 percent during the 1980s to 18 percent in the 1990s.

A large share in agricultural exports was caused by diversification of crop and livestock sectors. Diversification of agricultural commodities has promoted export of many nontraditional items. Historically, there was virtually no export of fruits, vegetables, and livestock and fish products. The exports of these commodities, as well as rice, increased during the 1990s. For example, exports of rice went up from 440,000 tons in TE 1981-1982 to 656,000 tons in TE 1991-1991 and reached to 3,145,000 tons in TE 1997-1998. India's share in world rice trade went up to more than 10 percent in the 1990s, up from a mere 3.7 percent in 1980-1981. Similarly, exports of fruits and vegetables more than doubled during the past two decades (from US$110 million in TE 1981-1982 to US$262 million in TE 1999-2000). The exports of fish increased from $320 million in TE 1981-1982 to $1,125 million by TE 1999-2000. More progress was registered in processed fruits and juices. Furthermore, exports of milk, milk products, eggs, and fish products have also made entry into the export markets. Exports of milk and milk products increased from US$1.1 million in TE 1981-1982 to US$1.7 million in TE 1991-1992 and reached a peak of US$3.2 million in TE 1999-2000. Exports of eggs, although too erratic, increased from a low of US$0.4 million in TE 1981-1982 to US$25.3 million in TE 1999-2000. The production of all these commodities increased substantially during the 1990s.

The progress in exports of nontraditional items was quite impressive during the 1990s as compared to the 1980s. This implies that diversification of agriculture can substantially contribute to exports provided a congenial environment through infrastructure development and institutional innovation is created. It must be supported by appropriate domestic policies and legal changes, which encourage development of new institutions for linking production, marketing, and processing.

CONCLUSIONS AND POLICY IMPLICATIONS

The study diagnosed the status of agricultural diversification in South Asian countries. Detailed investigations were carried out for India on the determinants of diversification. The results of the study revealed that the agricultural sector in South Asia is gradually diversifying in favor of high-value commodities, namely fruits, vegetables, livestock, and fish products. Much of the diversification came, if at all, with little support from the state governments. Food security issues are still critical in the subcontinent and government policy is still obsessed with self-sufficiency in cereals, which presumably contributes to a large share of area still being allocated to cereal crops. Countries such as Bangladesh, India, and Sri Lanka have achieved food self-sufficiency at the national level, but the emphasis is still focused toward increasing production of rice and wheat. Countries, such as Bhutan, Nepal, and Pakistan, which are still deficit in food-grain production, are making serious attempts to augment their production.

Despite efforts toward food-grain production, a silent revolution has occurred in high-value commodities. Production of fruits, vegetables, livestock, and fish products have increased remarkably in most South Asian countries. Due to their low share in gross value of agricultural output, the silent revolution was unnoticed. The production of these commodities was demand driven, which is unlike the supply-driven green revolution. During the 1980s, production increase was attributed to the rise in yield levels. During the 1990s, production increases came from area augmentation.

The determinants for high-value commodities (horticultural and livestock) were studied. Markets and roads were the key determinants to influencing the status of diversification. Another important determinant was technology absorption. The higher the technology adoption of cereals (particularly irrigation) the less was the diversification in favor of high-value commodities. Diversification in favor of horticultural and livestock commodities was more pronounced in rainfed areas, which were bypassed during the green revolution. The rainfed areas are becoming a hub of noncereals due to their low water requirement and abundant labor supply. The evidence confirmed that the regions that were diversifying in favor of noncereals have accomplished better growth performance as compared to those specializing in cereals. In addition, relative profitability of high-value commodities in relation to other crops also played an important role in determining status of diversification. This calls for further strengthening of R&D efforts for improving productivity in a sustainable manner. Most important, high-value commodities are usually produced by small farmers.

The study also highlighted the implications of diversification on food security, employment generation, and export earnings. The macro-level information showed that food security was not adversely affected as a consequence of agricultural diversification. Similarly, the high-value crops have substantial potential for generating employment opportunities. Most of the high-value crops are high-labor-requirement crops. Incidentally, the small and marginal farmers have abundant labor, which can be effectively utilized for production of high-value commodities. The high-value commodities have also witnessed good performance in the international trade.

To facilitate the process of agricultural diversification of high-value commodities, South Asian countries need to take series of measures to reform institutional arrangements, which can appropriately integrate production and markets. Among others, the immediate measures include ensuring markets, developing roads, creating appropriate infrastructure, and encouraging private-sector participation for value addition and processing. Domestic market reform is a precondition for agricultural diversification in favor of high-value enterprises. The most challenging issue is how to ensure greater participation of small and marginal landholders in the process of agricultural diversification for sharing benefits of globalization. These farmers are moving in favor of high-value commodities but they have high transaction costs due to tiny marketable surplus, which negates their higher production efficiency. Future research may be initiated to assess how appropriate institutional arrangements would convert weaknesses of small farmers into opportunities. Contract farming, cooperatives, and group actions may lead to better opportunities. These types of institutions may overcome risk and uncertainty and establish strong vertical linkage between production, marketing, and processing.

NOTES

1. Nineteen states in the country are major ones. The remaining nine states are small with respect to geographical area, production, and population.

2. It was computed as follows: Gains in Cropped Area (A) = Change in Gross Cropped Area (B) + Crop Substitution (C). Since (A) and (B) are known, (C) is residual.

REFERENCES

Birthal, P.S. and P. Parthasarthy Rao (Eds.). 2001. *Technology options for sustainable livestock production in India.* Proceedings of the workshop on documentation, adoption, and impact of livestock technologies in India held at ICRISAT-Patancheru, India, January 18-19. New Delhi: National Centre for Agricultural Economics and Policy Research; and Patancheru, India: ICRISAT.

Birthal, P.S., Kumar, A., and Tewari, L. (Eds.). 2002. *Livestock in different farming systems in India.* New Delhi: Agricultural Economics Research Association.

CMIE (Centre for Monitoring Indian Economy). 2001. Agriculture: Economic Intelligent Service, New Delhi: Center for Monitoring Indian Economy Pvt. Ltd.

Delgado, C.L. and Siamwalla, A. 1999. Rural economy and farm income diversification in developing countries. In *Food security, diversification and resource management: Refocusing the role of agriculture,* eds. G.H. Peters and J. Von Braun (pp. 126-143). Proceedings of Twenty-Third International Conference of Agricultural Economists. Brookfield, VT: Ashgate Publishing Company.

FAO (Food and Agricultural Organization of United Nations). 2001. *Contract farming: Partnerships for growth.* Rome: Author.

Government of India. 2000. *Cost of cultivation of principal crops in India.* Ministry of Agriculture, Department of Agriculture and Cooperation. New Delhi: Author.

Government of India. 2001. *Indian economic survey: 2000-2001.* New Delhi: Akalank Publications.

Government of India. 2002. *Agricultural statistics at a glance.* Directorate of Economics and Statistics, Ministry of Agriculture. New Delhi: Author.

Hayami, Y. and Otsuka, K. 1992. Beyond the green revolution: Agricultural development strategy into the new century. In *Agricultural technology: Policy issues for the international community,* eds. J.R. Anderson (p. 35). Washington, DC: The World Bank.

IASRI (Indian Agricultural Statistics Research Institute). 2001. *Agricultural research data book 2001.* New Delhi: Author.

Jha, D. 1996. Rapporteur's report on diversification of agriculture and food security in the context of new economic policy. *Indian Journal of Agricultural Economics,* 51(4): 829-832.

Kelley, T.G., Ryan, J.G., and Patel, B.K. 1995. Applied participatory priority setting in international agricultural research: Making trade-offs transparent and explicit. *Agricultural Systems,* 49: 177-216.

Krishnan, M. and Birthal, P.S. (Eds.) 2000. *Aquaculture in India: Retrospect and prospect.* New Delhi: National Centre for Agricultural Economics and Policy Research.

Kumar, A., Joshi, P.K., and Birthal, P.S. 2001. Fisheries sector in India: An overview of performance, policies and programs. In the International workshop on "Strategies and Options for Increasing and Sustaining Fisheries Production to Benefit Poor Households in Asia," from August 20-25. ICLARM, Penang, Malaysia. Published by the WorldFish Center.

Kumar, P., Joshi, P.K., Johansen, C., and Asokan, M. 1998. Sustainability of rice-wheat based cropping systems in India. *Economic and Political Weekly,* September 26: A-152-157.

Kumar, P. and Mathur, V.S. 1996. Structural changes in the demand for food in India. *Indian Journal of Agricultural Economics,* 51(4): 664-673.

National Dairy Development Board. 2002. *Annual Report 2001-02.* Anand, India: Author.

Pandey, V.K. and Sharma, K.C. 1996. Crop diversification and self sufficiency in food grains. *Indian Journal of Agricultural Economics,* 51(4): 644-651.

Pingali, P.L. and Rosegrant, M.W. 1995. Agricultural commercialization and diversification: Processes and policies. *Food Policy,* 20(3): 171-186.

Ramesh, C. 1996. Diversification through high value crops in western Himalayan region: Evidence from Himachal Pradesh. *Indian Journal of Agricultural Economics,* 41(4): 652-663.

Ryan, J.G. and Spencer, D.C. 2001. Future challenges and opportunities for agricultural R&D in the semi-arid tropics. Patancheru, Andhra Pradesh, India: International Crops Research Institute for the semi-Arid Tropics.

Subramanian, S.R., Varadarajan, S., and Asokan, M. 2000. India. In *Dynamics of vegetable production and consumption in Asia,* ed. M. Ali. Taiwan: Asian Vegetable Research and Development Center.

Tewari, L., Elumalai, K., Birthal, P.S., and Joshi, P.K. 2001. Implications of globalization on small farm holders: A SWOT analysis. In the Annual Conference of the Agricultural Economics Research Association, November 21-22 at the Indian Agricultural Research Institute, New Delhi.

von Braun, J. 1995. Agricultural commercialization: Impact on income and nutrition and implications for policy. *Food Policy*, 20(3): 187-202.

Vyas, V.S. 1996. Diversification in agriculture: Concept, rationale and approaches. *Indian Journal of Agricultural Economics*, 51(4): 636-643.

Chapter 13

Market Reforms in Agriculture:
An Indian Perspective

S. Mahendra Dev

INTRODUCTION

The Indian economy is predominantly agrarian as far as employment is concerned. Apart from the direct benefit that farmers gain from working in agriculture, the indirect benefits and linkages of the sector with the rest of the economy are also important. Thus, agriculture performance has an economywide impact and its growth can achieve equity if it is spread across impoverished regions. Several studies have even suggested that agricultural growth helps directly to reduce poverty (Ravaillon, 2000). A related issue is the enhancement of food security in India. Improving food security at the household level is an issue of great importance for a developing country such as India where millions of poor suffer from malnutrition and a lack of purchasing power.[1] Adequate purchasing power is important for the poor so that food can be bought. This purchasing power can be ensured in two ways. One is to supply food grains at subsidized prices such as India's public distribution system (PDS) does; the other is to raise the incomes of the poor through employment.

The economic reforms in India since 1991 have improved the incentive framework and agriculture has benefited from the reduced industry protection. The terms of trade for agriculture have improved and private investment has increased. The export of commodities, particularly cereals, has improved and there has been some progress on market reforms in terms of removing domestic and external controls.

However, there have also been concerns over agriculture and food security in the 1990s. The general concern is that the reforms in the 1990s have neglected agriculture. Agricultural growth during the Ninth-Plan period (1997-2002) has been only 2.1 percent per annum and the productivity growth rate has declined. Furthermore, food management has not been satisfactory. Currently, there are 60 to 65 million tons of grain held as buffer stock, with 260 million people living below the poverty line. In the 1990s diversification led to risk and uncertainty, subsidies continued, and public investment did not increase. Disparities between labor productivity in agriculture and nonagriculture have widened, and regional disparities also seem to have increased. Agricultural marketing problems for farmers constitute another set of issues. Finally, there are concerns regarding the environment and the sustainability of agriculture.

Regarding trade policy in India, in the past five decades since independence, the policies were highly interventionist and discriminated against agriculture. Basic foods such as cereals, in particular, were highly regulated and private trade was practically banned. Whatever little trade was permitted was restricted to basmati rice and maize imports for poultry. The year 1995 marked a new era for agriculture in external trade. For the first time the Indian agricultural sector was brought into the Uruguay Round. The nation's rules of trade in agriculture were thus rewritten under the Agreement on Agriculture (AOA).

The objective of this chapter is to examine the market reforms needed for achieving higher and sustainable growth in agriculture, including the "third revolution"[2] for diversification and processing, and food security for the poor. In the second section, it discusses the problem of buffer stock overaccumulation, and needed reforms. The third section examines other agricultural policies. Employment and rural transformation for food security and poverty alleviation are discussed in the fourth section.

THE PROBLEM OF GRAIN STOCK OVERACCUMULATION AND NEEDED REFORMS

Before addressing the problem of overaccumulation, a brief discussion of India's food policies is needed. Currently, the food security system and price policy together basically consist of three instruments: procurement prices/minimum support prices, buffer stocks, and the PDS.[3] In order to ensure remunerative prices to farmers, the government of India has been intervening in the food-grain market mainly for wheat, paddy, and coarse cereals through its minimum support price (MSP)/procurement policy. The importance of building up a buffer stock of food grains, generally of rice and wheat, is to provide food security to the country. The argument in favor of buffer stocking is that where variability of food-grain output is large due either to weather conditions or human factors, the state can ensure food security for the poor by building adequate buffer stocks from the surpluses of good production years and/or arranging to import the requisite food grains in times of need. The network of the public distribution system (PDS) has extended over the years since its inception during World War II. Currently, the food grains from the government stocks are being made available through the PDS to about eighteen crore[4] households through a country-wide network of about 4.6 lakh[5] fair price shops. Various alterations have been made over the years to the PDS to improve its performance, including expansion of the system in backward areas, the establishment of subsidized rates, etc. Food-for-work programs can also be considered as part of the provision of food security to the poor. In addition, we also have nutrition programs for children such as noon meal schemes and integrated child development services (ICDS).

Reasons for the Overaccumulation of Food Grains in Recent Years

For the past few years, public stocks of food grains have been rising to a disturbingly high level—62.55 million tons (May 1, 2002). The reasons for the overaccumulation are discussed as follows.

The continuous increase in the procurement prices/MSP has led to the accretion of stocks far above the levels required. The concept of MSP, which was originally based on the "paid out" or "variable cost of production," was enlarged to take into account the full cost of production (from A2 cost to cost C).[6] Due to pressures from the vocal farmers' lobby, all types of farm expenditure, incurred or imputed, were added to the cost of production for the purpose of fixing the MSP. This escalated the MSP, and it had to be increased each year. Now there is no distinction between procurement price and MSP and all quantities of food grains offered for sale by farmers were procured at the enhanced MSPs. Generally, the government-announced MSP is higher than the prices recommended by the Commission on Agricultural Costs and Prices. Moreover, some state governments give higher procurement prices than the central government-announced prices.

The continuous rise in procurement and issue prices, and the obligation to purchase all grains that farmers offered led to the growth of rice and wheat stocks to their present level. A significant increase in procurement and issue prices occurred in the 1990s as compared to the 1980s as shown in Table 13.1. The price system has led to the buildup of grain stocks for other reasons as well.

1. *Private trade is discouraged:* The higher levels of procurement and market prices seem to have discouraged private trade in recent years. Until the mid-1990s, private trade used to hold stocks of around 30 to 40 percent of the marketable surplus. Possibly, most of the stocks are

with Food Corporation of India (FCI) as high procurement prices make private trade unprofitable.

2. *Exports are uncompetitive:* Exports of food grains have become uncompetitive because of high procurement prices. The government has been exporting rice and wheat at subsidized prices in order to reduce the high stocks.

3. *Consumption was reduced due to low purchasing power:* In the 1990s, the relative prices of food grains were higher than those of the 1980s. This reduced the purchasing power of the population, particularly of the poor. Some simulations on the demand for food grains were conducted for this study. As shown in Figure 13.1, if relative prices in the 1990s were the same as those of the 1980s, the demand would have been higher by 2.62 million tons in 1991. In subsequent years, the estimated demand would have been higher by 3.0 to 6.0 million tons. Similarly, Figure 13.1 also shows the estimated demand if the growth of consumption in the 1990s is similar to that of the 1980s. In some years in the 1990s, the growth pattern was negative while in others it was positive. In 1998, the additional demand for food grains would have been around 7.3 million tons if the relative prices and growth pattern in the 1990s were the same as those for the 1980s.

4. *Offtake declined under PDS:* The increase in procurement prices led to an increase in issue prices for PDS. This led to reductions in offtake of food grains under PDS, particularly for wheat. In 1991-1992, the offtake of wheat was 86 percent of the quantity allocated for PDS while rice offtake was about 90 percent of the allocated quantity. However, in 2000-2001, wheat offtake fell to 32 percent while for rice it fell to 48 percent (GOI, 2000a). Offtake declined particularly after 1998-1999, indicating the introduction of targeted PDS (TPDS), which made a distinction between above poverty line (APL) and below poverty line (BPL) households.

A Long-Term Factor: Changing Consumption Patterns

Apart from the factors relating to procurement and issue prices, long-term factors such as changes in consumption patterns could also be responsible for the huge increase in stocks. It is now widely recognized that the food basket is more diversified and dramatic changes in food consumption patterns have taken place in India in the post-green revolution period. For example, at the national level, in rural areas, cereal consumption declined

TABLE 13.1. Growth of minimum support prices and issue prices

	1980-1981 to 1991-1992	1992-1993 to 1999-2000
Annual average inflation in MSP (%)		
Rice	7.5	10.0
Wheat	6.2	10.8
Average annual increase in issue price (%)		
Rice	8.4	9.3
Wheat	7.4	12.0

Source: GOI, 2002.

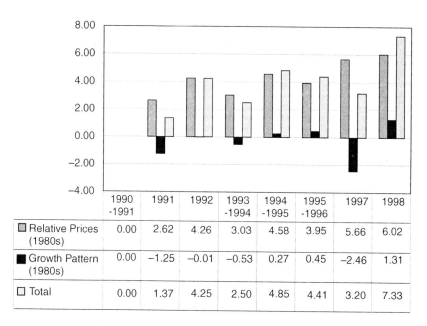

	1990-1991	1991	1992	1993-1994	1994-1995	1995-1996	1997	1998
▨ Relative Prices (1980s)	0.00	2.62	4.26	3.03	4.58	3.95	5.66	6.02
■ Growth Pattern (1980s)	0.00	−1.25	−0.01	−0.53	0.27	0.45	−2.46	1.31
☐ Total	0.00	1.37	4.25	2.50	4.85	4.41	3.20	7.33

FIGURE 13.1. Estimated additional demand for cereals in India.

from 15.3 kg per capita per month in 1972-1973 to 13.4 kg per capita per month in 1993-1994.[7] Recent National Statistical Survey data for 1999-2000 show a further decline in the per capita consumption of cereals. This is true for all classes, rich and poor. The cereal shares saw a dramatic decline of more than 10 percentage points between 1972-1973 to 1993-1994 in most regions, in both rural and urban India. Similarly, the share of meat and milk products, and vegetables and fruits has increased over time.

The Food Subsidy Increase in the Late 1990s

As a result of the accumulation of food grains, the food subsidy increased significantly in the late 1990s. Figure 13.2 shows that the food subsidy increased from Rs. 24.5 in 1990-1991 to Rs.212 billion (budgeted) in 2002-2003. It increased from Rs. 92.1 billion in 1999-2000 to Rs.120.1 billion in 2000-2001. It is expected to increase to Rs.212 billion in 2000-2003. As percent of GDP, the food subsidy increased from 0.43 in 1990-1991 to 0.48 in 1999-2000 (Figure 13.3). It increased significantly to 0.76 percent in 2001-2002 and to 0.83 percent in 2002-2003. Similarly, food subsidy as percent of total public expenditure also increased significantly from 2.3 percent in 1990-1991 to 5.17 percent in 2002-2003.

The related issue under food subsidy is that the producer subsidy has increased while the consumer subsidy has declined. Higher procurement with the reduced offtake resulted in the generation of larger stocks, which in turn led to higher carrying costs (comprised of freight, storage, interest charges, etc.). In 1997-1998, the buffer component of the food subsidy bill was only 9.37 percent. It increased to 42 percent in 2000-2001 (Figure 13.4).

Problems in the Public Distribution System

The PDS is one of the instruments for improving food security at the household level in India. PDS ensures availability of essential commodities such as rice, wheat, edible oils, and kerosene to the consumers through a

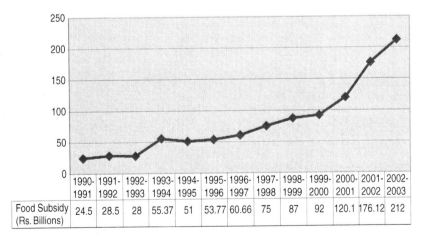

FIGURE 13.2. Food subsidy.

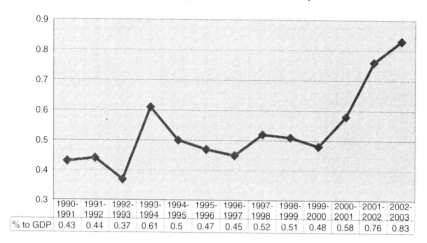

% to GDP	1990-1991	1991-1992	1992-1993	1993-1994	1994-1995	1995-1996	1996-1997	1997-1998	1998-1999	1999-2000	2000-2001	2001-2002	2002-2003
	0.43	0.44	0.37	0.61	0.5	0.47	0.45	0.52	0.51	0.48	0.58	0.76	0.83

FIGURE 13.3. Food subsidy as percent of GDP.

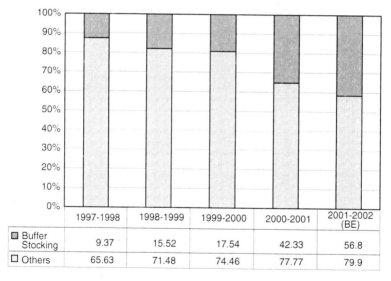

	1997-1998	1998-1999	1999-2000	2000-2001	2001-2002 (BE)
▨ Buffer Stocking	9.37	15.52	17.54	42.33	56.8
□ Others	65.63	71.48	74.46	77.77	79.9

FIGURE 13.4. Components of food subsidy.

network of outlets or fair price shops. These goods are supplied to consumers at below market prices. Access to the system was universal until 1997. During the first few decades of its existence, the PDS had actually never operated as an antipoverty program, but merely as an instrument of price stabilization. Before the late 1970s, the PDS was mainly restricted to urban areas and food-deficit regions. Since the Sixth Five-Year Plan, however, the importance of the PDS has been recognized. In the 1980s, the program covered rural areas in many states. In the 1990s, the government decided to restructure the PDS in the form of a revamped PDS (RPDS) and targeted PDS (TPDS). The PDS has been effective during drought years, for example in 1979-1980 and 1987-1988 (Tendulkar et al., 1993). It has also been effective in transferring food grains from surplus areas to deficit regions such as Kerala. However, the system has many problems.

A significant diversion of PDS commodities has occurred. A study conducted by the Tata Economic Consultancy Services to determine the quantity of PDS goods diverted from the system found that at the national level there was a diversion of 36 percent of wheat supplies, 31 percent of rice, and 23 percent of sugar. The diversion occurred more in the northern, eastern, and northeastern regions; it was comparatively less in the southern and western regions.

The overall impact of PDS on the poor seems to be less than assumed. Few studies have measured the welfare gains of the PDS. Parikh (1994) reports that for every rupee spent less than 22 paise are actually allotted to the poor. This is true of all states except Goa, Daman, and Diu. In these areas 28 paise reach the poor.

In a study conducted by the Bureau of Directorate of Economics and Statistics, Karnataka, revealed that only 13 paise out of a rupee spent on PDS reach the poor (Machaiah 1995). A study by Radhakrishna et al. (1997) concludes that the potential benefits from the PDS to the poor could not be realized cost effectively due to weak targeting and leakages. The cost of income transfer was high mainly because the program was open ended and never targeted. The study also reported that approaches other than quantity rationing, including self-targeting and food stamps, need to be considered in order to deliver food transfers to the needy in a cost-effective manner.

Poor states have not benefited much from PDS. The relationship between poverty and PDS offtake is weak across Indian states. Radhakrishna et al. (1997) provides the following conclusions on regional disparities:

1. Regional mistargeting exists in the distribution of food grains through PDS. The offtake by states such as Kerala and Andhra Pradesh, which are implementing a subsidized food scheme, is high. On the other hand, the offtake by the poorer states such as Bihar, Orissa, and Madhya Pradesh is low.

2. There has been poor targeting and access in the majority of states. The access of the poor to PDS is still limited even after four decades of op-

eration. Data for 1995, based on selected village surveys, confirm that in Bihar and Uttar Pradesh the poor have limited access to the PDS.
3. The empirical evidence for an urban bias seems to be weak. Urban bias was severe only in two states, Jammu and Kashmir and West Bengal.
4. In states such as Andhra Pradesh and Kerala, the access of the poor is much better. However, even in these states, per capita monthly PDS cereal purchases tended to be regressive.

The targeted public distribution system (TPDS) has undergone changes over time and has some weaknesses. TPDS was introduced in 1997 by issuing special cards to BPL families. A separate issue price is fixed for APL households. Under the scheme, each poor family was originally entitled to 10 kg of food grains per month at a subsidized price. To improve allocation to poor families, the government increased the allocation to BPL families from 10 kg to 20 kg of grain per family per month at 50 percent of economic cost from April 2000. The allocation for APL households was retained at the same level at the time of introduction of TPDS. However, the central issue prices (CIP) for APL families was fixed at 100 percent of economic cost from the date so that the entire consumer subsidy could be directed for the benefit of the BPL population. In order to reduce excess stocks, the central government has initiated the following measures under TPDS from July 2001:

1. The BPL allocation of food grains has been increased from 20 kg to 25 kg per family per month with effect from (w.e.f) July 2001; the CIP for BPL families at Rs. 4.15 per kg for wheat and Rs. 5.65 per kg for rice is 48 percent of the economic cost.
2. The government has decided to allocate food grains to APL families at the discounted rate of 70 percent of the economic cost. The CIP of APL wheat of Rs. 830 per quintal has been reduced to Rs. 610 per quintal, and the CIP of APL rice, which was at Rs. 1130 per quintal, has been reduced to Rs. 830 per quintal.

Furthermore, under the Antyodaya Anna Yojana program, 25 kg of food grain are provided to the poorest of the poor families at a highly subsidized rate of Rs. 2 per kg for wheat and Rs.3 per kg for rice.

Some problems still exist, however. The government's approach of TPDS is related to the income-based means test. However, identifying the poor through the income-based means test is difficult. Other methods should be used for targeting. Targeting also leads to exclusion and inclusion errors (Cornia and Stewart 1995; Sen 1995). The former error occurs when the poor are excluded while the latter occurs when the nonpoor are included. Both kinds of errors seem to be high under TPDS. The challenge is to minimize the errors with innovative programs.

Other associated problems with the scheme are: (1) the poor do not have cash to buy 20 kg at a time, and often they are not permitted to buy in installments; (2) the low quality of food grains—a World Bank report (1999) states that half of FCI's grain stocks are at least two years old, 30 percent between two and four years old, and some grain is as old as 16 years; and (3) weak monitoring, a lack of transparency, and inadequate accountability of officials implementing the scheme (GOI, 2000a).

In spite of the number of programs for the poor, concerns remain regarding calorie intake and malnutrition. For the bottom 30 percent of the population, there has hardly been any perceptible improvement in calorie intake in both rural and urban areas. In the 1990s, the per-capita calorie intake for the bottom 30 percent of the population was 1,678 calories per day in the rural areas and 1,682 calories per day in urban areas. This is a matter of concern because the bottom 30 percent of the population is deprived of the minimum required energy intake. The problem of malnutrition is acute and widespread, and it is extremely serious among women and children. The proportion of pregnant women with anemia is as high as 88 percent. At the national level, more than 50 percent of the children under five were underweight in 1998-1999. The percentage of malnutrition is much higher in poorer states such as Bihar, Uttar Pradesh, and Madhya Pradesh. Currently, the major nutrition-related public health problems are chronic energy deficiency and undernutrition, and deficiencies of micronutrients, such as iron (which leads to anemia), folic acid, vitamin A, and iodine (GOI, 2000b). In fact, nutritionists argue that energy intake is a poor measure of nutritional status, because nutritional status depends not only on nutrient intake, but also on nonnutrient food attributes, privately and publicly provided inputs, and health status (Martorel and Ho 1984, quoted in Radhakrishna 2001). The nonfood factors that influence biological absorption are also considered as equally important to food security.

What Are the Needed Reforms?

The food policy pursued since India gained independence could be considered a success story in some respects. Overall food-grain production increased from 50 million metric tons in 1950-1951 to 209 million metric tons in 1999-2000. The procurement and the reach of the PDS have also risen over time. Food/cash-for-work programs also helped to raise the purchasing power of the poor. However, as discussed, all is not well with India's food policy, particularly the procurement system and PDS. A review of past experience raises a number of questions about the working of the PDS and the associated policies. At the macro level, some of these questions refer to the fixing of the support and issue prices, the quantities to be procured and distributed, and the manner in which stocks should be managed. At the micro level, there are the obvious problems of extending the coverage of the PDS, ways of targeting, and managing the fair price shops to prevent leakages. There is lot of room for improvement in these policies.

Higher procurement prices invariably edge out private trade thus causing FCI to procure more than what is required for food security. Low volumes of food grains available through private trade generally lead to higher market prices during lean seasons while large stocks in storage go unutilized. The food security system has become expensive and unsustainable. It needs better options in terms of cost effectiveness, reaching targeted populations, and financial sustainability.

Procurement and Buffer Stock Policies: Needed Reforms

- *The private sector should be involved in the handling of grain to increase the efficiency of storage and distribution.* The inefficiency of the FCI has led to an increase in economic cost of buffer stocks, which in turn has led to an increase in the issue price. The private sector should be involved in the storage and distribution of food grains (World Bank, 1999). The inefficiency of the FCI is known (Gulati et al., 1996), and it has been responsible for increasing the economic cost of the food subsidy. Some studies have shown that private sector costs are lower than those of FCI in handling grains.
- *The MSP should be frozen for a few years.* In order to control the rise in minimum support prices, the Commission for Agricultural Costs and Prices (CACP) suggested freezing the MSP for a few years. Farmers may be interested in the total net income (yield × price) from crops rather than price only.
- *Procurement should be decentralized.* In an attempt to economize on procurement costs, the Union Budget 2001-2002 indicated an enlarged role for the state governments in both procurement and distribution of food grains for PDS. Instead of giving subsidized food grains, funds would be provided to the state governments to enable them to procure and distribute food grains. Decentralized procurement has already started in Uttar Pradesh, Madhya Pradesh, and West

Bengal. It is intended to benefit both farmers and consumers while simultaneously improving the financial position of the central government. However, state governments such as Punjab, Haryana, Andhra Pradesh, and Kerala opposed the idea because of the lack of infrastructure and resources within these states. These problems will have to be solved before decentralized procurement is implemented.

- *Food crops must be diversified.* The changing consumption patterns have implications for food-demand projections and there is a need for more focus on noncereal food crops and allied agricultural activities. Some forecasts on cereal demand show that the production of 224 million metric tons (193.5 for food and 30.1 for feed) by 2020 would be sufficient (Dyson and Hanchate, 2000). Other forecasts also predict around 250 million metric tons. Thus there appears to be no cause for concern over cereal demand outstripping cereal supply in India. A slowdown in population growth and a shift in consumption away from cereals would bring about a decrease in cereal demand in the future. As shown in Table 13.2, the share of expenditure on food grains in the total food expenditure of households has declined over time. The share of other foods, namely fruits/vegetables, milk, fish, eggs, etc., has relatively increased. It also indicates the need to take care of the supply-side factors for noncereal food crops and allied agricultural activities (e.g., dairy). Public and private investments in infrastructure and research are needed for these activities. The minimum support policy has to be modified to encourage changes in cropping patterns.

- *The new income policy suggested by the government must be reconsidered for viability.* The Department of Agriculture has been considering the adoption of a policy of compensating farmers directly for the difference between the market price and the MSP. This income policy may not be viable in a country like India where the government would have to deal with more than 100 million farm holdings.

- *Maintain the buffer stock at economically optimal levels.* As mentioned, there are about 65 million food grains with the government of India. It is very uneconomical to hold such huge stocks. Various committees have suggested the optimal size of a buffer stock, which varies from 15 to 25 million metric tons for a year, depending on the season. The Expenditure Reforms Commission recommended 17 million metric tons as the total average to be maintained for distribution and buffer stock. The FCI should procure only the required amounts from the open market instead of having an open-ended policy.

TABLE 13.2. Share of expenditure on food grains, fruits/vegetables, milk, meat, eggs, and fish in total food expenditure

	Expenditure on food grains		Expenditure on fruits and vegetables		Expenditure on milk, meat, eggs, and fish	
	Rural	Urban	Rural	Urban	Rural	Urban
1972-1973	63.10	42.02	10.15	9.92	13.44	19.53
1977-1978	58.01	40.83	9.95	10.67	16.17	21.67
1983-1984	55.34	38.75	11.43	12.01	16.01	21.66
1987-1988	47.81	33.16	12.66	13.83	18.59	23.23
1993-1994	44.78	31.63	14.40	14.99	20.25	24.13
1999-2000	44.11	31.81	14.48	15.59	20.37	24.53

Source: Government of India, 2002.

Reforms in the Public Distribution System

In spite of spending Rs.210 billion for food subsidization, many of the poor are not benefiting from the system. Methods need to be developed to reduce costs and improve targeting of the poor. Some of these could include:

- *Implement geographical targeting.* As previously mentioned, innovative programs are needed to minimize targeting errors. In this context, some studies have shown that geographical targeting is better than income-based targeting (Bigman and Srinivasan, 2001; Jha and Srinivasan, 1999). In fact, the limited evidence on RPDS (revamped PDS) in Andhra Pradesh and Madhya Pradesh suggests that targeting was more effective in RPDS villages than it was in non-RPDS villages with similar socioeconomic characteristics.
- *Conduct a trial with a food coupon or credit card system.* A food coupon system for the distribution of rice and kerosene through PDS was introduced in Andhra Pradesh in 1998-1999. The scheme was aimed at improving the delivery system of the two commodities. Under the scheme, mere possession of a card was not enough to obtain coupons. The physical presence of the cardholder, whose photo was on the card, was necessary. This has reduced the leakage of rice and kerosene to a great extent, if not totally eliminated it. Another alternative is a food credit card system that customers can use to buy subsidized food grains from the market and which allows retailers to claim the subsidy

from the government. However, the private sector has to be assured
that the government will reimburse it without delay.

- *Utilize the excess food grain stocks for food-for-work programs
  (FFW).* Four possible options exist for using excess stocks: sale on the
  domestic open market, export, issue price reduction, and use for
  FFWs. Selling in the open market would defeat the very purpose of
  price support operations, as traders would buy the grain at the depressed price and sell it back to the government at the procurement
  price. Exporting food grains might be feasible, and it has been successful recently in the case of rice. The stocks with FCI are 65 million
  tons and exports would not completely solve the excess stock problem. Moreover, in the case of wheat exports, the government has not
  been successful because international prices are lower than Indian
  prices. The third alternative of stimulating domestic demand through
  reducing the issue price would not be cost effective. The fourth alternative of using buffer stock for FFWs would be the best option. FFWs
  not only create employment and incomes for the poor workers but also
  provide rural infrastructure such as roads and improvements in land.

- *Improve public works programs so that food security may be
  achieved.* Public works programs provide many direct and indirect
  benefits to the poor (Dev, 1995). Rural public works programs have
  been in operation in India since the 1960s. Among the important ones
  are: the Employment Guarantee Scheme (EGS) of Maharashtra,
  Jawahar Rojgar Yojana (JRY), and the Employment Assurance Scheme
  (EAS). The EGS is a cash-for-work program. The JRY and EAS are
  food-for-work schemes, but the quantity of food supplied is very low.
  Two questions must be addressed in designing improved public works
  programs. One is how can the excess stocks be utilized effectively for
  food-for-work schemes? The second is how can cash-for-work programs be redesigned to increase the purchasing power of their unskilled participants? With an increase in purchasing power, workers
  could buy food grains from PDS or in the market. However, food security does not mean the ability to buy food grains alone. Increased
  purchasing power is important because it would allow program participants to afford other foods as well. Another question is how to link
  public works schemes with PDS. One option would be to ensure that
  subsidized food grains under PDS reach workers in FFWs. The participants would thus receive double benefits (under FFW and PDS). The
  prime minister's Sampoorna Gramin Rojgar Yojana program, to be
  funded at Rs.100 billion (10,000 crores), is supposed to provide full
  assistance for the states to implement FFWs.

- *Improve the effectiveness of ICDS and other nutritional programs.* Under- nutrition among children is a severe problem in India. One of the objectives of Integrated Child Development Services (ICDS) is to improve the health and nutrition status of children zero to six years by providing supplementary food and coordinating with state health departments to ensure the delivery of required health inputs. ICDS has been successful in some cases. For example, noon-meal programs for children in Tamil Nadu are effective in improving the nutrition of children and school enrollment. However, there are some gaps in implementation as shown by several evaluations. Some of these gaps are: irregular food supply; coverage of below three years under supplementary nutrition program has been relatively incomplete; there is little community participation in running ICDS; the Anganwadis do not have adequate buildings; interdepartmental coordination is poor.

OTHER AGRICULTURAL POLICIES

To revive agricultural growth, there is a need for increased public and private investment, greater availability of rural credit, infrastructure such as irrigation, better marketing systems, improvements in agricultural research and technology, and domestic and global trade liberalization. Also essential are cost-reducing technologies that would help make agriculture viable and enable India to compete with other countries in the WTO environment. Selected policy issues are discussed as follows.

Credit

The total credit flow to the agricultural sector through institutional channels is expected to reach the level of Rs. 750 billion (75,000 crores) in 2002-2003. However, the share of agriculture in the nonfood bank credit has been only 12 percent for the past few years.[8] Despite the banking system having a wide network of rural branches and the existence of many programs for the expansion of credit for agriculture and rural development, a large number of very poor people still remain outside the fold of the formal banking system. The credit system should reach marginal and small farmers. In fact, the growth rate of agricultural credit for small and marginal farmers declined in the 1990s from the 1980s (RBI, 2002). During the same period, there was no decline of growth in credit for large farmers. Some farmers' suicides in recent years have been due mainly to their being trapped in debt. Farmers take loans from moneylenders at high interest rates and fall into debt. Rural credit must also expand at a fast rate. Rising incomes cause increased demand for high-income elastic agricultural products such as dairy, livestock, horticulture, and those derived from agroforestry. The need to increase investment for the production of the goods will increase.

Public Investment

The decline in public investment in agriculture has been a cause for concern because it is crucial for the development of agricultural research and infrastructure, such as irrigation, electricity, roads, markets, and communications. In 2000-2001, public investment in agriculture was Rs. 40.07 billion (4,007 crore), which was less than that for 1999-2000 (GOI, 2000a). As a percent of agricultural GDP, public investment was between 7 and 8 percent in the 1990s. This percentage was less than that for the 1980s. Private investment has been increasing, however private investment often depends on public investment. For example, in West Bengal, private investment in irrigation increased only after rural electrification was established. Assam has also shown a significant increase in minor irrigation. Public investment has to be greater in rainfed areas and marginal areas where poverty is concentrated. Under the Rural Infrastructure Development Fund (RIDF), recent budgets have increased funds for this scheme. A closer look at the RIDF, however, shows that the utilization rate is low; around 40 percent. A few states such as Andhra Pradesh and Maharashtra utilize more funds than others. The low utilization of RIDF is attributed to bureaucratic delays in a number of areas, proposal submission, evaluation, the sanctioning of funds, and the disbursement of funds in installments. In March 2002's budget, the release of funds to the states was linked to reforms in agriculture and rural development. However, the criteria regarding this linkage are not clear.

Technology, Trade, and Institutions

Technology plays an important role in improving yields. Research and extension are important for improving and disseminating technology. A fresh look at the priorities of the Indian agricultural research system is necessary in light of emerging prospects in the open economy framework. States have to take a lead in research and extension. India spends only 0.5 percent of GDP on agricultural research as compared to more than 1 percent by other developing countries. The involvement of the private sector should be encouraged in seed research. Biotechnology offers many opportunities for increasing yields. The recent approval of the cultivation of *Bacillus thuringiensis* (Bt) cotton by the government of India is a step in the right direction.

Regarding trade, the WTO does not pose a threat to food security or any of the agricultural commodities. However, the expected gains since 1995 due to the reduction of agricultural subsidies by developing countries under the WTO have not materialized. Vigilance is needed so as to take timely measures within existing tariff bindings. Even if developed countries reduce subsidies the prices may not increase because they may be used in cost-reducing technologies. Therefore, the viability of Indian agricultural goods as trade commodities depends on cost reduction and not on higher prices. The domestic and trade liberalization measures have to be continued to give incentives to farmers.

Regarding institutional reforms, land reforms still have relevance. Apart from redistribution measures, focus should be on tenancy reforms and the consolidation of holdings. On irrigation and rainfed areas, water-user associations and watershed programs have to be strengthened.

ECONOMIC ACCESS, EMPLOYMENT, AND RURAL TRANSFORMATION

Food security is mainly a question of household economic access to food, as food availability at the national level is not a problem. One of the reasons for the accumulation of over 60 million metric tons of food stocks with FCI is the low purchasing power of the poor. Two factors could be responsible for the low level of purchasing power in the 1990s. First, high relative prices of food due to high procurement prices could have reduced the capacity of the poor to purchase more food grains. The second factor could be the decline in employment opportunities in spite of high GDP growth. The solution to food security and poverty is to improve economic access through employment. Diversification and the promotion of rural nonfarm employment are needed to create productive employment.

Employment creation in the 1990s was not very encouraging. The growth rate of rural employment was around 0.5 percent per annum between 1993-1994 and 1999-2000, as compared to 1.7 percent per annum between 1983 and 1993-1994. The daily status unemployment rate in rural areas has increased from 5.63 percent in 1993-1994 to 7.21 percent in 1999-2000. The overall employment growth declined from 2.04 percent during 1983-1994 to 0.98 percent during 1994-2000. Much of the decline in the growth was due to developments in two sectors viz, agriculture, and community, social, and personal services. These two sectors, which account for 70 percent of the total employment, did not show any growth during the 1990s.

How can more employment opportunities be created and the quality of life of the poor be improved? There are basically two approaches. One is through sectoral programs and the other is through direct employment programs. Sectoral programs refer to those policies and programs in agriculture, industry, and services. The direct programs are the antipoverty programs targeted to the poor. There is some overlap in the two approaches. In regard to the elasticity of employment to GDP, for the entire economy it was 0.53 in the period 1977-1983 and declined to 0.41 in the period 1983-1994. The elasticity declined sharply to 0.15 during 1993-2000. Employment elasticities in agriculture, and community, social, and personal services were zero during the latter period. In manufacturing it was 0.26 while in services it was more than 0.50 during this period.

Agriculture still contributes 60 percent of the total employment in the country. In the decade 1983-1994, agriculture contributed 50 percent of the additional employment. On the other hand, there was an absolute decline in agricultural employment between 1993-1994 and 1999-2000. In the process of economic development, workers in agriculture are supposed to shift to nonagriculture employment. However, underemployment can be removed with higher agricultural growth. Also, agriculture still has the potential to absorb workers in regions with higher incidences of rural poverty, such as Orissa, Bihar, Assam, Madhya Pradesh, West Bengal, and Uttar Pradesh.

Within agricultural and allied activities, there seems to be some diversification toward noncereal crops. Diversification to fruits and vegetables, fisheries, and animal husbandry is expected to promote employment. For example, in Maharashtra, the requirement of person days per hectare per crop season for wheat is 143 while for vegetables it is 200. The corresponding numbers for grapes and other fruits are 2,510 and 855, respectively. Thus fruits in general require nearly six times and grapes over seventeen times the person days required per hectare for wheat. However, risk and uncertainty are always associated with diversification. Technology, infrastructure, and markets have to be improved in order to shift the farmers to non-food-grain crops.

Development of wastelands would substantially increase employment opportunities. Currently, there are about 24 million hectares of land that are categorized as culturable wasteland and permanent fallows, and which could be developed and brought into cultivation. Two alternatives could be approached. First, these lands could be distributed to *panchayats* (local governments) and small and marginal farmers. Another alternative is to give them to the corporate sector on a long lease of approximately twenty years. *Panchayats* and the corporate sector can develop the wastelands by raising resources for the market. However, one has to make sure that the corporate sector does not occupy the fertile land.

Growth in rural nonfarm employment (RNFE) can improve rural wages and also be an escape route for agricultural workers, leading to an improvement in their purchasing power. The importance of the rural nonfarm sector in poverty alleviation and the improvement of livelihoods are being increasingly recognized. An increase in rural nonfarm employment is one of the main factors responsible for the reduction of poverty in the 1980s. The percentage of RNFE in total rural employment in India increased from 16.6 percent in 1977-1978 to 18.4 percent in 1983, to 21.6 percent in 1993-1994, and to 23.8 percent in 1999-2000. The compound growth rate in RNFE during 1977-1983 was 4.06 percent per annum; it was 3.28 percent per annum and 2.14 percent per annum during 1983-1994 and 1993-2000, respectively. In other words, during the reform period (1993-1994 to 1999-2000) the growth rate of employment in RNFE was lower than the prereform period. This is a matter of concern. However, it has to be seen in the context of the low overall growth of employment (around 1 percent per annum) during the reform period. At the same time, the expected growth in the nonagricultural sector due to economic reforms does not seem to have materialized. As compared to the East Asian experience, the growth in RNFE in India has been much slower.

One of the challenges of the reforms now is to improve the quality of employment and incomes in the rural nonfarm sector. A three-pronged strategy is needed for the enhancement of the livelihoods of the rural poor. First, the government should have policies to improve the education and skills of workers. Second, the government should have several policies to increase employment for unskilled workers. Third, the incomes of women have to be improved by creating opportunities in the higher-productivity sectors. Most women are confined to employment in agriculture. There was only a 0.7 percent increase in the share of RNFE during the reform period. To achieve these strategies, pro-poor growth engines have to be identified at the subsectoral level rather than at the broad sectoral level. Public investment in agriculture and rural nonagriculture has to be improved significantly. In order to develop the rural nonfarm sector, there is a need to look at issues such as rural-urban linkages, sectoral and subsector potentials, markets, regulations and promotional policies, human capital, training, entrepreneurship, skills, and finally infrastructure and technology. Rigidities in these factors have to be removed to promote the rural nonfarm sector. Allowing the poor to contribute to and benefit from increased growth rates will pose particular challenges as employment in India is largely in the unorganized sector.

CONCLUDING OBSERVATIONS

In this chapter, the market reforms needed for achieving higher and sustainable growth in agriculture and food security for the poor are examined. A paradoxical situation exists: 60 to 65 million metric tons of food grain are held as stock while 260 million people live below the poverty line. Continuous increases in the procurement price have led to the accumulation of stocks far above the levels required. The steep increases in the procurement price have discouraged the private sector from holding stocks, made exports uncompetitive, and reduced the domestic consumption of food grains, particularly by the poor. The performance of PDS is also not satisfactory. There is a need for reforms in procurement, buffer stock, and distribution policies.

The role of the FCI has to be restricted to price stabilization and the PDS should be left to the states. The procurement system has to be decentralized. The private sector may have to be involved in storage and handling the grains. In the case of PDS, cost-effective methods have to be tried in order to reach the poor effectively. In the case of agricultural policies, there is a need for faster reforms in agriculture in order to have win-win-situation for farmers, workers, and consumers. Supply-side factors such as irrigation, infrastructure, technology, research and extension, and marketing have to be improved to have higher and sustainable growth. A multidimensional reform agenda has to be developed by improving incentives, rationalizing subsidies, promoting investments, and protecting the poor. Domestic and external trade liberalization has to be faster than before. Commodity exchanges must be established in order to reduce the volatility of crop prices. The WTO provides opportunities and challenges for Indian agriculture. WTO poses no threat on food security and on other agricultural commodities in terms of a deluge of imports. The viability of Indian agriculture depends on internal factors such as public investment in infrastructure and research. Cost-reducing technology has to be promoted to compete in the world markets and to achieve food security for the poor.

The accumulation of food grains also indicates a need for crop diversification in the agriculture sector, which will also improve employment opportunities. Education, infrastructure, credit, etc., have to be improved in order to raise the productivity of rural nonfarm employment. The suggested reforms in the food and agricultural sector and in the rural nonfarm sector are expected to enhance the food security of the poor through rural transformation.

NOTES

1. Food security refers to availability, stability, access, and absorption. Many studies have also shown that improvements in nutrition are important even for increases in productivity of the workers. Thus food security has intrinsic (for its own sake) as well as instrumental (for increasing productivity) value.

2. In the 2002-2003 budget speech, the finance minister said, "having achieved great success with the green revolution and then the white revolution, the country is now ready for its third revolution of agricultural diversification and food processing" (Budget, 2002).

3. The Food Corporation of India (FCI) and the Commission on Agricultural Costs and Prices (CACP) were established in 1964-1965 to oversee the countrywide procurement, distribution, and stocking of food grains, and recommend minimum support prices of agricultural crops to the government, respectively. These food policy operations are aimed at stabilizing prices, protecting the interests of the producers, consumers, and the economy as a whole, and reducing interregional disparities in production and consumption of food grains.

4. One crore is equal to ten million.

5. Ten lakhas is one million.

6. The definitions of various costs given by the Commission for Agricultural Costs and Prices are as follows: Cost A1 = All actual expenses in cash and kind incurred in production by owner; Cost A2 = Cost A1 + rent paid for leased-in land; Cost B1 = Cost A1 + interest value of owned capital assets (excluding land); Cost C1 = Cost B1 + imputed value of family labor.

7. For more details on changing consumption patterns, see Rao (2000).

8. Bank credit is divided into food credit and nonfood credit. Food credit is given mainly for buying food grains, for bufferstock, and the public distribution system. Agricultural credit is part of the nonfood credit.

REFERENCES

Bigman, D. and P.V. Srinivasan (2001). *Geographical Targeting of Poverty Alleviation Programs: Methodology and Applications in Rural India.* Indira Gandhi Institute of Development Research, mimeo.

Budget (2002). Budget of finance minister. Available online at <http://216.122.48.63/budget2002FMSSpeech28Feb.pdf>.

Cornia, G.A. and F. Stewart (1995). Two Errors of Targeting. In Van de Walle, D. and K. Need (Eds.), *Public Spending and the Poor: Theory and Evidence*, Baltimore: Johns Hopkins University Press and World Bank.

Dev, S. Mahendra (1995). Alleviating Poverty: Maharastra Employment Guarantee Scheme. *Economic and Political Weekly*, 30(41/42), pp. 123-132.

Dyson, T. and A. Hanchate (2000). India's Demographic and Food Prospects: State Level Analysis. *Economic and Political Weekly*, November 11, pp. 402-403.

Government of India (2000a). *Mid-Term Appraisal of Ninth Five Year Plan (1997-2002).* New Delhi: Planning Commission.

Government of India (2000b). *Report on Food Security*. Chairman K.P. Geetha-krishnan, former Finance Secretary, Expenditure Reforms Commission. New Delhi: Government of India.

Government of India (2002). *Economic Survey of India, 2000-2001*, Ministry of Finance. New Delhi: GOI.

Gulati, A., P.K. Sharma, and S. Kähkönen (1996). *The Food Corporation of India: Successes and Failures in Indian Foodgrain Marketing*, August, IRIS-India Working paper No. 18, IRIS Center, College Park, MD: University of Maryland.

Jha, S. and P.V. Srinivasan (1999). Grain Price Stabilization in India: Evaluation of Policy Alternatives. *Agricultural Economics*, 21: 93-108.

Machaiah, M.G. (1995). 13 Paisa out of a Re Spent on PDS Reaching the Poor: Study. *Indian Express*, June 30.

Parikh, K.S. (1994). Who Gets How Much from PDS: How Effectively Does it Reach the Poor? *Sarvekshana*, 17(3): January-March.

Radhakrishna, R. (2001). Food Security: Emerging Concerns. In Dev, S.M., Piush, A., Gayathri, V., and R.P. Mamgain (Eds.), *Social and Economic Security in India*. New Delhi: Institute for Human Development.

Radhakrishna, R. and Subbarao, K., with S. Indrakant and C. Ravi (1997). *India's Public Distribution System: A National and International Perspective*. World Bank Discussion Paper No. 380. Washington, DC: The World Bank.

Ramaswamy, B. (2002). Efficiency and Equity of Food Market Interventions. *Economic and Political Weekly*, 37(12).

Rao, C.H.H. (2000). Declining Demand for Food grains in Rural India: Causes and Implications. *Economic and Political Weekly*, January 22.

Ravallion, M. (2000). What Is Needed for a More Pro-Poor Growth Process in India? *Economic and Political Weekly*, 35(13).

RBI (2002). *Report on Currency and Finance 2000-01*. Mumbai: Reserve Bank of India.

Sen, A. (1995). The Political Economy of Targeting. In van de Walle, D. and K. Nead (Eds.), *Public Spending and the Poor: Theory and Evidence* (pp. 11-24). Baltimore: Johns Hopkins University Press and World Bank.

Tendulkar, S.D., K. Sundaram, and L. R. Jain (1993). *Poverty in India, 1970-71 to 1988-89*. ARTEP Working Papers, ILO-ARTEP, New Delhi.

World Bank (1999). *India—Foodgrain Marketing Policies: Reforming to Meet Security Needs*. Rural Development Unit, South Asia Region, Working Paper, Report No. 18329, May 27. World Bank.

Chapter 14

Market Reform, Diversification, and Food Security in Sri Lanka

Saman Kelegama
Suresh Chandra Babu

INTRODUCTION

Sri Lankan policymakers are at the crossroads of furthering domestic market reforms and trade liberalization in the food and agricultural sectors. Due to increased agricultural and food production over the years, Sri Lanka has reduced its food imports substantially from 51 percent of total imports in 1975 to 9 percent in 1999. Yet the challenge of attaining food security for its growing population remains. Major tasks for planners and policymakers include understanding the role of technology in increasing agricultural productivity, designing and implementing policies that will facilitate increased employment in the rural sector, and identifying and nurturing institutional reforms that will increase the participation of agricultural producers both in regional and international markets. The major objective of this chapter is to provide an overview of the issues and challenges facing Sri Lankan policymakers in their efforts to introduce market reforms, liberalize external trade, and diversify the agricultural production base.

The chapter is organized as follows. An overview of Sri Lanka's agricultural sector is provided in the first part to illustrate the contribution it has made to rural incomes and food security, and the potential it still has. The issues related to post-WTO market reforms are then addressed in Part II. The challenges encountered in employing agricultural diversification as a means to boost rural employment and incomes are discussed in Part III. The policy and program constraints in attaining food security at the national and household levels are discussed in Part IV. The first part of the section looks at national food security from the broad perspective of the agricultural sector, and the second part examines the role of food and nutrition programs in achieving household food and nutrition security.

PART I: OVERVIEW OF SRI LANKAN AGRICULTURE

Agriculture is the mainstay of Sri Lanka's rural population. Yet in the past decade, that sector's growth has been around 2 percent per annum—the lowest in South Asia (RIS, 2002). Because the agricultural sector still accounts for 18 percent of GDP, and 65 percent of the population largely depends on it for their livelihood, the government is particularly concerned about its poor performance during the past decade.

Despite the significant structural transformation of the Sri Lankan economy over the past three decades, the agricultural sector still provides employment to about 32 percent of the labor force, surpassing the contribution of any other sector. Land for agriculture is mainly located within the dry zone where water availability limits land utilization to its full potential. Sri Lanka's land area under cultivation decreased significantly during the past decade, by about 18 percent.

The agricultural sector has two components, the plantation and the non-plantation. The three main plantation crops, tea, rubber, and coconut, account for 15.6 percent of the agricultural GDP and 3 percent of the GDP. About 40 percent of the lands in the tea and rubber sectors are in large plantations and the small landholders have the balance. The plantation sector accounts for 15.3 percent of total foreign exchange earnings. Total direct and indirect employment in the sector is estimated at around 1.5 million people. The plantation sector was under state control from the mid-1970s to 1992 after which it was privatized. In 1995, the land ownership lease of the management companies was extended to ninety-nine years.

The nonplantation sector, also referred to as the domestic food production sector, produces mainly paddy and subsidiary food crops. Paddy (rice), the main staple crop, dominates nonplantation agriculture. The area under paddy cultivation is 930,000 hectares or 45 percent of the total area under the agricultural sector. The majority of farmers in the paddy sector are small-scale producers; more than 70 percent of paddy holdings are less than one hectare and only about 5 percent have a holding size of greater than 2 hectares. The rice sector employs about half of the total agricultural labor force in Sri Lanka (about 20 percent of the total labor force). Rice accounts for approximately 25 percent of the consumer goods basket, about 75 percent of total grain consumption, and 45 percent of calorie intake in the country. Nearly 90 percent of the total domestic rice requirement is now met from local production (see Table 14.1).

The other nonplantation crops include subsidiary food crops such as maize, pulses, chilies, onions, and potatoes. With the emphasis on the promotion of domestic food production, most of the subsidiary food crops have also benefited from government investment in irrigation schemes, subsi-

TABLE 14.1. Value and share of food imports in 2001

Item	Value in 2001 (Rs Million)	Percentage contribution
Total food imports	48,683	100
Rice	969	1.99
Flour	122	0.25
Sugar	10,289	21.13
Milk and milk products	10,223	20.99
Fish and fish products	6,360	13.06
Other food items	20,720	42.56

Source: Central Bank of Sri Lanka, *Annual Report 2001.*

dized inputs, concessionary bank credit, and tariff protection/import restrictions aimed at maintaining domestic market prices above competitive world prices.[1] Despite this, production of potatoes, chilies, and big onions has been severely affected lately, as local producers whose costs are higher could not compete with cheap imports mainly from India and Pakistan. The paddy sector too is characterized by relatively low and stagnating yields, escalating costs of inputs, and thus slow growth. Land area under subsidiary food production declined by 40 percent and paddy production by 15 percent during the 1990-1997 period. In fact, the entire nonplantation sector has lost its momentum of growth over the past 10 to 15 years.

Agricultural policy since 1977 has been more liberal than what it was in the period prior to 1977. Although this has been the case, achieving self-sufficiency in rice has remained a primary goal among policymakers (Kelegama, 2000) (see Table 14.2). Although general tariffs were lowered from 1977 onward, those for the agricultural sector were still high. Quantitative restrictions were applied for some domestic food crop sectors in order to promote domestic production with less competition. A number of state-owned enterprises involved in purchasing and distribution of agricultural products operated actively until about the mid-1990s, but their role has gradually diminished over the past five years.

The land market was hardly liberalized, with restrictions on ownership and land sales still operating in many areas. Thus the first phase of policy reforms, begun in 1977 and continuing until the mid-1990s, did not go far enough to remove the distortions in the land market. This issue has been of concern to policymakers. We shall revisit this debate in Part III.

However, compared to the other South Asian countries, Sri Lanka, despite various distortions in the agricultural market, had the most liberal re-

TABLE 14.2. Paddy production and rice imports

Year	Gross extent sown (in thousand Ha)	Production (in thousand MT)	Yield (Kg/Ha)	Imports (in thousand MT)
1970	759	1,616	1,129	526
1975	696	1,154	2,270	457
1980	845	2,133	2,927	167
1985	882	2,661	3,465	182
1990	857	2,538	3,453	172
1995	915	2,810	3,535	9
1996	749	2,061	3,513	341
1997	730	2,239	3,618	306
1998	848	2,692	3,634	168
1999	896	2,868	3,672	214
2000	878	2,860	3,856	15
2001	798	2,695	3,954	52

Source: Central Bank of Sri Lanka. *Annual Report 2001.*

gime for agricultural (Athukorala, 2000). Just before the World Trade Organization Agreement on Agriculture (WTO-AOA) came into operation, Sri Lanka's situation was, briefly, as follows:

1. all agricultural export taxes were abolished,
2. all agricultural export and domestic production subsidies were within the AOA limits,
3. the tariff band for agriculture was 45 percent. Quantitative restrictions (QRs) were applicable on potatoes, chilies, big onions, wheat flour, and rice,
4. special tariffs applied to sugar and milk because the government assured foreign direct investment (FDI) in these two sectors and a special rate of return (Athukorala and Kelegama, 1998), and
5. state trading enterprises involved in importing agricultural items were still in operation with a reduced role.

PART II: POST-WTO MARKET REFORM

After considering various options available under the WTO-AOA, Sri Lanka decided to bind all its agricultural tariffs at 50 percent so that it could align it with the ongoing agricultural sector reforms and develop the sector according to its comparative advantage. However, QRs were retained on selected agricultural products after the AOA came into operation with a justification on balance of payment grounds (Athukorala and Kelegama, 1998).[2] This action was overruled by the WTO, and Sri Lanka, in July 1996, removed all QRs other than those on wheat. It was not possible in the case of wheat due to a condition with a Singaporean foreign direct investment firm in the wheat flour sector. The removal of QRs in the domestic food crop sector dealt a severe blow to the production of potatoes, onions, and chilies. It resulted in the disappearance of the price advantage enjoyed by local farmers in the protected market.

The situation compelled the policymakers to implement a seasonal tariff policy for the sectors, in which imports were liberalized (even with zero tariffs) when there was a shortage of domestic production and normal agricultural tariffs were imposed when there was a glut in domestic production. The seasonal tariffs had a number of problems associated with them: timing, hoarding, and unscrupulous traders making maximum use of the loopholes to gain profits. Various lobbies also influenced the seasonal tariffs, and thus ad hoc manipulations resulted.

Agricultural import duties were periodically waived for all essential food items that the government deemed were in the national interest. The periodic application and revocation of import duties gave confusing signals to the producers and contributed to price volatility, thereby reducing incentives for domestic storage and adding to the uncertainties faced by the farmers.

The producers constituted a strong lobby, and with the support of the minister of agriculture exerted pressure for more protectionist policies for their products during the post-1996 period. This was evident in early 1999 when they managed to convince the government to put the entire agricultural sector into the negative list of the Indo-Sri Lanka Bilateral Free Trade Agreement.

To accommodate the concerns of the protectionist lobby, various surcharges were implemented on the importation of food crops. Ad hoc tariff changes and imposition of surcharges raised the effective tariffs close to 50 percent in some years (IPS, 1999). Despite reducing tariffs to zero to please consumers, in certain seasons, protection to producers has come at the expense of consumers regarding all four items, chilies, potatoes, onions, and rice. Welivita et al. (2002) report that the average effective protection rates (EPRs) for the 1990-2000 period for rice, potatoes, chilies, and big onions were 59, 179, 109, and 39, respectively.

The case of paddy illustrates the ad hoc nature of the tariff policy. When there was a shortage of production the rice price increased. To lower the domestic prices, the government, in November 1997, liberalized rice imports by waiving the 35 percent duty effective at the time. Zero tariffs applied until February 1998 when the 35 percent duty was reimposed. In October 1999, the duty was reduced to 10 percent. Again in January 2000 the 35 percent duty was reintroduced. In mid-2000, the government of Sri Lanka introduced license requirements for rice trading.[3]

In early 2001, Sri Lanka indirectly protected the agricultural sector by imposing a 40 percent surcharge on imports to address a balance-of-payment problem. The 40 percent surcharge effectively made agricultural tariffs close to the bounded rate of 50 percent (Kelegama, 2001). This surcharge was reduced to 20 percent in March 2002 making the nominal agricultural tariffs close to 42 percent.

The country did not further trade policy reform and thus succeed in allowing the agricultural sector to develop according to its comparative advantage. The binding of tariffs at 50 percent in 1995 and the alignment of the agricultural sector reforms with trade liberalization proved to be an unsuccessful strategy. Perhaps Sri Lanka should have followed the strategy of other South Asian countries, which bound the tariffs at a high level to use as a bargaining chip in multilateral negotiations.[4] This strategy succeeded especially because developed countries did not reduce their subsidies. In Sri Lanka's situation, market prices could not be taken as a guide for comparative advantage. Using trade policy to drive agricultural development was not successful in Sri Lanka, inter alia, due to ad hoc tariff changes that resulted from binding agricultural tariffs at 50 percent.

As a result of the post-WTO reforms, the inefficiencies of the key parastatal agencies involved in food trade became apparent. Four key state-owned enterprises were involved in either importing or purchasing agricultural items, (1) Cooperative Wholesale Establishment (CWE), (2) State Trading Cooperation (STC), (3) Multipurpose Cooperative System (MPCS), and (4) Paddy Marketing Board (PMB). The PMB, which was involved in purchasing paddy from farmers at a guaranteed price, could not compete with private sector purchases and thus ran at a loss for a number of years before the government decided to close it down in 1997 (IPS, 1999).

However, the forward sales contract program, which was introduced by the Central Bank of Sri Lanka in 1999 to fill the void created by state purchasing institutions going out of operation, has not been effective as a purchasing instrument to assist the farmers (Kelegama, 2000; Welivita and Epaarchchi, 2002). Thus the CWE, which did not possess any expertise in paddy purchasing, had to step in. The CWE was stretched to its limits in the absence of alternative effective market instruments to ensure paddy purchases from the farmers. The forward sales contract scheme seems to be still in its infancy and needs to be popularized. Once the forward sales contract mechanism takes off, the public purchasing system will be redundant.

At least at this stage the government finds it very difficult to dismantle or privatize the CWE given the important role that it plays in the market.

PART III: AGRICULTURAL DIVERSIFICATION

Although the trade reforms were implemented (with some reversals later), because of a number of structural problems, significant diversification of agriculture has not taken place. In addition to the subsidies, which are within AOA regulations, there are a number of incentives for investment in the agricultural sector. There is an incentive package offered by the Sri Lanka Bureau of Investment and general incentives exist under the Inland Revenue Law for both foreign and domestic investment in the agricultural sector (Tabor et al., 2000; Welivita and Epaarachchi, 2002). However, as attractive as these packages may be, the incentives are nullified because of various impediments related to land acquisition, and the material that can be planted. Agrobased industries have been slow to emerge due to these impediments.

The state owns approximately 80 percent of the land. According to land reform legislation enacted in 1972, no individual can own more than 25 acres of paddy land and 50 acres of highland. This legislation remains intact even after 25 years of liberalization. Besides, it is difficult to acquire commercially viable land due to procedural delays, and many governmental agencies, such as the Department of Railways, have some ownership over land. Moreover, private land transactions are restricted by the absence of a unified transparent system of title registration. To make matters worse, approximately half of the privately owned land in Sri Lanka is subject to conflict-of-ownership claims (World Bank, 1996).

Diversification to higher-value crops requires significant areas of land to provide the critical volume to ensure continuous supply.[5] The World Bank (1996) argued that a major factor working against agricultural diversification is the built-in bias in favor of paddy cultivation in the agricultural incentive regime. In fact, there are a number of factors related to the land market that encourage rice production and discourage commercial crops. For example, in the government-irrigation areas, farmers are given land permits that severely limit tenure rights. In other words, there is no freeholding of land and farmers are virtually locked into these holdings. Besides, through land transfer to offspring, land fragmentation has occurred. There is also the problem of attachment to ancestral landholdings. All these factors have prevented land consolidation and promoted rice cultivation in the agricultural policy regime.

The Agricultural Development Authority Act No. 46 of 2000 was enacted to eliminate the tenant cultivation system and facilitate the sale of land to the private sector without any hindrance from cultivators. However, the Act requires the written approval of the commissioner general of the Agricultural Development Authority before paddy land can be used for the cultivation of any other crop. Such bureaucratic procedures have also acted as impediments for diversification.

Diversification takes place when farmers have the strength to bear risks (Dunham and Edwards, 1997). Because paddy farmers have low profit margins they are risk averse (Sanderatne, 2000). Other factors that have worked against agricultural diversification are the following:

1. A new seed and planting material policy was approved in 1996 by the government with the objective of promoting seed industry with the participation of the private sector to produce and market seeds. Although a Seeds Bill based on this policy was formulated in 1998, it has still not been enacted. In 1998, a seed paddy farm was privatized as a first step in the new policy setup. However, some seed and planting materials are banned from import. This is a major problem because a permit requirement to import agricultural raw material is a time-consuming exercise. Restrictive quarantine regulations have indirectly constrained potential exports because the inability to access improved varieties of seeds, for example, in ornamental flowers, has meant an inability to meet international market requirements.

2. R&D in agriculture for higher-value-added crops is not demand driven. United States Agency for International Development (USAID), jointly with the Ministry of Mahaweli Development supported gherkins, asparagus, and cantaloupe cultivation in the late 1980s on an experimental basis. Some blue chip companies also embarked on this exercise. However, the outgrower model did not work because of a number of factors centered on marketing (EDB, 1998).

3. Marketing is disorganized in Sri Lanka. No attempt has been made to set up a national private company with private traders and farmers' organizations as stakeholders to undertake the purchasing and marketing of agricultural items. In fact, only after the government lost two vital paddy cultivation areas in the October 2000 general elections did it propose to establish an Agricultural Purchasing Authority (APA).[6]

4. Postharvest losses of fruits and vegetables due to improper packing, handling, and transfer are significant. Some suggestions made to rectify the problem have still not caught the attention of policymakers (Bamunuarachchi, 2001).

5. High duties on agricultural raw material increase the costs of Sri Lanka's agro-industrial goods and render many of these products uncompetitive in global markets. Import duties on agricultural inputs also significantly increase the costs of agro-industrial goods sold in domestic markets. Higher domestic prices constrain domestic markets and limit the ability of the agro-industries to achieve economies of scale or scope (Tabor et al., 2000).
6. Much of the fertile topsoil layer on most of the lands has become eroded due to improper land management in the past.

Basically the supply potential of the agricultural sector has not been improved. Consequently, diversification of the nonplantation sector away from the focus on rice to other foods has not been accelerated. As argued, this is due to a plethora of domestic impediments and not due to any hindrances from the AOA. This points to the need for reform in the domestic agricultural sector so that it is aligned with the trade liberalization policies on which the Sri Lankan government has embarked.

PART IV: FOOD SECURITY

Sri Lanka has been well ahead not only of its South Asian neighbors, but also of most other countries with comparable (or even considerably higher) per capita incomes in terms of such indicators as human development and literacy levels, infant mortality, and life expectancy. However, food insecurity at the household level is still a major problem. For instance, average dietary energy intake falls below the minimum requirement of 2,200 calories, with 33 percent of women and 37 percent of men suffering chronic energy deficiency (CED) and 33 percent of children under five years being malnourished (with 13 percent actually malnourished) (FAO, 1999).

The food and nutritional inadequacies are concentrated among the poor, the children, and the population displaced by war for nearly two decades. The rural poor in rainfed agricultural regions suffer substantial food insecurity in low rainfall seasons; their energy intake may be only 60 percent of the energy requirement. Food insecurity is clearly an issue of pressing concern in Sri Lanka. Although its importance has been recognized by successive governments and nongovernmental and donor agencies, eliminating food insecurity has been a challenge. In fact, the nutritional welfare of the poor has deteriorated over the past one and a half years. In this chapter we look at the food security issues from the perspectives of rural agricultural development and food and nutrition intervention programs.

Agricultural Development and Food Security

Until about 1994, achieving self-sufficiency in various food items, in particular, rice, has preoccupied policymakers in Sri Lanka. Achieving self-sufficiency in rice was equated with achieving the glories of the ancient civilization in Sri Lanka.[7] However, since 1994, it has not been as valued a goal (Kelegama, 2000). This policy is subject to debate from time to time, however, there are no direct government measures in place to achieve self-sufficiency in any food crop at present. As is well known, income from the agricultural sector is the main source for food intake of rural households.

As stated in Part I, most domestic food crops have benefited from government investment and subsidies over the years. Despite this, Sri Lanka's domestic food crop sector has lost its momentum of growth during the past 10 to 15 years. Relatively slow and stagnating yields and escalating costs of inputs characterize this subsector. Domestic food crops cannot compete with cheap imports from India and Pakistan. In other words, food production at the margin is not internationally competitive. (Of course, Indian food grains are cheap because the Indian government provides large subsidies.) Thus the Sri Lankan domestic food sector is under pressure and the producer margin is eroding.

Rice yields have been stagnating and there are indications that paddy farming is becoming increasingly economically nonviable. Paddy prices have not increased with the increase in the costs of production. As stated earlier, over 70 percent of paddy holdings are below 1 hectare, which is inadequate to sustain a family of five (Sanderatne, 2000). According to impressionistic evidence, the stagnation of rice yields may be due to a combination of factors: problems in soil fertility, poor cultivation practices, poor land preparation, large postharvest losses, and other factors. The only way out of this situation is to: (1) increase yields by rectifying the various problems, (2) promote the cultivation of higher-value crops off-season, and (3) generate part-time off-farm employment. How the stagnation of yields and rural poverty and food insecurity will be addressed remains to be seen.

Kelegama (2002) reports that the production of chilies, gingelly, green gram, and soya beans has fallen by more than 50 percent while black gram, cowpea, groundnut, maize, and red onion declined by 25 to 50 percent during the 1991-2000 period. In volume terms, the production of all subsidiary food crops fell from 258,000 metric tons in 1991 to 184,000 metric tons in 2000.

Although the decline in production has led to a significant increase in food imports, this is not a cause for concern. As a percentage of overall imports, food imports declined from 43.7 percent in 1995 to 9.2 percent in 1999. Food imports amount to about 15 percent of export earnings, 20 percent of industrial export earnings, and 66 percent of agricultural export earnings (Sanderatne, 2001).

In spite of this, there is concern about food security at the household level. In addition to yield stagnation and the rise in production costs, farmers' incomes have declined rapidly. The classic answer to this problem is that surplus labor will increasingly be absorbed in industrial and service employment. Urban demand for agricultural products will rise and, as labor moves out, prices and wages will be bolstered and restructuring of the agricultural sector will become increasingly viable. The problem with the Sri Lankan case is that remunerative off-farm jobs in sufficient numbers are simply not available. A recent study shows that already approximately 30 percent of rural household incomes come from off-farm activities (De Silva et al., 1999).

Recent research has shown that the Sri Lankan rural economy is kept alive more by remittances from workers in the Middle East, from garment factories, and the armed forces, and from transfers from the poverty alleviation program, Samurdhi, than from income generated from the agricultural sector (Dunham and Edwards, 1997). These are clearly unstable sources of income considering the fact that income from the nonplantation sector is small. Agriculture has basically diminished as an income source.

In fact, Dunham and Edwards (1997, p. 39) state: "Agricultural policy— if it is serious about alleviating rural poverty—has to take a much longer and harder look at prevailing local conditions in Sri Lanka." A serious crisis in the agricultural sector needs to be addressed if it is to be viewed as a source of generating food security. This is all the more important as the poverty alleviation programs initiated by the government have not been very effective from the food security perspective.

Food and Nutrition Programs

The Sri Lankan government's food and nutrition policies generally reflect a high degree of concern for poor and vulnerable households. The food subsidy has been one of the major elements of the welfare-oriented strategies in Sri Lanka. In 1972, Sri Lanka introduced an income criterion to determine the beneficiaries of the food subsidy. A universal form of subsidy that was in existence was changed to a targeted one. In 1978, the income cutoff point was further revised, which resulted in the provision of food subsidies to only 50 percent of the population from a previously universal form. With the introduction of the food stamp scheme in 1979, Sri Lanka achieved further reductions in the costs of subsidizing food by improved targeting of the beneficiaries.

In order to address the high levels of poverty, the Janasaviya poverty alleviation program was introduced in 1989, and in 1995, this program was renamed the National Development Trust Fund (NDTF). The beneficiaries of the program were identified from the households originally entitled to receive food stamps. Four key programs were implemented under NDTF: the Human Resources and Institution Development (HRID) program, the Credit program, the Community Projects program, and the Nutrition program. The HRID program develops human resources of the poor by increasing awareness within poor households, promoting group action, and facilitating self-reliance through the provision of technical and financial assistance to partner organizations. The partner organizations in turn train the beneficiaries in income-generating skills. The Community Projects program addresses the "poverty pockets" by creating wage employment through short-term labor-intensive rural works. The Credit program supplies credit at the market rate to socially mobilized poor groups through partner organizations such as local financial institutions and NGOs. The poor, who have recently acquired organizational strengths and skills, are given credit to initiate self-employment activities and microenterprises. The Nutrition program includes a set of activities with specific nutritional goals.

In spite of a lack of rigorous evaluation on the impact of the NDTF program, there is a consensus that the program has not been effective in reducing poverty among the participating households (Hewavitharana, 1999). Several reasons are attributed to this poor performance. First, the poverty alleviation objective of the program has been sidelined by the emphasis given to transferring resources to the participating households. Second, the targeting mechanism has not been effective in keeping the nonpoor out and ensuring the inclusion of the ultrapoor. Third, the poor choice of partner organizations, mostly NGOs, and their strained relationships with government officials, have reduced the cooperation of these organizations in implementing the programs. Fourth, the lack of interdivisional collaboration and coordination among the four programs has resulted in poor follow-up and a patchy approach to poverty alleviation. Finally, the lack of a rigorous monitoring and evaluation system has resulted in poor feedback on the impact of the program on the beneficiaries and has reduced the opportunity to improve the program design and implementation.

The Samurdhi program is the current national program for the eradication of poverty. It replaced the food stamp scheme in 1995 and it covers roughly 1.5 million households, which is approximately 40 percent of the households in Sri Lanka. The main thrust of the Samurdhi program is the participation of the poor in the production process by ensuring their access to resources for self-employment opportunities. Development of village-level institutions and grassroots-level task forces *(balakaya)* are the key elements of the program that increase the participation of the rural poor. The Samurdhi program also includes a social security scheme to provide insurance for the participating families through their voluntary contribution to the Samurdhi Social Security Fund. In addition, a Samurdhi Banking Society was established to manage and operate savings and credit schemes for the beneficiaries of the Samurdhi program (Gunawardena, 1999).

The promotion of home gardens and the Nutrition Awareness Program, in collaboration with the NDTF, directly address the food security and nutritional status of the beneficiaries. Again, there is a lack of evaluation of the Samurdhi program for its impact on poverty alleviation (Divaratne, 1999).

Triposha is the main food-based nutrition intervention program in Sri Lanka. Introduced in 1973, it provides precooked food fortified with vitamins and minerals. The program is targeted to pregnant and lactating mothers and preschool children. Although the impact of the program on its beneficiaries has been positive, further refinement in identifying the malnourished and improved targeting will increase its effectiveness.

The Janasaviya and Samurdhi programs suffer from design and implementation weaknesses that have reduced their effectiveness in terms of allocative efficiency. The World Bank (2000, pp. 36-37) highlights this problem as follows:

> Both Janasaviya and Samurdhi have been large income transfer programs covering over 50 percent of the population or twice the percentage actually living in poverty. . . . Data show that transfers from poverty programs (Janasaviya, Samurdhi, Food Stamps) reach 66 percent of households in the poorest decile and 14 percent of the households in the top three deciles. . . . This inadequate targeting is also confirmed by the preliminary findings of the Sri Lanka Integrated Survey which reveals that only 60 percent of households in the lowest expenditure quintile receive Samurdhi transfers. . . . The thin spread of income transfers across the population and poor targeting divert resources away from the most needy segments of the population.

Looking at the poverty alleviation programs from the perspective of agricultural development, De Silva et al. (1999, p. 3) argue:

> A rethinking in the design of the poverty alleviation programs is necessary as some of the anti-poverty strategies may merely serve to perpetuate unproductive small farms. . . . The effectiveness of the safety nets must be enhanced through better targeting and increasing the volume of support to the needy while development programs must stimulate growth by concentrating on potential growth centres and

commercially oriented farmers. . . . Also new forms of organizations such as commercially organized farmer companies, management firms in charge of irrigation systems for instance, consultancy and financial services may be beneficial for promotion of farmer interest in a competitive market environment.

CONCLUDING REMARKS

The Sri Lankan agricultural sector has no doubt been subject to reform, but ad hoc policy changes have prevented more investment in the agricultural sector. Internal agricultural markets have been substantially liberalized from numerous state controls, export taxes have been eliminated, and import restrictions have been reduced. However, the import agricultural goods remain relatively protected. Agriculture remains an area of continuing state intervention with regard to trade policy.

A steady decrease has occurred in both real producer and consumer prices for several food crops, which undermines the profitability of these products. These trends have not been compensated by adequate productivity improvements or favorable trends in input prices. When considered together with the level of protection, these trends confirm that Sri Lanka is a high-cost producer of rice and of many subsidiary food crops. Thus continuation of, and increase in further production, unless brought about by productivity improvements, will be at a very high cost to the economy.

Given this situation, agricultural diversification has been slow to emerge. A number of restrictions in the land market created a bias toward rice production and posed other obstacles to diversification. The environment was not conducive to commercialized agriculture. Institutional constraints and tariff policy on agricultural inputs also acted as impediments to agricultural diversification.

As a result, agriculture showed a slow growth rate during the 1990s—at 2 percent, the lowest in South Asia. If not for the plantation-sector restructuring via the privatization program, it would have been difficult to achieve even this growth rate. Due to the poor performance of the non-plantation sector, food imports increased over the past decade. Overall food imports accounted for less than 10 percent by value of total imports, having being above 40 percent from the 1950s to the late 1970s. Of course, the ability to finance food imports depends on overall export performance.

Arguably, in the long run, sustained economic growth does reduce absolute poverty and the pervasiveness of food insecurity. In the context of ongoing globalization and policy liberalization, it is the world market prices and trends that will increasingly set the level, direction, and volatility of domestic prices. The way these are transmitted to domestic food markets depends on the degree of regional food market integration and the transmission of external prices to domestic prices—the latter depending on the structure of domestic markets. How does internal volatility of food prices compare with that in global markets? Though there is some limited information on these issues, they have not been studied rigorously in Sri Lanka.

Import liberalization—so long as export growth is maintained to finance it—would tend to lower food prices if domestic food markets function well. The urban poor who currently suffer from food insecurity should be clear gainers. But how would the food security of some of the poorest households living in relatively remote areas, and dependant on rainfall for their agricultural production, or those living in the tea plantations be affected by increased globalization?

Cheaper competing imports can undercut the already precarious livelihoods of rainfed-area farmers and worsen their food insecurity. Similarly, plantation-sector restructuring brought on by more market-oriented policies may lead to reductions in employment. Members of these households need to be able to tap into the gains of cheaper food by enhancing cash-generating activities, including paid employment. Many other issues will present themselves as the forces of globalization penetrate ever deeper into the Sri Lankan economy. On the other hand, the prospect of a shrinking—or at least not significantly expanding—staple-food crop sector is consistent with an expansion of other food industries, both for export and domestic consumption, and could be made possible with foreign investment. All of these possibilities raise issues of short-term assistance and long-term labor mobility. More broadly, they point to the need for safety nets, as these are needed in periods of rapid change. Research suggests that the safety net programs that do exist, such as Samurdhi, fail to direct resources to the most needy.

Although there will be no major direct impact from WTO commitments in the immediate future on food security, the world is clearly entering into a period of heightened structural changes, unanticipated shocks, and greater volatility. It is likely to be a period in which Sri Lanka becomes more reliant on staple-food imports, barring major technological advances that can enhance productivity in the domestic food sector. This means that the way domestic food markets function, how they are linked to global markets and experience external shocks, and how they can be accessed by the poor will determine both what happens to food security and, because of the politics of food, also the sustainability of market-oriented reform processes. The continuing policy liberalization process in Sri Lanka, the increasingly important role played by the WTO in the global trading system, and overarching globalization pose important challenges for policymakers as well as researchers and analysts.

NOTES

1. Irrigation is still supplied free of charge by the government.
2. The items were chilies, big onions, potatoes, wheat, and rice.
3. See Table 1 in Welvita and Epaarachchi, 2002.
4. See Athukorala (2000) for a discussion on the agricultural trade reforms in other South Asian countries.
5. Duhnam and Edwards (1997) report that in certain areas in northwestern Sri Lanka even in small plots of land the productivity has been very high, but these are rare exceptions.
6. See 2002 Budget Speech (Minister of Finance, 2001). Before the authority could be established the government was defeated in the polls in December 2001.
7. According to historical texts, Sri Lanka was not only self-sufficient, but it also exported rice to countries such as Burma during the twelfth century.

REFERENCES

Athukorala, P. (2000). Agricultural Trade Policy Reform in South Asia: The Role of the Uruguay Round and Policy Options for the Future WTO Agenda. *Journal of Asian Economics*, 11: 169-193.

Athukorala, P. and S. Kelegama (1998). The Political Economy of Agricultural Trade Policy: Sri Lanka in the Uruguay Round. *Contemporary South Asia*, 7(1): 7-26.

Bamunuarchchi, A. (2001). *Minimizing of Post Harvest Losses in Fruits and Vegetables*, Colombo, Sri Lanka: National Task Force for Minimizing of Post Harvest Losses.

Central Bank of Sri Lanka. Annual Report 2001. Available online at: <http://www.lanka.net/centralbank/Ar_Index.html>.

De Silva, K.T., S.B. de Silva, and S. Kodituwakku. (1999). *No Future for Farming?: The Potential Impact of Non-Plantation Agriculture on Rural Poverty in Sri Lanka*. Kandy, Sri Lanka: Centre for Intersectoral Community Health Studies.

Divaratne, S.B. (1999). Sri Lanka (1). Country Paper in *Rural Poverty Alleviation in Asia*. Tokyo: Asian Productivity Organization.

Dunham, D. and C. Edwards (1997). *Rural Poverty and an Agrarian Crisis in Sri Lanka, 1988-1995: Making Sense of the Picture*. Institute of Policy Studies, Poverty and Income Distribution Series, No. 1.

EDB (1998). Export Development: Concepts and the Role of the Participants in the Process. Study prepared for the Export Development Board of Sri Lanka.

FAO (1999). Food Security and Nutrition: Sri Lanka. Colombo: FAO Office.

Gunawardena, P. K. (1999). Sri Lanka (2). Country Paper in *Rural Poverty Alleviation in Asia*. Tokyo: Asian Productivity Organization.

Hewavitharana, B. (1999). The two leading meso policy interventions for rural poverty alleviation—Sri Lanka: A case study. *Rural Poverty Alleviation in Asia and the Pacific*. Tokyo: Asian Productivity Organization.

IPS (1999). Agriculture and the New Trade Agenda in the WTO 200 Negotiations: Economic Analysis of Interests and Policy Options for Sri Lanka. Report prepared for the World Bank by the Institute of Policy Studies of Sri Lanka.

Kelegama, J.B. (2002). What Has Happened to Our Agriculture? *The Island*, April 1, 2002.

Kelegama, S. (2000). Food Security in Sri Lanka. In S.G. Samarasinghe (Ed.), *Hector Kobbekaduwa Felicitation Volume*. Colombo: HARTI.

Kelegama, S. (2001). Sri Lankan Economy in Turbulent Times: Budget 2001 and IMF Package. *Economic and Political Weekly*, 36(28).

Minister of Finance (2001). Budget Speech: 2001. Government of Sri Lanka.

RIS (2002). South Asia: Development and Cooperation Report-2001/02. Research and Information Systems for Non-Aligned and Other Developing Countries, New Delhi: RIS.

Sanderatne, N. (2000). The Rice Economy: Challenges in the Next Decade. Inaugural Address at the Rice Symposium, 2000, Postgraduate Institute of Agriculture, University of Peradeniya, Sri Lanka.

Sanderatne, N. (2001). Food Security: Concepts, Situation, and Policy Perspectives. *Sri Lanka Economic Journal*, 2(2).

Tabor, S. et al. (2000). An Agro-Industry Strategy and Policy Reform Action Plan for Sri Lanka. Report prepared by agent for Ministry of Industrial Development, Sri Lanka.

Welivita, A. and R. Epaarachchi (2002). Agricultural Policies and their Implications for the Non-Plantation Agriculture Sector, 1995-2000. Report prepared for the World Bank Resident Mission by the IPS.

Welivita, A., Jayanetti, S., and R. Epaarachchi (2002). The Level and Structure of Trade Protection in Domestic Agriculture Sector of Sri Lanka: 1995-2000. *Sri Lanka Economic Journal*, 3(1).

World Bank (1996). *Sri Lanka: Non Plantation Crop Sector Alternatives*. Agriculture and Natural Resources Division, Country Department I, South Asia Region. Washington, DC: The World Bank.

World Bank (2000). *Sri Lanka: Recapturing Missed Opportunities*. World Bank Country Report. Washington, DC: The World Bank.

Chapter 15

Food Security in Bhutan: Policy Challenges and Research Needs

Suresh Chandra Babu
Choni Dendup
Deki Pema

INTRODUCTION

Bhutan has made considerable progress in the past two decades toward achieving food security for its population. Although landlocked by two large neighbors, India and China, Bhutan has been successful in using its land, water, and other natural resources to increase food security and reduce poverty. This chapter seeks to describe the policies and strategies related to food security that are in place in Bhutan and identify the policy challenges that must be considered for the future. It also discusses some of the policy research needs that should be addressed for designing future food, agriculture, and natural resource policies and programs. The chapter is organized as follows: the characteristics of the agricultural sector are described in section two. In section three, the policies undertaken to achieve food security are discussed. Section four addresses the opportunities related to trade and diversification. The final section identifies the policy challenges and policy research needs for the country based on the ongoing discussions within its food and agriculture sector.

THE ROLE OF AGRICULTURE IN BHUTAN

Agriculture plays a major role in the economy of Bhutan, having accounted for 35.9 percent of GDP in 2000. Maize and rice are the major domestic crops, contributing 49 and 43 percent, respectively, to total cereal production in the country. Wheat and other minor cereal crops make up the remainder. Rice is the main and preferred staple crop. About 56 percent of the Bhutanese diet consists of rice, and maize constitutes about 29 percent.

The demand for rice is expected to rise as the population of Bhutan continues to increase and the incomes of Bhutanese households rise. The population of Bhutan is an estimated 677,934, 70 percent of which lives in scattered rural villages and homes. Although preferences will shift from coarse cereals, such as maize and millet, to finer cereals, such as rice and wheat, minor cereals cultivated on small acreage would continue to play an important role for household food security in many remote areas of Bhutan. Table 15.1 presents figures on the productivity and production of cereal crops in Bhutan.

Production systems in Bhutan are determined by the diverse agroecological conditions including the alpine zone, in which pastoralists rear yak; the cool temperate zone where livestock are reared, and potatoes, buckwheat, and barley are grown; the warm temperate zone, where a wide range of crops are produced including fruits and vegetables destined for the export market; the dry subtropical zone, where maize and other cereal crops are produced; and the humid and wet subtropical zone, where rice is cultivated. Farm production, particularly in the warm temperate zone, depends highly

TABLE 15.1. Productivity and production of cereal crops

Total arable land (acre) as per cadastral survey	Crops	Net area cultivated in 1999	Yield in MT/acre 1999	Production (rough grains) (MT) 1999	Production (milled grains) (MT)	Net available food grains (minus feed and other purposes) (MT)	
Wet Land = 71,849	Rice	64,664	0.923	59,685	35,810	35,810	—
	Maize	90,362	0.634	57,290	45,832	36,667	less 20%
	Wheat	17,109	0.398	6,809	6,809	5,788	less 15%
	Millet	16,414	0.323	5,302	5,302	4,507	less 15%
	Buckwheat	14,366	0.394	5,660	4,528	4,528	—
	Barley	7,352	0.492	−3,617	3,617	3,617	—
					Total domestic	90,917	
					Import	38,000	Rice
					Import	3,500	Wheat flour, White flour
					Total consumption	132,417	
					Per capita	195 kgs	Population 680,000

Source: Minsitry of Agriculture, 2002. *Agricultural Yearbook 2001.* Thimphu, Bhutan: Ministry of Agriculture, Royal Government of Bhutan.

on the availability of leaf litter from forests, which is the main ingredient for maintaining soil fertility. Furthermore, farm mechanization is very limited, and exists mostly in the broader valleys where rice is grown.

The conservation and management of natural resources are also important objectives of the renewable natural resources sector. For this reason, much of the country's land is under forest cover and therefore not cultivable. About 72.5 percent of the land area is covered by forest, of which 26 percent consists of reserves and protected areas, and 9 percent of biodiversity corridors. Only 7.8 percent of the country's land is used for agriculture, with the highest concentrations of population and agricultural activity found in the southern foothills, lower valleys of the inner Himalayas, and the temperate-climate valleys at elevations of 2,000-3,000 meters above sea level. Given the small amount of land that can be cultivated, although population density is low, farm sizes are very small. Around 56 percent of the farm households have land holdings sized between one and five acres, 22 percent between five and ten acres, 8 percent between ten and twenty-five acres with 14 percent having less than one acre. Generally, it is estimated that the average farm size is about two acres in the north and seven acres in the south.

In fact, the Royal Government of Bhutan (RGB) does not consider the maximization of gross domestic product as its goal. Rather, the official policy is to maximize gross national happiness, which consists of five objectives: human development, maintaining culture and heritage, balanced and equitable development, good governance, and environmental conservation.

Between 1996 and 2000, a key trend in Bhutanese agriculture was the decline of the share of agricultural GDP in national income from 39.1 percent to 35.9 percent. A breakdown of the agricultural share in 2000 shows that crop products constituted about 17.9 percent, livestock products 7.6 percent, and forestry products 10.4 percent. Although agriculture's contribution to national income is decreasing, rural Bhutan accounts for 79 percent of employment. In different terms, 75 percent of the active labor force is engaged in agricultural production.

FOOD SECURITY IN BHUTAN

The major challenge for the Bhutanese food and agriculture sector is to increase the availability of food in the country. It can be met through either an increase in the production of food crops or imports, or both. However, one of the primary goals of the country's food security policy is achieving self-sufficiency in food grains. The target in terms of self-sufficiency has been 70 percent of the total food requirement, as identified under the eighth five-year plan, and currently the country is at the level of 69 percent (some official sources cite 65 percent). Since the fifth five-year plan, implemented in the 1970s, the RGB has pursued food grains self-sufficiency through various development policies and programs, particularly subsidies for agricultural inputs, land development, small farms, and irrigation development. Self-sufficiency in staple food, namely rice, became the major focus of the sixth five-year plan. In the seventh five-year plan, the major aim was to sustain food production at existing levels for the purpose of achieving food security at both the national and household levels. In the eighth five-year plan, the goal of reaching a minimum level of 70 percent in food grain self-sufficiency was conceived. The high level of political commitment to providing enough food for its citizens has resulted in the RGB having essentially achieved the goal of self-sufficiency in rice.

The food availability gap of 31 percent is being met through the import of rice. The present level of rice self-sufficiency through domestic production stands at around only 49 percent, which points to the need for increasing rice output on existing cropland if local production is to meet the major part of local demand. The present production of maize, at 57,000 metric tons, almost meets domestic maize demand. Bhutan imports about 2,159 metric tons of whole maize grain annually, mainly to support livestock production. To meet the food availability gap the RGB has undertaken food imports through the Food Corporation of Bhutan. Currently, it is estimated that about 52 percent of the rice requirement is imported from India. In 1999, for example, 38,800 tons of rice was imported.

The constant policy dilemma is whether to increase imports or boost local production. Increasing local production without high levels of productivity per unit of land will involve bringing additional land, which is limited in Bhutan, under cultivation. Because the RGB is committed to maintaining at least 60 percent of its land under forests at all times, increasing self-sufficiency in rice will have to come mainly from the generation of higher yields through improved technology, and better policies for marketing and distribution of food grains in the country. Improved farm management practices that reduce the need for land expansion, and increasing the opportunities for bioprospecting and ecotourism are the main avenues through which a balance between farming and forest conservation can be achieved.

Another problem the country faces is that if it imports rice from India, which is cheaper than locally produced rice, domestic rice farmers will have a reduced incentive to grow rice. Thus, there is a need for policies in Bhutan that reduce the cost of production and improve the competitiveness of domestic rice.

Three components make up the government's food security strategy: (1) increasing foreign exchange revenues through the promotion of export crops to cover the import of food grain, (2) attaining 70 percent self-sufficiency in food grains, and (3) strengthening household food security by increasing access to food and improving nutritional standards. Raising rural income and thus increasing the standard of living requires improving the productivity of the agricultural systems. This can be done through using appropriate and viable technologies, making the benefits of the market, both domestic and global, more accessible to farming households, and enabling a better regulatory and policy framework that supports increased food and agricultural production.

One of the goals behind the policy of increasing rural employment is to keep rural households in rural areas. Such a policy requires making agricultural activities economically viable and attractive to the rural population. Toward this end, training in new farming technology and agricultural business skills will have to be imparted to rural communities.

NEW OPPORTUNITIES: DIVERSIFICATION AND TRADE

Agriculture constitutes 17 percent of the total exports from Bhutan; the other major export, which is nonagricultural, is electricity, which accounts for about 47 percent of total exports. In 2000, agricultural exports earned US$17.62 million. Yet Bhutan is a net agricultural importer. In 2000, about 6.7 percent of total imports were agricultural products, with a value of US$25.59 million. The Food Corporation of Bhutan, through its network of auction yards, handled 76 percent of the agricultural exports in 2000.

Agriculture has become increasingly diversified in the past few decades. In horticulture the major products are oranges, potatoes, and apples. In 1999, these goods generated a gross revenue of Nu. 86, 82, and 66 million, respectively. Other fruits and vegetables together earned Nu. 55 million. Rice, potato, and apple production have rapidly expanded. These goods have supplemented the more traditional wheat, maize, and livestock products. However, extremely high transportation costs continue to make it difficult for farmers to take advantage of international trade opportunities to increase their incomes.

Bhutan has a general free trade agreement with India, its major trading partner. About 88 percent of Bhutan's trade is with India. Bhutan also has a preferential trade agreement with Bangladesh. It is currently trying to diversify its markets and thus find export opportunities for its fruits and vegetables. For trading in agricultural commodities, traders require export-import licenses although the formalities are simple and usually the licenses are issued within a short time. The tariffs on agricultural imports have also been lowered. For example, prior to 1996 the customs duty was 100 percent on products imported into Bhutan. In 1996 and again in 2001, the tariffs were lowered. Currently the tariff rates range from 0 to 30 percent with some exceptions. The taxes on exports have been eliminated and goods are free from export duties.

Furthermore, inspection and certification systems have recently been introduced to improve export performance. There has also been improved marketing support in the form of research, information, and the promotion of agricultural exports. While the Food Corporation of Bhutan imports and manages the quarantine measures for essential commodities, inspection and certification processes and quarantine measures for both the import and export of other agricultural commodities are being strengthened by the RGB through the Ministry of Agriculture.

FOOD POLICY CHALLENGES AND RESEARCH NEEDS

In its effort to formulate policies that will together achieve food security for all of its citizens, the RGB faces a number of challenges. These can be addressed through research and information generation. First, how the different and, in a sense, conflicting policies of food self-sufficiency, which involve increasing yields, and environmental conservation can both be achieved needs to be fully understood at all levels of government. Achieving self-sufficiency will, moreover, require a thorough understanding of the extent and causes of food insecurity in Bhutan. This understanding will help to determine what the self-sufficiency target should be and provide the economic, social, and political justification for it. Furthermore, the comparative advantage of rice production over other commercial crops that could be exported should be understood. Improving food security in Bhutan depends on increasing the productivity and income of rural households, increasing rural employment, improving household nutrition, and managing the environment for conservation.

Research will help to solve the problem of low agricultural productivity due to input unavailability, farm pricing, and the lack of technological options and irrigation facilities. However, making investments for the improvement of agricultural production will require policy analysis of the challenges facing the agricultural research, extension, and irrigation systems. It is important to understand how research, extension, and irrigation programs implemented in districts are complementing and contributing to self-sufficiency policies. It is also important to analyze the input supply policies to increase the availability of inputs to farmers. Agricultural extension systems need to be improved through an increase in the role of specialists on each commodity's production. Investment should also be made for research on crops other than rice since most of the research conducted in agriculture has focused on rice-based farming systems.

Increasing agricultural yields will also require the provision of new or improved technologies. The production and distribution of quality seeds needs to be given due consideration. Agricultural mechanization policy also needs a closer look. Since labor comprises about 83 percent of the cost of rice production, making farm machinery available to farmers will help to reduce production costs and make rice a competitive crop. Soil fertility management issues also need the attention of policymakers. Although the use of organic fertilizers, such as farmyard manure is encouraged, depending on organic fertilizers for increasing productivity may not be possible in the long run. The pricing and availability of fertilizers also needs to be given due consideration.

The consumption patterns for rice show that currently about 45,000 tons of rice are consumed. Rice consumption per capita is highest in the western part of the country, which consumes nearly all of the high-quality local rice. The urban consumers also prefer rice to other cereal crops, which increases the future demand for rice. Estimates show that in 2020 the demand for imported rice and high-quality local rice will be 70,500 tons and 60,700 tons, respectively. Given the increasing demand for rice it is imperative to analyze the rice self-sufficiency policy and develop policy guidelines for improving the rice availability in the country through both importation and increasing local production. The domestic production of rice can be increased if additional wetlands are brought into cultivation. A ban on the conversion of wetlands to other forms of land use is in place to protect the existing wetlands although there is a mechanism through which conversion can be approved on a case-by-case basis. Nevertheless, the policy of using wetlands for rice alone instead of other commercial and more remunerative crops should be thoroughly studied. There is a need to design a comprehensive food policy strategy that is supported by sound empirical research and to develop sustainable capacity for implementing future research on emerging policy issues that affect the food, agriculture, and natural resource sectors in Bhutan.

Although it is expected that the recent publication of the first agriculture census data will address some of the requirements, relevant and reliable data for conducting policy analysis will continue to be a major challenge in Bhutan. Therefore, an immediate research need will be to generate empirical data from rural household surveys. These surveys could be conducted in the major agricultural regions of the country, and could collect information on agricultural production, land use, labor and other inputs, sources of household income, and household consumption. Information is also needed to conduct market analyses. This information, which could be obtained through surveys of traders, can be used to understand market flows, assess how well markets are functioning for farmers, and investigate how transaction costs in the agricultural marketing sector can be lowered. Such research is needed for not only the major food crops such as wheat, maize, and rice, but for the commercial crops, such as potatoes, apples, and other fruits and vegetables as well, which have potentially large export markets. The market surveys should also collect information on the number of traders, trading volumes, sources of credit for trading, how trading contacts are established, costs of transport, storage, and other aspects of agricultural trade.

Studies are also needed to measure the comparative advantage of agricultural products over other goods, and understand the impacts of trade and exchange rate policies on production, consumption, and trade within the country. In order to understand the land use options for meeting food, agricultural, and environmental goals, a major effort is needed to map out various agroecological zones and their implications for production of various food and nonfood commodities. The capacity for undertaking food policy analysis in Bhutan continues to be limited, and it needs to be enhanced through collaborative research programs with partners within and outside the region.

Policymakers in Bhutan face complex challenges and promising opportunities in their efforts to achieve these objectives. Meeting these challenges and taking advantage of the opportunities will require improving the institutional capacity and the availability of policy analysis and information to support complex policy decision making. Not taking these steps will have major implications for economic growth, poverty reduction, and environmental sustainability.

PART VI:
FOOD SECURITY INTERVENTION
IN SOUTH ASIA

Chapter 16

Feeding Minds While Fighting Poverty: Food for Education in Bangladesh

Akhter U. Ahmed
Carlo del Ninno

INTRODUCTION

Bangladesh has led the world in creating innovative development programs that can be replicated successfully in other developing countries. The Grameen Bank microcredit program for the poor and the Comilla Model for rural development are notable examples. Bangladesh has also implemented the first-ever Food for Education (FFE) program, which may soon be added to the list of successful antipoverty interventions.

The Government of the People's Republic of Bangladesh launched FFE in 1993 on a large-scale pilot basis. FFE was designed to develop long-term human capital through education by making the transfer of food resources to poor families contingent upon enrollment of their children in primary school.

Pervasive poverty and undernutrition persist in Bangladesh. About half the country's 130 million people cannot afford an adequate diet. Poverty has kept generations of families from sending their children to school, and without education, their children's future will be a distressing echo of their own. Furthermore, from birth, children from poor families are often deprived of the basic nutritional building blocks that they need to learn easily. Consequently, the pathway out of poverty is restricted for the children from poor families.

Many children from poor families in Bangladesh do not attend school, either because their families cannot afford books, other school materials, or clothes, or because the children contribute to their family's livelihood and cannot be spared. Children often have to work in the fields, sell various products, or care for younger siblings so that their parents can earn an income away from home. Thus, these children bring direct or indirect income into the household—income that can make a difference between one or two meals a day for the family.

The FFE program provides a free monthly ration of food grains to poor families if their children attend primary school. Thus, the FFE food-grain ration becomes an income entitlement enabling children from poor families to go to school. The families can consume the grain, thus reducing their food budget, or they can sell the grain and use the cash to meet other expenses. FFE provides immediate sustenance for the poor, but perhaps more important, it has the potential to empower future generations by educating today's children. Education would equip children from poor families to improve their productivity, thereby expanding their future income-earning opportunities.

This chapter describes the main features of the FFE program and evaluates its performance in fulfilling its official objectives: to increase school enrollment, promote school attendance, prevent dropouts, and improve the quality of education. This study also examines the targeting effectiveness of the program, its impact on food consumption and nutrition, and the efficiency of the food-grain distribution system. After evaluating program performance, the study presents conclusions for policy.

OVERVIEW OF THE FFE PROGRAM

Origin of FFE

In 1992, the government of Bangladesh closed down the *palli* (rural) rationing program, one of the largest channels in the public food distribution system (PFDS). The government was providing subsidies of $60 million per year to run the program, but about 70 percent of the subsidized food was going to those who were not poor, i.e., ineligible to receive the subsidy (Ahmed, 1992). The high cost of the subsidy and heavy leakage to the nonpoor motivated its abolition.

Following the demise of *palli* rationing, the government commissioned a working group, chaired by International Food Policy Research Institute (IFPRI), to review the options for developing food programs that would reach the neediest people in a cost-effective manner. In 1993, drawing on suggestions of the review, the government launched a large-scale pilot test of the innovative FFE program (WGTFI, 1994).

Expansion of FFE in Relation to Overall Primary Education

Table 16.1 shows the trends in primary education in Bangladesh during the ten years from 1988/1989 to 1997/1998. Over this period, the number of primary schools increased by 46 percent, teachers employed in primary schools by 30 percent, and students in primary schools by 50 percent. Table 16.1 also indicates that almost the entire expansion in primary education during the period was due to the growth in private-sector schools. A sudden and big surge occurred in the number of nongovernment primary schools, which increased from 13,043 in 1992/1993 to 28,640 in 1993/1994. This increase was response to a new government directive that provided incentives to rural communities to build new schools.

The average number of students per teacher in all primary schools increased from sixty-one in 1988/1989 to seventy in 1997/1998. There are more students per teacher in government schools than in nongovernment schools. In 1988/1989, government schools had a student/teacher ratio of sixty-five, while in nongovernment schools the ratio was fifty. This ratio increased to seventy-seven for government schools and sixty-five for nongovernment schools in 1997/1998.

The FFE program started in 1993 in 460 unions, one union in each of the 460 rural *thanas* in Bangladesh.[1] The program expanded to 1,247 unions by 2000. From 1993/1994 to 1999/2000, the number of primary schools covered by the program increased by 262 percent and the number of students in the program schools by 245 percent. About 40 percent of the students in FFE schools receive FFE food grains. Hence, out of the 5.2 million students enrolled in schools with the FFE programin 2000, 2.1 million students and about 2 million families benefited from the program. Currently, FFE covers about 27 percent of all primary schools and enrolls about one-third of all primary school students in Bangladesh. FFE beneficiary students account for about 13 percent of all students in primary schools.

In 1993/1994, the FFE program started at a cost of Tk. 683 million ($17 million),[2] involving distribution of 79,553 metric tons of food grains. By 1999/2000, the annual cost increased to Tk. 3.94 billion ($77 million), and the distribution of food grains to 285,973 metric tons. The cost of the program in 2000 translates into Tk. 5.20 ($0.10) per beneficiary student per day. The share of FFE food grains in the total PFDS was about six percent in 1993/1994, which increased to15 percent in 1999/2000. The share of the FFE program in total expenditure for primary education in the country increased from 4.7 percent in 1993/1994 to 19.9 percent in 1997/ 1998. In 1997/1998, expenditure on FFE accounted for about 1.5 percent of the total government expenditures.

TABLE 16.1. Number of government and nongovernment primary schools, teachers, and students in Bangladesh

Year	Number of schools			Number of teachers			Number of students (thousand)		
	Govern-ment	Nongovern-ment	Total	Govern-ment	Nongovern-ment	Total	Govern-ment	Nongovern-ment	Total
1988/1989	37,910	7,429	45,339	154,814	34,402	192,816	10,053	1,721	11,774
1989/1990	37,760	8,023	45,783	162,237	37,819	200,056	10,494	1,851	12,345
1990/1991	37,659	10,487	48,146	160,744	42,103	202,847	10,722	2,313	13,035
1991/1992	38,097	11,867	49,964	158,180	50,091	208,271	11,157	2,560	13,717
1992/1993	37,855	13,043	50,898	160,497	54,282	214,779	11,239	2,963	14,202
1993/1994	37,528	28,640	66,168	159,538	82,714	242,252	11,266	3,919	15,185
1994/1995	37,717	24,900	62,617	161,251	87,532	248,783	11,826	4,603	16,429
1995/1996	37,752	23,831	61,583	161,026	88,689	249,715	12,026	5,042	17,068
1996/1997	37,348	24,290	61,638	161,597	88,331	249,928	12,248	5,071	17,319
1997/1998	41,248	24,987	66,235	160,677	90,313	250,990	12,423	5,206	17,629

Source: Bangladesh Bureau of Statistics (BBS), 2000. *Statistical Yearbook of Bangladesh.*
Note: Nongovernment schools include the following: (1) registered nongovernment primary school, (2) high school attached primary school, (3) experimental school, (4) Ebtadayee Madrasa (EM), (5) high madrasa attached EM, (6) kindergarten school, (7) satellite school, and (8) community school.

Salient Features of FFE

The FFE is one of the food-grain distribution channels of PFDS of the Ministry of Food. The program is administered and funded by the Primary and Mass Education Division (PMED), even though the food grain in the PFDS system comes from different sources.[3]

The FFE program uses a two-step targeting mechanism. First, two to three unions that are economically backward and have a low literacy rate are selected from each of the 460 rural *thanas*. Second, within each union, the program covers all government, registered nongovernment, community (low-cost), and satellite primary schools, and one *Ebtedayee Madrasha* (religion-based primary school) in these selected unions. Households with primary-school-age children become eligible for FFE benefits if they meet at least one of the following four targeting criteria:

1. a landless or near-landless household that owns less than half an acre of land;
2. the household head's principal occupation is day laborer;
3. the head of household is a female (widowed, separated from husband, divorced, or having a disabled husband); or
4. the household earns its living from low-income professions (such as fishing, pottery, weaving, blacksmithing, and cobbling).

A household that meets the targeting criteria, but is covered under another targeted food-based program of the government (such as the Vulnerable Group Development (VGD) program or the Rural Development program), is not eligible to receive FFE food grains.

If a household is selected to participate in the FFE program, it is entitled to receive a maximum free ration of 20 kilograms (kg) of wheat or 16 kg of rice per month for sending all its primary-school-age children to a primary school.[4] If a household has only one primary-school-age child (six to ten years) who attends school, then that household is entitled to receive 15 kg of wheat or 12 kg of rice per month. To maintain their eligibility, children must attend 85 percent of classes each month. Thus, the total food-grain allotment to a school may vary from month to month depending on the variation in the number of students who meet the attendance requirement.

Based on the targeting criteria, a school managing committee (SMC) and a compulsory primary education ward committee jointly prepare a list of FFE beneficiary households in every union at the beginning of each year. Due to resource constraints, the total numbers of beneficiary households are identified so that no more than 40 percent of students receive FFE rations.

The beneficiary list is recorded in a registry book. The headmaster of the school, who is a member and secretary of the SMC, is the custodian of this registry book. Each FFE-enlisted household gets a ration card that entitles it to receive the monthly free food-grain ration.

At the beginning of each month, the headmaster prepares a list of students from beneficiary households who met the 85 percent attendance requirement in the previous month. Based on this list, the SMC calculates the food-grain requirement for the school for that month and prepares a procurement request. The *thana* education officer certifies the procurement request and then forwards it to the *thana* controller of food, an official of the Ministry of Food. Each union has a designated private grain dealer who distributes FFE food grains to all beneficiary households in that union. Based on the procurement requests, the *thana* controller of food issues a delivery order to the Ministry of Food's local supply depot to provide all grain dealers in the *thana* with monthly supplies of FFE food grains for distribution to all beneficiary households living in that *thana*. Each beneficiary student's parent or guardian holding the FFE ration card picks up the monthly ration from the dealer on a day specified by the school. Designated officials are responsible to supervise the food-grain distribution (PMED, 2000).

ANALYSIS OF PROGRAM EFFECTS

In September and October 2000, IFPRI collected primary data from multiple surveys covering primary schools with and without FFE, households including program beneficiaries and nonbeneficiaries, communities, and food-grain dealers. In addition to the surveys, academic achievement tests designed to assess the quality of education received by students were given to students enrolled in both FFE and non-FFE schools. The evaluation of the FFE program reported in this section is based on these data. Based on these data, IFPRI researchers used a variety of quantitative and qualitative methods to evaluate the FFE program.

Data Sources

The sample includes 600 households in 60 villages in 30 unions in ten *thanas,* and 110 schools in the same 30 unions from which the household sample was drawn. First, ten *thanas* were selected with probability proportional to size (PPS), based on *thana*-level population data from the 1991 census. Second, in each *thana* two FFE unions and one non-FFE union were randomly selected. Third, two villages from each union were randomly selected with PPS using village-level population data from the 1991 census.

Fourth, a complete census of the households was carried out in each of the selected villages. Then, ten households that had at least one primary-school-age child (six to twelve years old) were randomly selected in each village from the census list of households. Only those schools attended by the children in the sample households were selected for the school survey. A total of 110 primary schools (70 FFE and 40 non-FFE schools) were surveyed.

Several questionnaires were used in the surveys. The village census questionnaire collected information on household demography, school enrollment, literacy, and FFE participation from 17,134 households. A team of male and female interviewers who completed separate male and female questionnaires for each household administered the household survey. The household questionnaires collected information on a wide variety of topics, such as household composition, occupation, education, school participation, dwelling characteristics, assets, expenditures, food consumption, anthropometric measurements of women and children, and use of the FFE system. The school questionnaire collected information on student enrollment, class attendance, dropout rates, teacher qualification, school facilities, school expenditures, and FFE program participation. Questionnaires to food-grain dealers and program-implementing officials captured various operational aspects of the FFE program. A community survey was conducted in all sample villages to collect primary data on union-level and village-level variables.

In addition to the surveys, a standard academic achievement test, designed to assess the quality of education received by students, was given to 3,369 fourth-grade students enrolled in both FFE and non-FFE schools.

Performance of the FFE at School Level

General Information on Schools

Observations during the school survey suggest that, in general, the conditions of nongovernment primary school buildings in rural Bangladesh are much poorer than those of government primary schools. Only about 11 percent of the total sample of nongovernment schools have permanent building structures of concrete or tin roofs, brick walls, and cement floors compared to 45 percent of all surveyed government schools that have such structures.

Table 16.2 indicates that the average size of FFE schools (in terms of number of students per school) is about 27 percent larger than that of non-FFE schools because the FFE program entices more children to attend schools. Overall, about half of all students are girls. The proportion of girls to total students is slightly higher in nongovernment FFE schools than in nongovernment non-FFE schools.

Table 16.2 also shows that average annual school operating expenses per student (excluding teacher salaries) are generally low (around Tk. 40 per student a year), or very low (only Tk. 27 per student a year) for nongovernment FFE schools.[5] Both government and nongovernment schools under the FFE program are more intensively inspected than schools that are not in the program. Teachers of over 90 percent of the FFE as well as non-FFE schools receive training. More teachers in nongovernment schools are engaged in private tutoring compared to government schools, and this is true for both FFE and non-FFE schools.

The number of teachers per school (FFE and non-FFE, government and nongovernment) ranges from 3.9 to 4.8 and these numbers have remained virtually the same since 1992. In FFE schools, female teachers as a percentage of all teachers increased from 20 percent in 1992 to 29 percent in 2000.

TABLE 16.2. General information about schools by type

Information	FFE schools			Non-FFE schools		
	Government	Nongovernment	All	Government	Nongovernment	All
Average number of students per school in 2000	350	315	343	286	162	270
Proportion of girls (% of total)	50.0	50.0	50.0	50.0	48.3	49.9
Average operating expenses per student (taka/year)*	43	27	40	41	—	—
Inspection made by school inspectors in 1999 (% of schools)	100.0	92.9	98.6	88.6	80.0	87.5
Number of inspections in 1999	5.7	3.4	5.2	5.1	2.4	4.8
Fully follow curriculum (% of schools)	94.6	92.9	94.3	91.4	100.0	92.5
Teachers who received subcluster training (% of schools)	94.3	90.9	93.7	98.1	100.0	98.3
Teachers engaged in private tutoring (% of teachers)	14.3	50.0	21.4	25.7	80.0	32.5

Source: Based on data from IFPRI's 2000 *Food for Education Evaluation Survey, 2000: School Survey, Bangladesh.*
School operating expenses exclude teacher salaries, and include the costs of stationary and supplies, repair and maintenance, utilities, and communication.

In non-FFE schools, 23 percent of all teachers in 1992 and 33 percent in 2000 were female.

The educational qualifications of teachers in FFE and non-FFE schools are about the same. However, teachers in government schools have higher education levels than nongovernment schoolteachers. About 32 percent of government schoolteachers have a bachelor's degree or above. In contrast, only 9.3 percent of all nongovernment schoolteachers have a bachelor's degree. In all types of schools, each teacher teaches around four classes per day and five subjects per week.

There is almost no difference in teacher salaries between FFE and non-FFE schools. However, the average salary of a government schoolteacher (Tk. 4,439 per month) is about 3.5 times higher than that of a nongovernment schoolteacher (Tk. 1,285 per month). Furthermore, most nongovernment schoolteachers are not paid regularly. Mainly due to much higher salaries, government schoolteachers are better off than nongovernment schoolteachers. Average monthly household expenditures of a government schoolteacher are about 72 percent higher (Tk. 6,991) than that of a nongovernment schoolteacher (Tk. 4,072). School salary accounts for about three-fourths of total income for the government schoolteachers, while it accounts for only 27 percent of total income for the nongovernment schoolteachers. Nongovernment schoolteachers mainly depend on agriculture for their livelihood. This indicates that nongovernment schoolteachers are less likely to devote themselves to teaching full time.

Enrollment Level in Schools

The school survey results presented in Table 16.3 show that student enrollment in FFE schools increased by 35 percent per school over the two-year period from the year before the program to the year after the introduction of the program.[6] Enrollment of girls increased by a remarkable 44 percent, and for boys, the increase was 28 percent. In contrast, per-school enrollment in non-FFE government primary schools at the national level increased by only 2.5 percent—0.1 percent for boys and 5.4 percent for girls—over a two-year period from 1992 (the year before FFE was introduced) to 1994 (the year after FFE). Nongovernment schools had a higher increase in enrollment than government schools in the initial year of the introduction of the FFE program.

Table 16.3 also shows that the per-school rate of increase in enrollment in the surveyed FFE schools declined significantly in the years following the introduction of the program, largely due to capacity constraints in the

TABLE 16.3. Changes in enrollment rate and dropout rate by type of schools

Information	FFE schools			Non-FFE schools		
	Government	Nongovern-ment	All	Government	Nongovern-ment	All
			Percentage change			
*Before FFE to after FFE (over a two-year period)**						
All students	33.7	43.0	35.2	2.5	—	—
Boys	27.1	32.9	28.1	0.1	—	—
Girls	41.3	55.3	43.6	5.4	—	—
Average annual rate during the period 1997 to 2000						
All students	2.1	2.2	2.2	1.5	0.8	1.1
Boys	2.2	2.1	2.2	1.3	0.7	1.0
Girls	2.0	2.2	2.2	1.6	0.9	1.3
			Dropout rates from 1999 to 2000*			
All schools						
All students	10.4	12.5	10.9	11.2	8.3	10.8
Boys	9.6	13.5	10.5	10.9	7.5	10.8
Girls	11.1	11.6	11.2	11.4	9.8	11.3
FFE schools (FFE beneficiary students)						
All students	5.3	10.1	6.3	—	—	—
Boys	4.5	7.7	5.2	—	—	—
Girls	6.1	12.2	7.4	—	—	—
FFE schools (non-FFE beneficiary students)						
All students	15.0	14.6	14.9	—	—	—
Boys	13.9	18.3	14.9	—	—	—
Girls	16.2	11.1	14.9	—	—	—

Source: Based on data from IFPRI's 2000 *Food for Education Evaluation Survey, 2000: School Survey, Bangladesh.*
Note: For non-FFE schools, the percentage change in enrollment per school is calculated at the national level from 1992 (the year before FFE) to 1994 (the year after FFE).
Dropout rates are computed using the following formula: Dropouts from class i in year t = enrolled students in class i in year t − promotees from class i in year t + 1 − repeaters in class i in year t + 1 where, promotees from class i, year t + 1 = enrolled students in class i + 1 in year t + 1 − new entrants in class i + 1 in year t + 1 − repeaters of class i + 1 + transfers out from class i + 1 in year t + 1.

same schools. Nevertheless, year-to-year increases in the rate of enrollment in the sample schools remained somewhat higher in FFE schools than in non-FFE schools.

School Attendance and Dropout Rates

The school survey collected information on attendance of total enrolled students on the day of the survey. As recorded in the attendance register, the overall rate of attendance was 70 percent in FFE schools and only 58 percent in non-FFE schools. In order to check the validity of attendance recorded in the school attendance register, survey enumerators counted all students in each class in surprise visits to the schools. The difference in attendance between head count and official record was fairly small. This suggests that the attendance information from school records is quite reliable.

Table 16.3 indicates that FFE encourages children to stay in school. About 40 percent of the students in FFE schools receive FFE food grains. From 1999 to 2000, only about 6 percent of the FFE beneficiary students dropped out, compared with 15 percent of the students in FFE schools that did not receive benefits.

Quality of Education

The quality of education in FFE and non-FFE schools is judged on the basis of student/teacher ratio, use of classroom seating capacity, and students' academic achievement test results. A large student/teacher ratio is often seen as detrimental to the quality of education. In this regard, by encouraging children to attend school, FFE has become a victim of its own success. There are more students per teacher in FFE schools than in non-FFE schools. Although there were sixty-two students per teacher in non-FFE schools, on average, FFE schools had seventy-six students per teacher in 2000. In fact, nongovernment FFE schools had eighty students per teacher, while those without FFE had only forty-one students per teacher in 2000.

Because of increased enrollment and class attendance rates, classrooms of FFE schools are more crowded than non-FFE school classrooms. FFE schools in general utilize about 98 percent of their classroom seating capacity. Indeed, nongovernment FFE schools exceed the capacity. In contrast, non-FFE schools use about 79 percent of their seating capacity.

For this evaluation of FFE, a standard academic achievement test was administered to students. This test was given to all fourth-grade students in FFE and non-FFE schools. The average test scores are lower in FFE schools (49.3 percent of total points) than in non-FFE schools (53.0 percent of total points), and this difference is statistically significant. Within FFE schools, the average test score of FFE beneficiary students (46.0 percent of total points) is less than that of the nonbeneficiary students (53.3 percent of total points), which brings down the aggregate score in FFE schools. FFE beneficiaries score lower than nonbeneficiaries probably because of their relatively lower socioeconomic status.

FFE schools have class sizes of 31 percent more students than classrooms in non-FFE schools. About 40 percent of the students in FFE classrooms are FFE beneficiaries and the rest are nonbeneficiaries. If a larger class size leads to adverse effects on the quality of education (as measured in terms of students' test achievements), then this should be true for both FFE beneficiary and nonbeneficiary students in the same classroom. Analysis of the achievement test scores shows that, on the average, the non-beneficiary students in FFE schools scored (53.3 percent of total points) about the same as the students in non-FFE schools (53.0 percent of total points), despite a significantly larger class size in FFE schools.[7] Therefore, it is likely that larger class size in FFE schools does not necessarily cause lower test scores.

Furthermore, the difference in test scores is larger between government and nongovernment schools than that between FFE and non-FFE schools, with government-school students performing better than nongovernment-school students (government, 51.0 percent versus nongovernment, 40.0 percent of total points in FFE schools; and government, 53.3 percent versus nongovernment, 45.7 percent of total points in non-FFE schools). The school survey findings suggest that government primary schools have better facilities, have more qualified teachers, and pay much higher salaries to teachers compared to nongovernment primary schools.

Impact of FFE at Household Level

Most of the comparative analyses that are based on household survey data classify the sample households into five categories:

1. Households living in FFE unions
   A. FFE beneficiary households.
   B. Nonbeneficiary households with primary-school-age children who attend FFE schools.
   C. Nonbeneficiary households with primary-school-age children who do not attend any school.
2. Households living in non-FFE unions
   D. Households with primary-school-age children who attend school.
   E. Households with primary-school-age children who do not attend school.

Household Characteristics

Table 16.4 presents the characteristics of A, B, C, D, and E household categories. The average sizes of the sample households are slightly larger than the national average rural family size because the sample purposely in-

TABLE 16.4. Characteristics of respondent households (in averages)

	FFE Unions				Non-FFE Unions		
	(A) FFE beneficiary households	(B) Nonbeneficiary households with children attending FFE school	(C) Households with children not attending school	All	(D) Households with children attending school	(E) Households with children not attending school	All
Household size (persons)	5.4	5.4	6.4	5.5	5.5	6.1	5.6
Years of schooling, father	2.2	3.1	0.6	2.3	3.0	1.6	2.7
Years of schooling, mother	1.1	1.6	0.7	1.2	1.8	0.6	1.6
No schooling, adult male (percent)	49.8	47.9	84.3	53.5	45.7	73.7	51.0
No schooling, adult female (percent)	73.0	69.0	80.4	72.5	67.3	86.8	71.0
Less than 0.5 acre of land owned (percent)	68.1	43.7	60.8	58.5	54.3	50.0	53.5
Per capita monthly expenditure (taka)*	607.9	973.7	575.9	733.7	843.3	617.4	800.4
Share of all households (percent)	51.8	35.5	12.7	100.0	81.0	19.0	100.0
Principal occupation of household head (percent)							
Farmer	14.5	28.9	29.4	21.5	22.8	15.8	21.5
Business/trade	21.7	23.2	21.6	22.3	19.1	13.2	18.0
Salaried, service	5.3	5.6	2.0	5.0	9.3	7.9	9.0
Salaried, professional	1.5	2.8	0.0	1.8	4.3	0.0	3.5
Day laborer	28.5	12	29.4	22.8	17.3	36.8	21.0
Fisherman	4.4	1.4	3.9	3.3	3.1	0.0	2.5
Rickshaw puller	5.8	4.2	7.8	5.5	4.9	15.8	7.0
Other	18.3	21.8	5.9	18.0	19.1	10.5	17.5
Share of all households (percent)	51.8	35.5	12.7	100.0	81.0	19.0	100.0
Number	207	142	51	400	162	38	200

Source: Based on data from IFPRI's 2000 Food for Education Evaluation Survey, 2000: Household Survey, Bangladesh.
*Per capita monthly expenditures of FFE beneficiary households exclude income transfer from FFE program.

cluded only those households that had at least one primary-school-age child. The Household Income and Expenditure Survey (HIES) of 2000 reports the average rural household size of 5.2 persons at the national level (BBS, 2001).

The average years of schooling of parents are very low in general, and extremely low for mothers, and for C- and E-category households. Among the adult household members, over half of all adult males and almost three-quarters of all adult females never attended school. Indeed, these proportions are very high for the C and E categories that do not send their children to school.

In FFE unions, per capita monthly expenditure (as a proxy for monthly income)[8] is higher for B-category households than A-category households, but A-category households have higher incomes than C-category households.[9] In non-FFE unions, households belonging to D category have higher incomes than those belonging to E-category households.

Targeting Effectiveness

The results reported in Table 16.4 suggest that the average monthly per capita income (expenditure) of B-category households (nonbeneficiary households with children attending an FFE school) is 60 percent higher than that of A-category households (FFE beneficiaries). This income difference between A- and B-category households is statistically significant. This finding implies that FFE is effectively targeted to low-income households.

However, there are still some households who have primary-school-age children who do not attend any school (C-category households). The household survey reveals that many households in this category are extremely poor, and their children contribute directly or indirectly to household income. As a result, the opportunity cost of attending school for some of these children is higher than their expected income transfers from FFE. For other poor households in this category, the net income transfer (that is, net of opportunity cost of children to attend school) would not be enough to afford even the bare minimum of clothing and supplies needed to send their children to school. As a group, these nonbeneficiaries constitute about 13 percent of all households in FFE unions and are somewhat poorer than the households receiving FFE benefits. The average income of C-category households is 5.3 percent lower than that of B-category households (FFE beneficiaries). However, this difference is not statistically significant.

The FFE program is also designed to target the most "economically backward" unions in each *thana*. A comparison of average incomes between FFE unions and non-FFE unions suggests that FFE unions are poorer than non-FFE unions. The average income of households living in FFE unions is 8.3 percent lower than the average income of households who live in non-FFE unions, and this difference is statistically significant.

As described previously, a household is required to meet at least one of the four selection criteria to be eligible for the FFE program. Table 16.5 shows that about 44 percent of B-category households (nonbeneficiaries whose children attend FFE school) meet at least one criterion—owning less than half an acre of land—yet are not in the program. The results of the analysis also suggest that 21.3 percent of the FFE beneficiary households do not meet any criteria. Nevertheless, 57 percent of these households have incomes less than the average income of the beneficiary households who meet the criteria. These findings suggest that the official targeting criteria need to be improved for better identification of the needy households.

Effects on Food Consumption

Table 16.6 presents household expenditures on various food items. For the entire sample, rice accounts for about 35 percent of the total food budget. Household budget allocations for various foods across the five categories indicate similar patterns, except for wheat. Since FFE beneficiaries receive their ration mostly in wheat, the imputed expenditure on wheat for A-category households is higher than that in other groups.

FFE beneficiaries consume 10 percent more calories than do the C-category households. One-third of the program beneficiary households are calorie deficient, while as many as 60 percent of the C-category households consume fewer calories than they require.

TABLE 16.5. Households in FFE unions who fulfill the official targeting criteria (percent of all households)

Targeting criteria	FFE beneficiary households	Nonbeneficiary households with children attending FFE schools
Female-headed household	14.0	12.7
Less than 0.5 acres of land owned	68.1	43.7
Day laborer	28.5	12.0
Low-level profession	10.2	5.6

Source: Based on data from IFPRI's 2000 *Food for Education Evaluation Survey, 2000: Household Survey, Bangladesh.*
Note: 21.3 percent of FFE beneficiary households do not meet any of the criteria.

TABLE 16.6. Budget shares of food items, expenditures, and calorie consumption

Food items and expenditure categories	FFE unions			Non-FFE unions		
	FFE beneficiary households	Nonbeneficiary households with children attending FFE schools	Households with children not attending schools	Households with children attending schools	Households with children not attending schools	All
	Percent of total food expenditure					
Rice	35.21	32.55	41.24	35.01	39.22	35.18
Wheat	4.16	0.92	0.48	0.51	0.31	1.74
Bread/other cereal	0.48	0.62	0.51	0.58	0.41	0.55
Pulses	2.55	2.05	1.96	2.54	2.17	2.34
Oil	2.58	2.65	2.46	2.84	2.42	2.65
Vegetables	13.19	12.51	11.58	11.79	10.77	12.34
Meat	6.39	7.76	5.80	6.61	7.00	6.80
Eggs	1.02	1.25	1.09	1.39	0.95	1.19
Milk	1.67	3.15	2.33	3.11	2.50	2.58
Fruits	5.92	7.39	5.20	6.53	5.30	6.39
Fish	13.89	14.36	15.01	14.74	11.60	14.21
Spices	4.49	4.26	4.51	4.40	4.73	4.42
Sugar	5.18	6.22	4.56	5.43	6.10	5.53
Beverage	2.72	3.21	2.96	3.67	4.31	3.23
Prepared food	0.55	1.09	0.31	0.87	2.21	0.86
Total	100.00	100.00	100.00	100.00	100.00	100.00
Household food expenditure (taka/month)	2,182	2,788	2,363	2,570	2,309	2,452
Expenditure	3,372	5,272	3,590	4,569	3,752	4,188
Share of food in total expenditure (percent)	71.00	66.07	70.77	65.95	69.22	68.34
Per capita calorie consumption (kcal/day)	2,376	2,651	2,154	2,480	2,234	2,446
Calorie-deficient households (percent)	33.3	26.1	56.9	30.9	44.7	33.7

Source: Based on data from IFPRI's 2000 Food for Education Evaluation Survey, 2000: Household Survey, Bangladesh.
Note: Households have been defined calorie deficient if they consume less than 2,122 kilocalories (kcal) per capita per day.

The results of the descriptive analyses do not permit the separation of program effects on food consumption from the effects of other factors. In a multivariate analysis, the effects of income and other factors can be isolated from those of the FFE program. Here, the static comparisons are supplemented by the results of the multivariate analysis. All econometric models in this study appropriately control for the endogeniety of program placement. The average effects of FFE participation on food consumption at the household level have been assessed by following the difference-in-differences approach.[10] In this approach, the within-union difference in any outcome variable (e.g., food consumption) between eligible and noneligible in a "program union" (i.e., a union where the FFE program is available) is compared with the within-union difference between would-be-eligible and noneligible in the "non-program union." The difference between these within-union differences yields an estimate of the average impact of the program.

In a multivariate framework, the difference-in-differences comparison lies behind the instrumental variable estimates of average impact of participation in the FFE program on the outcome of interest (here, food consumption), which can be specified linearly as:

$$C_i = \alpha + \beta FFE_i + \delta e_i + \sum_{k=1}^{K} \gamma_{k,i} X + \varepsilon_i \qquad (1)$$

where C_i denotes an indicator of food consumption of the ith household, FFE_i denotes participation of the ith household in the program, e_i equals 1 if the household is eligible (i.e., meets at least one of the four eligibility criteria), zero otherwise; X is a set of explanatory variables (for example, income, household size, and *thana*-level dummy variables that capture fixed effects) indexed by $k = 1, \ldots, K$; α is a scalar; β and δ are parameters of FFE_i and e_i, respectively; γ is a $K \times 1$ vector of parameters; and ε_i is an error term.

Program participation (FFE_i) is instrumented with $e_i b_i$, where eligibility status (e_i) is interacted with b_i, an indicator of FFE program availability in the union where the ith household resides. The estimates of the model are obtained using two-stage least squares (2SLS) regression.

Table 16.7 presents the 2SLS results of determinants of calorie consumption. The FFE program transfers income to participating households. The particular interest here is to assess whether this extra income increases food consumption (in terms of calorie) of the beneficiaries at the household level. The dependent variable in estimating the regression equation is calorie consumption per adult equivalent units (AEU). The corresponding right-hand side variables include program participation, natural logarithm of

TABLE 16.7. Effects of FFE participation on calorie consumption, 2SLS regression results

Variable name	Coefficient	t-statistic
Log daily household expenditure per AEU	1,346.77	6.66**
Dummy: FFE beneficiary household = 1	351.20	2.53*
Log adult equivalent household size	−226.29	−1.36
Number of male with primary education	16.98	0.43
Number of male above primary education	−79.67	−1.72
Number of female with primary education	19.13	0.46
Number of female above primary education	19.90	0.36
Dummy: Meets eligibility criteria = 1	19.34	0.20
Rice price	−72.09	−2.49*
Wheat price	55.52	2.13*
Dummy: Living in *thana* 1 = 1	769.52	5.28**
Dummy: Living in *thana* 2 = 1	−152.93	−1.16
Dummy: Living in *thana* 3 = 1	396.16	2.92**
Dummy: Living in *thana* 4 = 1	539.04	4.15**
Dummy: Living in *thana* 5 = 1	−69.77	−0.44
Dummy: Living in *thana* 6 = 1	374.02	2.65**
Dummy: Living in *thana* 7 = 1	653.14	4.34**
Dummy: Living in *thana* 8 = 1	222.25	1.64
Dummy: Living in *thana* 9 = 1	−472.86	−3.25**
Constant	−5,566.42	−3.89**
F- statistic 18.03**		
Adjusted R-squared 0.42		

Note: Dependent variable is daily household calorie consumption per adult equivalent unit (AEU).
*Significant at the 5 percent level
**Significant at the 1 percent level

daily household total expenditures per AEU (instrumented with the total owned arable land, and the total value of productive assets), natural logarithm of household size in AEU, education levels of male and female household members, FFE eligibility status (e_i), rice and wheat prices, and *thana*-level fixed effects.

The results suggest that participation in FFE leads to increased calorie consumption in the beneficiary households, and this is statistically significant. Other statistically significant determinants of calorie consumption are income (positive), rice price (negative), wheat price (positive), and a number of location-specific fixed effects. The positive coefficient of wheat price reflects cross-price substitution effects among different food items and variations in food-to-calorie conversion factors.

Impact on Enrollment and School Participation

The data from the census of all households carried out in all sixty sample villages and covering 17,134 households, were used to select the sample households and schools and to estimate the level of enrollment and literacy at union level in 2000. The data show that, at the aggregate level, the enrollment rates are higher in FFE unions compared to non-FFE unions. There is a large difference in school enrollment levels between survey *thanas*, ranging from a low of less than 70 percent to over 90 percent.

The descriptive statistics may not be able to determine the effect of the FFE program on schooling. Therefore, we used an appropriately formulated multivariate analysis to isolate the effects of income and other factors to capture the true effect of FFE on enrollment and assess its impact on the probability of a child's school attendance.

A number of studies have estimated the impact of FFE program on enrollment. They all suggest that FFE has resulted in increased primary school enrollment (Ahmed and Billah, 1994; BIDS, 1997; DPC, 2000; Khandker, 1996; Ravallion and Wodon, 1997). The analysis presented here is based on individual observations of 930 primary school-age children from the sample of households in the household survey data, regardless of their school attendance status and FFE program participation. In particular, the sample includes all children between six and thirteen years of age that have not completed primary school.

We employ a regression model to isolate the effect of FFE program from the effects of other factors on school enrollment. Given that there are several unobservable factors that might have determined program placement at the union and village levels, we have specified a model of analysis, following the basic model structure of Ravallion and Wodon (2000) that considers FFE participation as endogenous.

The schooling of a child i (SC_i) is determined by the participation of the ith child of a household receiving the FFE ration (FFE_i), and a set of other explanatory variables (child, household, and community characteristics) denoted by X, and indexed by $k=1, \ldots, K$. The model takes the form

$$SC_i = \alpha + \beta FFE_i + \sum_{k=1}^{K} \delta_k X_{k,i} + \varepsilon_i \qquad (16.2)$$

where α is a scalar, β is a parameter of FFE_i, δ is a $K \times 1$ vector of parameters, and ε_i is an error term. Individual program participation FFE_i is instrumented by

$$FFE_i = \gamma FFEV_i + \sum_{k=1}^{K} \zeta_k X_{k.i} + \mu_i \qquad (16.3)$$

where $FFEV_i$ is equal to one if the child is a resident in a union that has the FFE program, zero otherwise; and μ_i is an error term.

The model has been estimated in two stages. In the first stage, we explain program participation using equation (16.3), and then we use the resulting predicted values of program participation in equation (16.2) to measure the impact of program participation on the probability of a child to attend school. We use two different approaches to estimate program participation. In the first approach, following the Ravallion and Wodon (2000) model, we use a Tobit specification in the first stage regression where the dependent variable is zero for nonparticipants, and is equal to the amount of grain actually received by participants. In the second approach, we use a simple probit model in the first stage where the dependent variable is equal to 1 if the household participates in the program, zero otherwise.

The estimation of the second stage equation, specified in equation (16.4), is virtually the same in both model specifications, even though the value of the coefficient of the FFE variable will have a different meaning depending on the specification used in the first stage. To control for the correlation between instruments used in the first stage and the error term in the second stage, we added to the estimation the predicted value of the residuals from the estimation of the first stage regression $(FFER_i)$ (Ravallion and Wodon, 2000; Datt and Ravallion, 1994; Rivers and Vuong, 1998), as follows

$$SC_i = \alpha + \beta FFE_i + \delta FFER_i + \sum_{k=1}^{K} \delta_k X_{k.i} + \varepsilon_i \qquad (16.4)$$

The results of the models are reported in Table 16.8. The first four columns report the results of the model in which the first stage is estimated using a Tobit. The last four columns report the results in which the first stage is estimated using a probit.

The results of the first stage regressions of the two models (columns one and two for Model 1 and columns four and five for Model 2) are quite similar. Among the four official criteria for FFE beneficiary selection, only one criterion—the household head's principal occupation is a daily laborer—is a statistical determinant of program participation. The values of "housing"

TABLE 16.8. Impact of FFE on school enrollment, econometric model results

| | Model 1 with Tobit first stage | | | | Model 2 with Probit first stage | | | |
| | Dependent variable: Grain Received | | Dependent variable: Going to school | | Dependent variable: Receiving FFE | | Dependent variable: Going to school | |
Variable name	Coeff. (1)	t-test (2)	dF/dX (3)	z-test (4)	Coeff. (5)	z-test (6)	dF/dX (7)	z-test (8)
Dummy: FFE union = 1	1.574	6.58[c]			3.950	5.17[c]		
Predicted FFE beneficiary household							0.084	1.81[a]
Residual from FFE beneficiary model							−0.135	4.71[c]
Predicted amount of FFE transfer			0.113	1.74[a]				
Residual from FFE model			−0.155	4.23[c]				
Age of child in years	−0.010	0.78	−0.003	0.62	−0.037	0.99	−0.003	0.67
Dummy: Sex, female = 1	−0.010	0.16	0.048	1.94[a]	−0.045	0.26	0.045	1.85[a]
Number of female younger siblings	−0.193	3.29[c]	−0.002	0.09	−0.619	3.37[c]	0.001	0.02
Number of male younger siblings	−0.069	1.22	−0.028	1.13	−0.218	1.22	−0.027	1.07
Number of female older siblings	−0.223	3.70[c]	−0.024	0.94	−0.660	3.59[c]	−0.025	0.95
Number of male older siblings	−0.129	2.17[b]	−0.024	0.97	−0.423	2.31[b]	−0.020	0.82
Household size	0.012	0.30	−0.036	1.96[a]	0.087	0.68	−0.038	2.08[b]
Percent members 19 to 34 years of age	−0.642	2.28[b]	0.272	2.34[b]	1.820	2.20[b]	0.273	2.35[b]
Percent members 35 to 65 years of age	−0.324	1.04	0.032	0.25	−0.905	0.99	0.025	0.20
Number males with primary education	0.200	5.52[c]	0.101	6.58[c]	0.600	5.44[c]	0.101	6.57[c]
Number males with education above primary	0.152	3.53[c]	0.062	3.45[c]	0.289	2.27[c]	0.062	3.50[c]

TABLE 16.8 (continued)

Variable name	Model 1 with Tobit first stage						Model 2 with Probit first stage					
	Dependent variable: Grain Received		Dependent variable: Going to school				Dependent variable: Receiving FFE		Dependent variable: Going to school			
	Coeff. (1)	t-test (2)	dF/dX (3)	z-test (4)			Coeff. (5)	z-test (6)	dF/dX (7)	z-test (8)		
Number females with primary education	0.240	5.82^c	0.092	5.64^c			0.758	6.10^c	0.092	5.55^c		
Number females with education above primary	0.176	3.87^c	0.106	4.63^c			0.531	3.76^c	0.107	4.68^c		
Dummy: Household head is farmer = 1	0.002	0.03	-0.087	1.88^a			-0.043	0.15	-0.082	1.80^a		
Dummy: Household head has business = 1	0.166	1.90^a	-0.020	0.52			0.419	1.64	-0.018	0.46		
Dummy: Household head is salaried = 1	0.016	0.12	-0.036	0.57			0.211	0.53	-0.033	0.53		
Dummy: Household head rickshaw puller = 1	0.201	0.93	-0.034	0.38			0.392	0.64	-0.044	0.48		
Value of housing (10,000 Tk)	-0.082	3.91^c	-0.006	1.23			-0.227	3.70^c	-0.006	1.24		
Value of consumable assets (10,000 Tk)	0.006	0.07	-0.025	1.12			0.025	0.10	-0.027	1.24		
Value of domestic assets (10,000 Tk)	-0.104	0.77	0.008	0.23			-0.152	0.41	0.009	0.25		
Value of liquid assets (10,000 Tk)	-0.079	0.83	-0.008	0.25			-0.170	0.66	-0.008	0.23		
Value of productive assets (10,000 Tk)	-0.109	2.99^c	0.026	2.26^b			-0.307	3.04^c	0.028	2.36^b		
Value of other assets (10,000 Tk)	-0.529	3.82^c	0.005	0.31			-1.352	3.29^c	0.005	0.26		
Dummy: Female head = 1	0.150	1.29	0.010	0.21			0.305	0.88	0.008	0.16		
Dummy: Head daily laborer = 1	0.185	2.17^b	-0.025	0.69			0.620	2.41^b	-0.026	0.72		

| | Model 1 with Tobit first stage | | | | Model 2 with Probit first stage | | | |
| Variable name | Dependent variable: Grain Received | | Dependent variable: Going to school | | Dependent variable: Receiving FFE | | Dependent variable: Going to school | |
	Coeff. (1)	t-test (2)	dF/dX (3)	z-test (4)	Coeff. (5)	z-test (6)	dF/dX (7)	z-test (8)
Dummy: Head low-level laborer = 1	−0.009	0.05	0.006	0.08	0.107	0.19	0.009	0.12
Dummy: Own land less than 50 decimal = 1	0.040	0.68	−0.036	1.48	0.046	0.27	−0.036	1.48
Dummy: Government school = 1	0.622	5.62c	−0.174	5.58c	1.531	5.06c	−0.172	5.52c
Dummy: Nongovernment school = 1	0.600	4.76c	0.008	0.09	1.626	4.55c	0.008	0.10
Village wage rate	−0.000	0.04	−0.003	1.81a	0.004	0.24	−0.003	1.70a
Village percent people with < = 0.5 acre	−0.549	1.02	−0.348	1.70a	−0.703	0.45	−0.335	1.64
Village percent people with 0.5 to 2 acres	−0.958	1.32	−0.239	0.97	−1.810	0.92	−0.219	0.89
Village Dummy: Grows two crops = 1	0.560	3.38c	−0.009	0.19	1.667	3.09c	−0.003	0.06
Village electrical connections	−0.000	0.58	0.000	0.45	−0.002	1.00	0.000	0.52
Village government schools	−0.018	1.14	0.002	0.48	−0.011	0.22	0.002	0.36
Village registered schools	0.006	0.20	0.018	2.17b	0.010	0.12	0.018	2.16b
Village Madrasa schools	−0.025	2.35b	0.001	0.28	−0.067	2.24b	0.001	0.40
Village NGO schools	−0.044	2.03b	0.010	2.59b	−0.153	2.29b	0.009	2.53b
Price of rice	0.135	2.02b	0.023	0.96	0.377	1.92a	0.020	0.87
Price of wheat	−0.216	1.99b	0.012	0.31	−0.596	1.88a	0.015	0.40
Price of atta	−0.101	1.20	0.025	1.43	−0.067	0.27	0.023	1.32
Price of onions	0.035	1.95a	−0.011	2.22b	0.121	2.19b	−0.011	2.19b
Price of potatoes	0.031	0.48	−0.030	1.18	0.032	0.15	−0.029	1.16
Price of eggplants	0.128	3.10c	−0.026	2.17b	0.349	2.92c	−0.025	2.09b
Price of mustard oil	−0.000	0.04	0.002	0.65	−0.010	0.63	0.001	0.63
Price of soybean oil	0.011	0.42	−0.013	1.92a	0.046	0.62	−0.013	1.85a
Price of pulses	0.041	0.81	−0.002	0.18	0.047	0.31	−0.002	0.14

TABLE 16.8 (continued)

Variable name	Model 1 with Tobit first stage				Model 2 with Probit first stage			
	Dependent variable: Grain Received		Dependent variable: Going to school		Dependent variable: Receiving FFE		Dependent variable: Going to school	
	Coeff. (1)	t-test (2)	dF/dX (3)	z-test (4)	Coeff. (5)	z-test (6)	dF/dX (7)	z-test (8)
Dummy: Living in *thana* 2 = 1	1.337	5.36c	-0.421	2.47b	3.789	5.15b	-0.424	2.49b
Dummy: Living in *thana* 3 = 1	1.289	4.97c	-0.614	3.22c	3.155	4.30c	-0.611	3.21c
Dummy: Living in *thana* 4 = 1	1.501	4.57c	-0.631	3.24c	3.964	4.21c	-0.631	3.23c
Dummy: Living in *thana* 5 = 1	0.948	3.82c	-0.043	0.42	2.966	3.94c	-0.071	0.63
Dummy: Living in *thana* 6 = 1	0.518	1.94a	-0.001	0.01	1.559	2.02b	-0.013	0.13
Dummy: Living in *thana* 7 = 1	0.798	2.80c	-0.021	0.24	2.189	2.70c	-0.031	0.34
Dummy: Living in *thana* 8 = 1	0.907	3.61c	-0.067	0.69	2.825	3.74c	-0.075	0.75
Dummy: Living in *thana* 9 = 1	2.302	1.65a	-0.056	0.13	4.477	1.09	-0.047	0.11
Dummy: Living in *thana* 10 = 1	1.969	1.34	-0.034	0.08	2.848	0.66	-0.018	0.05
Constant	-5.689	3.38c			-6.212	3.27c		
R-squared	0.50		0.41		0.56		0.43	

Note: dF/dX represents the change in probability for an infinitesimal change in each in dependent, continuous variable and, by default, the discrete change in the probability for the dummy variables. The equations have been estimated using the *"dprobit"* command of the Stata statistical software.
[a]Significant at the 10 percent level
[b]Significant at the 5 percent level
[c]Significant at the 1 percent level

and "other assets." are strongly and negatively correlated with program participation, which indicates that the program is effectively targeted to poorer households.

Columns three and four for Model 1 and columns seven and eight for Model 2 present the results of the probit model on the determinants of school enrollment. The estimated coefficient of the amount of FFE transfer in Model 1 (with the Tobit first stage specification) implies that, at the sample mean transfer of 70 kg of FFE grain over the five-month period of survey recall, the probability of a household's child going to school increases by 7.9 percent. The results of Model 2 (with the probit first stage specification) show this increase in probability to be 8.4 percent. [11]

The results of the models also show that girls have a higher probability of being enrolled compared to boys. Although household size is negatively correlated with school enrollment, the number of siblings of any gender does not appear to affect whether a child goes to school.

All four variables representing the levels of education of male and female household members have a strong and positive impact on child primary school enrollment. On the other hand, the probability of a primary school-age child's school enrollment decreases if the household head is engaged in farming activities (which may increase the demand for child labor, as was found by Ravallion and Wodon, 2000). The probability also decreases if the wage rate in the village increases, indicating the effect of an increased opportunity cost of the child's school attendance.

The total value of productive assets has a positive impact on enrollment. This result may be interpreted as follows. An increase in the value of productive assets is likely to increase income of the household as well as the marginal productivity of family labor. Although an increase in the marginal productivity of labor is expected to increase the opportunity cost of sending a child to school, the increased income can also be expected to reduce demand for a household's own child labor and to increase demand for the child's education. The regression result in this case shows that the income effect outweighs the substitution effect.

Finally, the results indicate that the presence of a registered nongovernment school or an NGO school in the village increases the probability of a child's enrollment.

Efficiency of FFE Food-Grain Distribution

From July 1993 to January 1999, the School Management Committee (SMC) distributed food grains to FFE beneficiary households at the school premises once a month. However, there were concerns that teachers were spending too much of their time in food-grain distribution, and that the quality of education in FFE-supported schools had deteriorated as a result. These concerns led to a government decision that the SMC would no longer distribute food grains. Instead, private dealers (one dealer per union) have been appointed for FFE food-grain distribution since February 1999.

All grain dealers in the thirty sample unions were interviewed for this evaluation to estimate costs and returns of their operations. On average, a dealer covers 1,534 FFE card-holding beneficiary households, and distributes 21.15 metric tons of food grains per month.

The estimates for dealers' profitability provided in Table 16.9 suggest that, on the average, a dealer earns a profit of Tk. 2,356 per month from FFE food-grain distribution. The return on dealers' investments are determined by dividing the profit (or net income) by the operating expenses. Interest on operating expenses is subtracted from profit at this point.[12] The average return on investment is calculated to be 27.3 percent per year. This is a conservative estimate of return on investment, because it is based on an assumption that the turnover of operating capital requires one year. However, since the dealers lift their quota of food grains twelve times per year, the rate of turnover of operating capital should be much quicker than what is assumed

TABLE 16.9. Average profitability to dealers of FFE food-grain distribution

Item	Per metric ton of food-grain distributed (taka)	Per dealer (taka per month)
Total cost	267	5,643
Food-grain loading cost	34	719
Food-grain carrying cost	124	2,613
Food-grain unloading cost	20	425
Staff salary	52	1,106
Other costs	37	780
Interest charges imputed at 14 percent per year	3	66
Total operating expenses	270	5,709
Total revenue	381	8,065
Commission	250	5,288
Sales proceeds of sacks	131	2,777
Profit	114	2,356

Source: Based on data from IFPRI's 2000 *Food for Education Evaluation Survey: Foodgrain Dealer Survey, Bangladesh.*

in this analysis. Even the conservative estimates of annual return on investment for the dealers are quite high (27.3 percent) compared to the 14 percent interest rate on borrowed capital. Although most dealers complained about high transport costs and labor wages, this analysis suggests that FFE food-grain dealership is a profitable enterprise.

Despite the fact that their dealerships are profitable, dealers often divert FFE food grains to the black market for extra profit. In the household survey, 71 percent of FFE beneficiaries reported that the quantity of FFE food grains they actually received from dealers was less than what they were entitled to. Reportedly, a number of dealers sold FFE food grains to private traders, sometimes even at the distribution centers. For instance, some of the extremely poor participants of FFE in a highly distressed union reported that the dealer had lent money to them at exorbitant interest rates. Subsequently, the dealer took their FFE wheat entitlements because they could not repay the loan with interest.

The average distance of dealers' food-grain distribution centers from beneficiaries' homes is 5.1 kilometers, ranging from 1.5 to 11.2 kilometers.[13] Most beneficiaries reported that the transaction costs to collect their FFE rations from distribution centers were higher than the costs under the old SMC distribution system when food grains were distributed at the school premises. Most schools are within one kilometer from beneficiaries' homes.

Mainly due to these reasons, the household survey results suggest that 92 percent of the FFE beneficiary households prefer SMC to dealers for food-grain distribution. The rationale for changing the distribution system from SMC to dealers was to improve the quality of education by eliminating teachers' involvement in food-grain distribution. However, 82 percent of the FFE participants opined that there has been no improvement in the quality of education with the change.

CONCLUSIONS FOR POLICY

This chapter presents an evaluation of the FFE program in Bangladesh, based on primary data collected from multiple surveys covering schools, households, communities, and food-grain dealers. The school survey results suggest that FFE has been highly successful in increasing primary school enrollment, promoting school attendance, and reducing dropout rates. Furthermore, the enrollment increase is greater for girls than for boys.

During the past seven years the number of teachers per school has remained virtually constant in all schools and student enrollment has increased significantly in FFE schools. As a result, there are more students per teacher in FFE schools than in non-FFE schools. Moreover, because of increased enrollment and class attendance rates, FFE school classrooms are more crowded than non-FFE school classrooms. Consequently, there have been concerns about the deterioration of the quality of education in FFE schools.

The student academic achievement test scores, on average, are lower in FFE schools than in non-FFE schools. However, further analyses reveal that, within FFE schools, the average test score of FFE beneficiary students is less than that of the nonbeneficiary students, which brings down the aggregate score in FFE schools. In fact, the nonbeneficiary students in FFE schools scored about the same as the students in non-FFE schools on the average, despite a significantly larger class size in FFE schools. Therefore, larger class size in FFE schools does not necessarily cause lower test scores. Hence, there is a caution against drawing conclusions regarding the success of the FFE program based upon lower achievement test scores in FFE classrooms. Follow-up research on FFE could focus further on this important issue.

Students in government primary schools performed better in the achievement tests than the students in nongovernment primary schools, and this is true for both FFE and non-FFE schools. Government primary schools have better facilities, more qualified teachers, and pay higher salaries to teachers compared to nongovernment primary schools, indicating that the quality of primary education is directly related to the characteristics of primary schools. Therefore, in order to improve the quality of education in FFE schools in general and in nongovernment FFE schools in particular, the program would need complementary financial assistance to improve school facilities, hire better-qualified teachers, and provide training as well as adequate incentives to teachers.

The household-level analysis suggests that, in general, the FFE program is targeted to low-income households. However, considerable scope exists for improving targeting, as a sizable number of poor households remain excluded from the program even while many nonpoor households are included. A more accurate yet low-cost means testing method, such as the indicator-based proxy means tests to predict household income and welfare, needs to be considered to improve targeting (see Ahmed and Bouis 2002, for example).

FFE has a positive impact on household food security. The program significantly increases overall calorie consumption in beneficiary households, even after controlling for effects of income and other factors.

The village census findings suggest that, given the large regional disparity in the rates of enrollment, FFE could have a much larger impact on enrollment (and consequently on the literacy rates) if larger shares of program resources were targeted to areas with relatively lower rates of enrollment.[14]

The results of the multivariate analysis suggest that the availability of an FFE program for a household increases the probability of its child going to school by 8.4 percent. Although this average impact may seem to be rather small, this represents the situation in 2000—seven years after the introduction of the program.

The evaluation of the new food-grain distribution system through private dealers rather than through the SMC, as was previously done, found that the new system is far from satisfactory. Individual FFE beneficiaries have difficulty claiming their free and full ration from powerful and profit-minded private dealers, and hence they experience losses in their food-grain entitlement due to dealer malpractice. Also, a great deal of their time and money is spent on traveling to dealers' distribution centers to collect their FFE ration.

Alternatively, the FFE program could lower leakage by modifying the distribution system by having either a local NGO, or a youth club, or even a private dealer, deliver food grains to the beneficiaries in the schools on a set day each month. This system would empower beneficiaries by establishing a sense of group solidarity among recipients, assisting them in clarifying the exact amounts of rations to which they are entitled, and facilitating collective action against pilferage. This would also reduce inconvenience and transaction costs to beneficiaries in collecting their FFE rations.[15]

NOTES

1. The administrative structure of Bangladesh consists of divisions, districts, *thanas,* and unions, in decreasing order by size. There are five divisions, sixty-four districts, 489 *thanas* (of which twenty-nine are in four city corporations), and 4,451 unions (all rural). The FFE program is implemented in all 460 rural *thanas.*

2. The official exchange rate for the taka (Tk), the currency of Bangladesh, was Tk. 40.25 per US$1.00 in June 1994. The exchange rate was Tk. 51.00 per $1.00 in June 2000.

3. From 1997/1998 to1999/2000, food aid from donor countries accounted for 44 percent; domestic procurement, 39 percent; and government commercial imports, 17 percent of the total PFDS food grains.

4. Of the total quantity of FFE food grain distributed from 1997/1998 to1999/2000, wheat accounted for about 64 percent, and rice about 36 percent.

5. School operating expenses exclude teacher salaries, and include the costs of stationery and supplies, repairs and maintenance, utilities, and communication. Information on school expenses was not available for the non-FFE, nongovernment schools.

6. Half of the sample FFE schools were brought under the FFE program in 1993 and the other half in 1995. The change in enrollment is calculated from 1992 to 1994 for the schools that entered the program in 1993, and from 1994 to 1996 for the schools entering the program in 1995.

7. The FFE program targets children from poor households, most of who would not have attended school without the program. The socioeconomic status of the nonbeneficiary students in FFE schools, therefore, can roughly be compared with that of the students in non-FFE schools, although several factors need to be controlled for in order to make a sound comparison.

8. In this study, per capita expenditure is used as a proxy for income for two reasons. First, expenditures are likely to reflect permanent income and are, hence, a better indicator of consumption behavior (Friedman 1957). Second, data on expenditures are generally more reliable and stable than income data. Because expenditures are intended to proxy for income, the terms "expenditure" and "income" will be used interchangeably.

9. Per capita monthly expenditures of FFE beneficiary households (A-category) exclude the income transfer from FFE programs.

10. See Morduch (1998) for a detailed description of the approach.

11. This result may be compared with the preceding result of a descriptive analysis that shows about 15 percent higher enrollment for beneficiary households than for nonbeneficiary households. Note also that the enrollment rate calculated at the household level cannot be compared to the large increase in enrollment observed at the school level. The increase in enrollment at the school level reflects the initial impact of the program over a two-year period, while results at the household level represent the marginal increase in enrollment in the same schools seven years after the beginning of the program. The result presented here shows a smaller impact of the FFE program on school enrollment than that reported by Ravallion and Wodon (2000). Ravallion and Wodon used a nationally representative multipurpose data set of 1995/1996, while the present analysis is based on data collected in 2000. As was pointed out earlier, the impact of the program was larger at the time of introduction of the FFE program.

12. The bank-lending rate for commercial activities was 14 percent per year in 2000. The dealers are assumed to receive credit at an annual interest rate of 14 percent, and they are to repay the loan at the end of every year.

13. IFPRI survey enumerators measured this distance using a global positioning system (GPS).

14. In cases of local shortage in schools and teachers, a higher concentration of FFE program resources should be considered first for those areas where low rates of enrollment are related to higher rates of poverty, rather than lack of school capacity.

15. This proposal is based on past IFPRI studies on the public food distribution system in Bangladesh. See Ahmed, 2000; Ahmed and Billah, 1994; Haggblade et al., 1993; and WGTFI, 1994.

REFERENCES

Ahmed, A.U. 1992. *Operational performance of the rural rationing program in Bangladesh*. Working paper on Bangladesh No. 5. Washington, DC: International Food Policy Research Institute.

Ahmed, A.U. 2000. Targeted distribution. In R. Ahmed, S. Haggblade, and T.E. Chowdhury (Eds.), *Out of the shadow of famine: Evolving food markets and food policy in Bangladesh* (pp. 213-231). Baltimore: The Johns Hopkins University Press.

Ahmed, A.U. and K. Billah. 1994. *Food for education program in Bangladesh: An early assessment*. Bangladesh Food Policy Project Manuscript 62. Washington, DC: International Food Policy Research Institute.

Ahmed, A.U. and H.E. Bouis. 2002. *Weighing what's practical: Proxy means tests for targeting food subsidies in Egypt*. Food Consumption and Nutrition Division Discussion Paper 132. Washington, DC: International Food Policy Research Institute.

Ahmed, A.U., del Ninno, C., and O.H. Chowdhury. 2000. *Food for Education Evaluation Survey, 2000: School Survey, Bangladesh*. Washington, DC: IFPRI.

Bangladesh Bureau of Statistics (BBS). 2000. *Statistical Yearbook of Bangladesh*. Dhaka: BBS.

Bangladesh Bureau of Statistics (BBS). 2001. *Preliminary report of household income and expenditure survey, 2000*. Statistics Division, Ministry Planning, Dhaka: BBS.

Bangladesh Institute of Development Studies (BIDS). 1997. *An evaluation of the food for education program: Enhancing accessibility to and retention in primary education for the rural poor in Bangladesh*. Dhaka: Government of the People's Republic of Bangladesh.

Datt, G. and M. Ravallion. 1994. *Income gains for the poor from public works employment: Evidence from two Indian villages*. Living Standards Measurement Study Working Paper 100. Washington, DC: World Bank.

Development Planners and Consultants (DPC). 2000. *Comprehensive assessment/ evaluation of the food for education programme in Bangladesh*. A report prepared for the Primary and Mass Education Division, Food for Education Programme Project Implementation Unit, Dhaka: DPC.

Friedman, M. 1957. *A theory of the consumption function*. Princeton: Princeton University Press.

Haggblade, S., S.A. Rahman, and S. Rashid. 1993. *Statutory rationing: Performance and prospects*. Mimeo. Bangladesh Food Policy Project. Dhaka: International Food Policy Research Institute.

IFPRI. 2000. *Food for Education Evaluation Survey, 2000: Foodgrain Dealer Survey, Bangladesh*. IFPRI.

IFPRI. 2000. *Food for Education Evaluation Survey, 2000: Household Survey, Bangladesh*. IFPRI.

IFPRI. 2000. *Food for Education Evaluation Survey, 2000: School Survey, Bangladesh*. IFPRI.

Khandker, S.R. 1996. *Education achievements and school efficiency in rural Bangladesh.* World Bank Discussion Paper 319, Washington, DC: World Bank.

Morduch, J. 1998. *Does microfinance really help the poor? Unobserved heterogeneity and average impacts of credit in Bangladesh.* Draft report. Stanford: Hoover Institute, Stanford University.

Primary and Mass Education Division (PMED). 2000. *Project report: Food for Education program* (in Bangla). Project implementation unit, Food for Education Program. Dhaka: PMED.

Ravallion, M. and Q. Wodon. 1997. *Evaluating a targeted social program when placement is decentralized.* Washington, DC: World Bank.

Ravallion, M. and Q. Wodon. 2000. Does child labour displace schooling? Evidence on behavioural response to an enrollment subsidy. *The Economic Journal,* 110, C158-C175.

Rivers, D. and Q. H. Vuong. 1988. Limited information estimators and exogeneity tests for simultaneous probit models. *Journal of Econometrics,* 39, 347-366.

Working Group on Targeted Food Interventions (WGTFI). 1994. *Options for targeting food interventions in Bangladesh.* Washington, DC: International Food Policy Research Institute.

Chapter 17

Agricultural Trade Policy Issues for Pakistan in the Context of the AOA

Sarfraz Khan Qureshi

INTRODUCTION

Agriculture plays an important role in Pakistan's economy, accounting for about 25 percent of the GDP and almost half of the country's labor force. Agricultural output has increased at an average annual rate of over 4 percent in the past two decades, contributing significantly to overall economic growth, food supplies, and nutrition and exports. Crop production accounts for the largest share of agricultural GDP (63 percent in 1995), followed by livestock (32 percent), and fishery and forestry (5 percent).

Despite the rising trend of agricultural output, the country faces a number of challenges in this sector. One is to reduce food imports, which have been growing steadily, especially in recent years. According to Food and Agriculture Organization (FAO) projections, food demand will rise substantially by the year 2010, with the share of imports in domestic consumption likely to go up further. With limited scope for the expansion of cropped land, higher crop output will have to come essentially from higher yields, which requires investments in agricultural research and irrigation, among other areas. Yet, investment in agriculture has been on the decline.

The continued high incidence of poverty is also a major challenge. According to FAO estimates, 20 percent of the population was undernourished in 1995/1997. One response of the government in the area of food security has been to keep consumer prices of basic foods down through a publicly rationed food distribution program and more recently through open-market operations. However, the program has been an expensive one, costing almost 1.5 percent of agricultural GDP in 1995/1996. Broad-based agricultural growth provides the best chance for reducing poverty in general and rural poverty in particular.

Agricultural exports have contributed significantly to overall export growth. Pakistan has had a strong comparative advantage in production for years and horticultural exports have increased rapidly with little support from subsidies. The country's climate and location give it an advantage for accessing a number of niche markets and it is generally considered that, given an enabling environment, horticultural exports could grow further.

Until the mid-1980s, the government pursued an economic policy that was strongly interventionist. One of the consequences was price discrimination against agriculture. According to one study, the taxation of the sector amounted to 39 percent during 1960-1985, through both the direct and indirect effects of policy.[1] As a result, resource transfers from agriculture to industry were substantial, about 156 billion rupees from 1980/1981 to 1989/1990, according to some estimates.

However, because of substantial policy reforms the structure of incentives within the agricultural sector is now generally neutral.

PAKISTAN'S EXPERIENCE IN IMPLEMENTING THE AOA

Market Access

Pakistan's trade policies have undergone significant change over time toward greater trade liberalization, involving both the dismantling of various nontariff barriers (NTBs) and the reduction of ordinary tariffs. The NTBs had included outright import bans, special dispensation and licensing, quotas, negative lists, and parastatal monopolies. Import surcharges were removed in 1992/1993. License fees and the *eqra* surcharge were abolished in 1994/1995. The maximum applied rate of ordinary tariffs has also been reduced substantially in phases. It was reduced from 225 percent in 1987/1988 to 65 percent in 1996 and, over the following two years, to 35 percent, where it has been since the end of March 1999 (Table 17.1). As a result, in 1999, applied rates varied from 0 to 35 percent. Appendix Table 17A.1 presents information on tariff and nontariff barriers in agriculture.

In the Uruguay Round (UR), Pakistan committed itself to bind more than 90 percent of the agricultural tariff lines. Products for which tariffs have not been bound include alcoholic beverages, swine, pig meat, and others, for religious reasons. As in many other developing countries, tariffs were bound at relatively high levels—at 100 percent for virtually all products. For ten commodities, however, the binding was 150 percent and for ten others it was understood that these high levels of binding were adopted in order to safeguard import-competing agricultural sectors in the short run from possible disruption as NTBs were removed. As Pakistan offered "ceiling bindings," no commitment was required to reduce the tariffs during the UR implementation period.

TABLE 17.1. WTO tariff bindings and applied rates for selected major products (percentage, ad valorem)

Product	Bound rate	Range of applied rates				
		1995	1996	1997	1998	1999
Cereals	100-150[1]	0-6	0-65	0-25	0-25	0-15
Oilseeds	100	10-70	10-65	0-65	0-45	0-35
Vegetable oils[2]	100	25-70	25-65	25-65	15-45	10-35
Live animals	100	15-65	15-65	15-65	10-45	10-35
Meat	100	35-70	35-65	15-65	15-45	10-35
Dairy products	100	25-70	25-65	25-65	25-45	10-35
Sugar	100	35-70	35-65	45-65	25-45	25-35
Coffee and tea[3]	100-150	15-70	15-65	0-65	15-45	25-35
Simple average[4]	100.5	—	—	—	—	—

Source: Government of Pakistan 2002.
[1]100 percent for rice and wheat flour and 150 percent for wheat.
[2]There were also specific rates of duty for some oils during 1995-1997.
[3]Bound rate for coffee 100 percent and tea 150 percent. Applied rates are also typically high for tea.
[4]Average of roughly 670 tariff lines.

From 1995 to 1999, applied tariffs were even lower than these statutory maximum rates, even on foodstuffs that are produced locally. In fact, the government often waives tariffs completely, for food security reasons, on products such as wheat and sugar. In addition to maintaining low prices through a public distribution program, Pakistan has had a long tradition of subsidizing wheat imports. Occasionally, duty-free imports, or imports at very low tariffs, are decreed. For example, tariffs on a large number of high-value products, e.g., cereal preparations and fruit and vegetable products, were reduced in 1999 to levels much below the statutory maximum of 35 percent.

This observation is confirmed by the estimates of the rate of protection as measured by the nominal protection coefficient (NPC), i.e., the ratio of the domestic to the world market price, or the import parity price. Although the NPCs fluctuate from year to year with changes in world market prices (since domestic prices are typically more stable), they were generally lower than unity. For example, for wheat it was 0.77 during 1990/1991-1994/1995, 0.63 in 1997/1998, and 0.85 in 1998/1999.

A key policy challenge has been setting "optimum" applied tariffs that address different policy objectives, notably those of safeguarding the interest of import-competing domestic sectors, maintaining full capacity of domestic processing industries, raising the value-added content of output, and ensuring the food security of the large urban population and the rural poor. The difficulty in achieving this was most evident in the case of vegetable oils. At nearly US$800 million, the cost of edible oil imports is second only to that of petroleum and projections point to an increase. The government has undertaken measures to raise the degree of self-sufficiency in edible oils through the introduction of canola oil production and minimum support prices for other oilseeds. Border policies have been used to support these measures. A related concern has been ensuring capacity utilization of the edible oil industry and thus adding to domestic value. In support of these objectives, the government allowed duty-free imports of oilseeds, for example, in 1998 and 1999, and raised tariffs on vegetable oils. However, this reduced the incentive for domestic oilseed producers to increase production.

In sum, the import regime has been fairly liberal in recent years with applied tariffs generally far below the WTO bound rates. However, implementing these reforms has not been smooth. In some cases, tariffs had to be lowered substantially for food security reasons, even though that undermined incentives to domestic producers and to longer-term growth of the sector. In many other cases, tariff rates had to be varied, often in response to swings in world market prices, in order to stabilize domestic markets. Appendix Table 17A.2 presents changes in the quantitative restrictions and maximum tariff. Tariff changes often have economywide consequences, and there is no easy solution to setting appropriate tariff rates. This is obviously an area in which in-depth policy analysis is needed.

Domestic Support

In the UR, Pakistan provided details of its domestic support measures in the various areas as specified in the AOA. As may be seen from Table 17.2, Pakistan's outlays on green box measures (minimal trade-distorting or production effects) rose from US$230 million in 1986/1988 (base period) to US$440 million in 1995/1996—an increase of 92 percent. In rupee terms, the increase amounted to 258 percent, or by about 32 percent per year, which was higher than the rate of inflation. Nevertheless, the total outlay in 1995/1996 was only about 3 percent of agricultural GDP, and the outlay fell by 21 percent by 1997/1998. In the base period, infrastructural services accounted for 64 percent of total outlays, followed by irrigation (flood protection) (14 percent), extension services (10 percent), and research (6 percent). By 1997/1998, the share of infrastructural services had increased to 85 per-

TABLE 17.2. Green box outlays, in US$ million

Type of measures	1986-1988	1995-1996	1996-1997	1997-1998	1998-1999	1999-2000
General services on research	14.5	12.8	7.7	7.6	—	—
Storage facilities	4.8	0.8	0.3	0.2	—	—
Marketing services	0.1	0.1	0.1	0.0	—	—
Extension services	22.1	2.4	2.2	1.6	—	—
General services	0.3	0.5	0.0	0.0	—	—
Infrastructural services	147.5	335.0	312.6	266.1	—	—
Flood protection services	7.9	34.6	15.9	22.8	—	—
Water supply services	31.3	53.7	53.9	14.1	—	—
Total	229	440	393	312	264	238

Source: Government of Pakistan 1989-2000; WTO 1995.

cent, at the expense of other categories. Flood protection services have increased considerably in recent years. Because green box outlays are exempt from reduction commitments there are no AOA-related issues here.

Table 17.3 shows the product-specific aggregate measurement of support (AMS) for the base period (1986-1988) and current years. Pakistan's data covered eleven crops for which there were market price support programs in the base period. The total product-specific AMS in that period was a negative 11,524 million rupees (US$640 million), with a positive AMS for sugarcane only. The total AMS amounted to a negative 7.6 percent of the total value of agricultural production, with wheat, cotton, and rice together accounting for most of the negative support.

In dollar terms[2], the product-specific AMS levels have fallen sharply. For example, the level in 1996/1997 was only 11 percent that of 1986/1988, and 22 percent in 1997/1998.[3] This was mainly because Pakistan reported the eligible production for most crops, and hence the AMS, to be zero. For example, of the eleven original crops for which the 1986/1988 AMS was computed, eligible production was assumed to be zero for seven in 1995/1996, nine in 1996/1997, and nine in 1997/1998, for which year the AMS for wheat only was computed. This assumption about the definition of eligible production has been questioned in WTO and has not been fully resolved. Wheat also accounted for part of the sharp decline in total AMS, for which the support became less negative.

TABLE 17.3. Product-specific aggregate measurement of support (AMS) in US$ million

Product	Base period (1986-1987)	1995-1996[1]	1996-1997[1]	1997-1998[1]	1998-1999	1999-2000
Wheat	−251.8	−172.0	−72.4	−143.4	191	257
Seed cotton	−187.1	0	0	0	—	—
Rice, basmati	−117.4	−20.1	0	0	—	—
Rice, corase	−48.6	−10.1	0	0	—	—
Sugarcane	+24.2	0	0	0	—	—
Onions[2]					—	—
Potatoes	−0.2	0	0	0	—	—
Gram	−57.8	0	0	0	—	—
Soybean	−1.1	0	0	0	—	—
Sunflower[3]	−0.4	0	0	0	—	—
Safflower	−0.1	0	0	0	—	—
Total	−640	−203	−72	−143	191	257

Source: Government of Pakistan 1989-2000; WTO 1995.
Note: AMS includes the market price support component only. A negative value indicates that the administered price was lower than the fixed (1986-1988) external reference price.
[1]Where the AMS is shown to be zero, this was the result of "zero"-eligible production, although administered prices were set (except for soybean and safflower in 1996/1997).
[2]Negligible support in the base period and 1995/1996.
[3]Negligible support in 1996/1997.

The total subsidy on farm inputs in the base period was reported as having been 3,652 million rupees (US$203 million), 43 percent of which fell within nonproduct-specific AMS and 57 percent under special and differential treatment (SDT). The AMS-related expenditures amounted to US$87 million in 1986/1988, but have been much lower in the implementation period (Table 17.4). Electricity subsidies originally accounted for roughly two-thirds, followed by subsidies on fertilizers (32 percent), and very little on credit. Only the electricity subsidy was reduced from 1996/1997. The total nonproduct specific AMS, which amounted to less than 1 percent of the value of agricultural production in 1986/1988, has since fallen even further, to around or below 0.1 percent. The level of support is thus minimal, compared to the 10 percent de minimis level specified in the AOA.

In regard to the SDT subsidies, in which the total outlay was 2,085 million rupees (US$116 million) in 1986/1988, 67 percent was on fertilizer, 32 percent on credit, and 1 percent on tubewells. In Pakistan's report, tubewell

TABLE 17.4. Nonproduct-specific aggregate measurement of support (AMS)

Type of measure	1986-1988	1995-1996	1996-1997	1997-1998	1998-1999	1999-2000
Fertilizer subsidy	27.4	0.4	0.0	0.0	—	—
Electricity subsidy	1.6	1.0	0.0	0.0	26.7	12.6
Credit subsidy	0.0	1.0	0.0	0.0	0.0	0.0
Total	87.1	10.8	15.5	22.5	26.7	12.6
As percent of product value status	0.80	0.06	0.08	0.12	0.04	0.03

Source: Government of Pakistan 1989-2000; WTO 1995.

subsidies were justified under the SDT category as being part of a national strategy for agricultural and rural development. The 74 percent of the total subsidy on fertilizers that was reported under the SDT category benefited poor farmers with land holdings of less than five hectares. Credit subsidies for interest-free loans and subsidized credits were estimated separately. The former was shown under the SDT category and the latter under the AMS, again using the five-hectare criterion. In the 1995/1996 notification, the total SDT outlay was reported as only one million rupees (US$55,500), a drastic cut from the base period, and exclusively on fertilizers. No SDT outlays were reported in the subsequent two years, essentially indicating that all forms of SDT subsidies had been eliminated.

Thus, on the whole, there were hardly any consequences for Pakistan as a result of reduction commitments. The base period product-specific AMS was negative (or zero, technically, from the AOA viewpoint) for all crops except sugarcane, which was within the de minimis level. Nonproduct-specific AMS was also within the de minimis level. The AMS outlays in the implementation period were also negative or very small. Consequently, the policy changes and/or reductions in support outlays from 1995 to 1999 could not have been due to the AOA.

Notwithstanding this conclusion, Pakistan's domestic support notifications for current years triggered much discussion in the Committee on Agriculture of the WTO and raised a number of issues. Resolving these issues is important for both Pakistan and other WTO members:

- Why were support outlays notified in dollar terms instead of in rupees as was done originally? Pakistan's response was that due to high inflation and currency depreciation, calculating the AMS in rupee terms would have presented a distorted view of the support levels. This issue has not been resolved.

- Pakistan defined eligible production as the quantity actually procured, hence the many zero values in Table 17.3 (indicating that there were no procurements). Some argue that in cases in which support prices were not effective, there was no support. This issue of what constitutes eligible production remains unresolved.
- Pakistan's decision to deduct from the total AMS the outlays that constituted negative support (e.g., on wheat): Pakistan's position was that the negative support, brought about the price support program, constituted an implicit tax on wheat producers. Since this was a form of distortion that the AOA was meant to address, the negative support should be included in the computation of total AMS. Canada submitted a detailed discussion paper on the question.[4]
- The full recovery of the operation and maintenance (O&M) costs of public irrigation schemes: Some WTO members argued that what was not recovered amounted to a subsidy.
- The noninclusion of the price support for canola in the AMS calculation. The response was that the necessary statistics were not yet available, since canola was a new crop that had only recently been introduced.

Export Subsidies, Taxes, and Restrictions

Pakistan did not report the existence of any export subsidies on agricultural products in the base period and accordingly cannot resort to them in the future. In regard to the entitlement to provide subsidies to reduce the costs of marketing exports, and internal transport and freight charges on export shipments, it reported relevant expenditures for 1995/1996 to 1997/1998. These consisted of a 25 percent subsidy on the actual freight paid on export consignments of fresh fruits and vegetables, amounting to US$1.7 million in 1995/1996, US$2.3 million for 1996/1997, and US$3.9 million for 1997/1998.

Prior to the establishment of WTO, Pakistan provided occasional direct export subsidies. Exports of rice and cotton were subsidized when the export trade was a monopoly of the public sector, but the subsidies were abolished when the private sector was permitted to trade in these products.

As in many other developing countries, some incentive measures are in force to promote export. For example, duty rebates are granted to some exports as well as for the import of raw materials and machinery for export-oriented industries. However, agriculture has on the whole not benefited much from this assistance, perhaps with some exceptions, such as the import of agricultural equipment and machinery. An export refinance program provides credit at concessional rates for the export of high-value-added products, including some agricultural products such as fish and packed rice. Higher and specific credit limits are also allowed for cotton and sugar, which ultimately are exported. Furthermore, in the late 1990s, the importation of meat, poultry goods, and sugar for the purposes of adding value and export was given support.

Pakistan has a long history of imposing quantitative export restrictions on a number of major agricultural products in the form of outright export bans, export quotas, and the channelling of rice and cotton through export consortia. Minimum export prices were fixed for some commodities, such as breeding animals, and for agricultural raw materials, such as hides, skins, and cotton; the main objective of instituting export controls and taxes was to ensure an adequate supply for local industries and to promote value-added exports. Most of these restrictions, as well as export taxes, have been lifted as a matter of general policy.

Despite the general policy, export duties and restrictions are occasionally reinstated on a temporary basis, essentially for two types of commodities. One group includes essential foodstuffs, in which exports are regulated to ensure, for food security reasons, that domestic supplies are available at fair prices. The other group includes agricultural raw materials such as hides, skins, and cotton, for the same reasons.

In the WTO, Pakistan has been designated as a net exporter of cotton and rice. Although the AOA has very little input on rules governing the export of nonfood products such as cotton, and other agricultural products like it, its Article 12 disciplines export prohibitions and restrictions on foodstuffs. A country prohibiting or restricting exports has to give due consideration to its effects on the food security of importing countries. Developing countries are not exempt from this provision, unless they are net exporters of the specific foodstuff concerned. No issues have arisen so far in the WTO with respect to Pakistan's exports of rice and cotton.

Other Experiences

Sanitary/Phytosanitary and Tariff Barriers to Trade Agreements

Pakistan has a number of regulations and standards to prevent food adulteration and to ensure hygienic and quality standards on both domestic production and imports. In 1995 the WTO documented the country's efforts to base its standards on international norms. The review also noted that national standards on a number of items were inferior to international norms due mainly to the lack of required technology and resources. It also noted that, overall, the standards did not seem to constitute a major impediment to exports from other countries. In some cases, Pakistan relaxed its standards on imports, such as on the shelf life of edible oils.

Exports of fruits and vegetables suffered considerably due to the limited ability for the enforcement of sanitary and phytosanitary standards (SPS) standards. The volume of exports to Europe, North America, Japan, and China has remained minimal despite the huge export potential. The general feeling in Pakistan is that the UR has hardly changed the situation since the standards and the inspection systems of the major importing developed countries are considered to be too strict for Pakistan to comply with. Efforts are being made to improve the national standards. For example, in 1998/1999, a scheme was announced for the inspection of all rice shipments by the Export Promotion Bureau in consultation with the Rice Exporters Association of Pakistan, so as to ensure quality exports. A similar system has been in place for the past two years for the exports of basmati rice to the European Union (EU).

Dispute Settlement

In another dispute, concerning export measures affecting hides and skins, the European Commission (EC) was the complainant against Pakistan. The issue was Pakistan's prohibition on the export of hides and skins and wet blue leather made from cowhides and cow calf hides, among other products. The EC contended that the measure limited access of EU industries to competitive sourcing of raw and semi-finished materials.

Pakistan was a joint complainant with India, Malaysia, and Thailand against a United States ban on the importation of shrimp and shrimp products from these countries. The experience of the complaining countries was positive, as the United States agreed to implement the rulings and recommendations of the Dispute Settlement Body (branch of WTO).

EXPERIENCE WITH FOOD AND AGRICULTURAL TRADE

Agricultural Trade

Food products accounted for roughly 70 percent of total agricultural imports during most of 1985-1998, while their share in total agricultural exports has been on the rise, from 45 percent in 1985/1987 to 70 percent in 1996/1998. The main export products are cotton, rice, and fruits and vegetables, and the principal imports are wheat and wheat flour, vegetable oils, pulses, tea, and (occasionally) refined sugar. Vegetable oils, together with wheat and flour, have constituted as much as 80 percent of agricultural imports in some years.

Total agricultural imports increased rapidly during 1985-1994, at the linear rate of US$49 million per year. In the post-1994 period, they rose sharply in 1995 (by 87 percent), fell slightly in the next two years, and rose by 12 percent in 1998. As a result, the average value of agricultural imports in 1995-1998 was 50 percent higher than in 1990-1994 and still 30 percent higher than the extrapolated value for 1995-1998 (Table 17.5).

In contrast, agricultural exports, which are much lower in value than imports, went slightly down during 1985-1994, a linear decline of about US $6 million per annum. Performance was better in the later years: during 1995-1998 exports increased in all years. Their average value in that period (US$1,100 million) was 14 percent higher than that in 1990-1994[5] and, since the decline was only slight, almost as high, at 12 percent, when measured against the trend.

As a result of the trends in imports and exports, the import surplus in 1995-1998 was 127 percent (an annual US$62 million) above the already high level that existed in 1990-1994. However, since the net deficits were strongly rising during the entire period 1985-1994, the 1995-1998 performance appeared better (only 58 percent higher) when measured against that trend.

Table 17.6 shows that for the major export products, the quantities exported have generally been higher in the post-UR years than those prior to 1995. Cotton enjoys favorable access terms in world markets, with few tariff and nontariff barriers, but the persistent decline in domestic production as a result of pest attacks and bad weather has reduced exports to below pre-UR levels. In recent years, export prices on cotton have also been depressed.

TABLE 17.5. Agricultural trade in 1990-1994 and 1995-1998 (average annual value, in millions [US$], and percentage change)

Period	Imports	Exports	Net imports
1990-1994 actual (a)	1,404	962	441
1995-1998 actual (b)	2,104	1,101	1,003
1995-1998 extrapolated (c)[1]	1,617	981	636
(b) − (a)[2]	700 (50%)	139 (14%)	562 (127%)
(b) − (c)[2]	487 (30%)	120 (12%)	367 (58%)

Source: Computed from FAOSTAT online database. Agriculture excludes fishery and forestry products.
[1]Extrapolated value based on 1985-1994 trend.
[2]Numbers in parentheses are percentage changes over (a) and (c), respectively.

TABLE 17.6. Exports of major agricultural products, 1995-1998 (in 000 tonnes)

Period	Cotton	Rice	Primary fruits	Primary vegetables	Nonprimary fruits and vegetables	All fruits and vegetables
1990-1994 actual (a)	359	1,095	111	28	41	180
1995-1998 actual (b)	186	1,798	168	28	53	249
1995-1998 extrapolated (c)[1]	295	1,148	123	7	41	171
percentage						
(b) / (a)	−48	64	51	0	29	38
(b) / (c)	−37	27	36	293	31	46

Source: Computed from FOASTAT online database.
[1]Extrapolated value based on 1985-1994 trend.

As a result of all these factors, exports earnings from cotton fell sharply, at the linear rate of US$26 million per year from 1985 to 1994, and continued to suffer thereafter.

Rice exports have increased steadily since 1985, and have accelerated in recent years: the tonnage exported in 1995-1998 was 64 percent higher than in 1990-1994. Of the two types exported, aromatic basmati rice faces competition only in its traditional market (the Persian Gulf countries and the United Kingdom), while the coarse variety faces stiff competition in most markets in Asia and Africa. It had been hoped that the UR would improve access terms for rice, but that did not turn out to be the case, especially in several developed-country markets.

Experience has been mixed in regard to the export of fruits and vegetables, but positive on the whole. Notable progress has been recorded in the post-UR period, particularly in 1996/1997 and 1997/1998, but it is not clear whether it can be sustained. Market access terms are difficult and the UR made very little difference in most developed-country markets. Moreover, the level of SPS standards in developed countries had been a major impediment. Although Pakistan has not, so far, taken any formal action in the WTO, it did bilaterally approach some of its developed trading partners, such as the EU and Japan, requesting them to relax their standards. In response, Japan offered a food-processing plant, but the vast potential for fruit exports remains to be exploited.

Food Trade

As noted, food products predominate in agricultural imports and are becoming increasingly important in exports.[6] The trends in food imports are thus quite safe and similar to those for agricultural imports as a whole (Table 17.7). In contrast, food exports followed a pattern that was different from total agricultural exports. They rose modestly during 1985-1994, at the linear rate of US$13 million per year. They surged by 68 percent in 1995, fell by 5 to 10 percent in the following two years, and rose again by 46 percent in 1998. As a result, the average values of food exports in 1995-1998 exceed that of 1990-1994 by 64 percent and the trend average by 45 percent.

In spite of this impressive export performance, because food exports are small relative to imports the deficit in 1995-1998 was 43 percent higher than in 1990-1994, and 19 percent more when measured against the trend.

Table 17.8 compares the imported quantities of selected major food products in 1995-1998 and 1990-1994. The volumes imported in 1995-1998 exceeded those of the preceding period in wheat and wheat flour, vegetable oils, fruits and vegetables, and pulses, but were below the earlier levels for dairy products and tea. On the other hand, the 1995-1998 imports were lower than the trend values, with the exception of fruits and vegetables, and pulses. In other words, there was a slowdown in the rate of imports of several food products.

Most of these food products are essential and their import demand is inelastic with respect to the world market price. Moreover, for food security reasons, the government refrains from excessive regulation of the imports, often preferring to reduce tariffs or allow duty-free entry. At the same time, most of these products are also produced domestically, and accordingly Pakistan faces the usual policy dilemma of reconciling the interests of producers with those of consumers. Import trends are consequently closely moni-

TABLE 17.7. Food trade in 1990-1994 and 1995-1998 (annual average value, in millions [US$], and percentage change)

Period	Imports	Exports	Net imports
1990-1994 actual (a)	999	480	519
1995-1998 actual (b)	1,527	785	741
1995-1998 extrapolated (c)[1]	1,166	543	623
percentage			
(b) / (a)	527 (53%)	305 (64%)	222 (43%)
(b) / (c)	360 (31%)	242 (45%)	118 (19%)

Source: Computed from FOASTAT online database. Food excludes fishery products.
Note: Numbers in parentheses are percentage changes over (a) and (c), respectively.
[1]Extrapolated value based on 1985-1994 trend.

TABLE 17.8. Imports of major food products during 1995-1998 (in thousand tonnes)

Period	Wheat and wheat flour	Vegetable oils	Fruits and vegetables	Dairy products	Pulses	Tea
1990-1994 actual (a)	1,966	1,074	258	168	157	113
1995-1998 actual (b)	2,429	1,248	410	93	205	107
1995-1998 extrapolated (c)[1]	2,619	1,322	376	142	261	132
percentage						
(b) / (a)	24	16	59	−45	31	−5
(b) / (c)	−7	−6	9	−34	22	−19

Source: Computed from FOASTAT online database. Food excludes fishery products.
[1]Extrapolated value based on 1985-1994 trend.

tored, and applied duties are adjusted in response to changes in domestic supply conditions and world market prices.

In 1985-1987, food imports were 90 percent of agricultural exports (import-export ratio was 0.9). This ratio fell to a low of 0.6 in 1988 and increased thereafter, fluctuating from year to year. The average value for 1995-1998 was 1.42, some 31 percent higher than in 1990-1994 (when it was 1.1), and still 11 percent higher than the extrapolated trend value. Thus, there has been clear and market deterioration in the balance of food imports and agricultural exports.

ISSUES OF CONCERN IN FURTHER NEGOTIATIONS ON AGRICULTURE

The Main AOA Provisions and Pakistan's Commitments

In market access, Pakistan's only commitment was on bound tariffs. Most nontariff barriers were lifted and applied tariffs reduced in stages to a maximum of 35 percent in 1999. These were unilateral initiatives as the AOA did not require Pakistan to reduce the bound tariffs. In section II it was concluded that Pakistan should, on the whole, be able to live rather comfortably with its AOA commitment (FAO document Repository, 2004).

This study does not analyze the full consequences of tariffication and tariff reduction and the impact of imports on the domestic import-competing sector. However, such analysis is important for any economy in which a large segment of the population depends on agriculture, and it should be undertaken as a matter of priority before further tariff reductions or similar measures are pursued at the WTO. A review of the experience during 1999-

2003 shows that it was not always easy for the Pakistani government to rely on low applied rates. For many products, tariffs often had to be revised, upward or downward, in response to developments such as local crop conditions and world market prices. This flexibility was important, but more analysis is required to fine-tune the process and to determine the range of the bound tariffs that could help achieve agricultural development objectives. Pakistan also needs to determine a negotiating position on agricultural special safeguards, that is, whether to press for access to such measures or for the complete abolition of agricultural safeguards for all WTO members, keeping in mind their importance when bound tariffs are very low.

Regarding domestic support, the AOA imposes no constraints on policy. Pakistan's product-specific AMS was negative and nonproduct-specific AMS was less than 1 percent of the value of agricultural production. Accordingly, ample scope exists for increasing support without contravening the AOA. The AOA also sets no ceiling on green box and SDT expenditures. Rather, the main problem is very low levels of public support for agriculture, given the important role of this sector in the economy.

Nevertheless, Pakistan needs to follow the debate in the Committee on Agriculture (COA) closely. First, a number of definitions (e.g., eligible production) and methods (e.g., measuring AMS) need clarification. Second, it is important to ensure that the current exemptions granted to developing countries under the AOA, such as the de minimis threshold and SDT, are protected. Third, the government needs to prepare itself for negotiations aimed at reducing domestic support in its major export markets and enlarging access to them. These issues are addressed as follows.

Access to Export Markets

Although Pakistan has broadly fulfilled its contractual obligations under the AOA, there is a general feeling that the country has not yet realized the promised benefits of trade. Although a number of problems on the domestic side are recognized, receiving the benefits also depends upon the extent to which the country's major trading partners undertake policy reforms, in both the spirit and the letter of the various WTO Agreements.[7] Much still remains to be done in this respect. A case in point is the continued high level of protection of and support to agriculture in the Organization for Economic Cooperation and Development (OECD) countries, a major market for Pakistan. The commodities that receive high support and protection in these countries are also the main export products of Pakistan, namely rice, sugar, fruits and vegetables, and meat. Policy in those countries has not only undermined Pakistan's access to their markets directly, indirectly, it has also undermined exports to other markets. For example, as a nonsubsidizing, low-cost rice exporter, Pakistan has been hurt by the very high domestic support to rice in the OECD countries, where the producer support estimate amounts to 74 percent of the value of rice production. It has also been hurt by the nominal rate of border protection of 3.8 for rice, unusually high among all commodities. Nominal coefficients for various crops are given in Appendix Table 17A.3.

A major related area of concern is the current tariff rate quote (TRQ) system. More than 70 percent of all tariff quotas under the AOA are on products of export interest to Pakistan: fruits and vegetables (26 percent), and meat products (13 percent). Overall, tariff quota fill rates have been less than two thirds. Perhaps even more important, the administration of the quotas is far from transparent; access is largely limited to traditional, specified suppliers, to the detriment of countries such as Pakistan. The administration of the quotas is also complex, requiring a great deal of effort and resources to understand the system. There is a need for a general overhaul of the current TRQ system.

One of the problems of the AOA for Pakistan is that it allows countries to liberalize their imports selectively. Except for cotton, Pakistan's principal exports, rice, sugar, fruits and vegetables, and processed hides and leather, face difficult market access terms in all major markets because importing countries have either not liberalized the entry of these goods, or done so only to a small extent. Rice is one of the most protected and supported commodities, and processed hides and leather products are subject to tariff peaks and considerable tariff escalation. The UR did very little to reform the world sugar market, leaving the old quota-ridden import regimes intact. Import regimes for fruits and vegetables also continue to be complicated, notably in the EU. There is much at stake for Pakistan in the new round of negotiations, but in order to achieve concrete results in terms of the export commodities of interest, Pakistan needs to prepare itself by analyzing the issues and putting forward, together with like-minded WTO members, concrete proposals for reform.

SPS/TBT Agreements

Although the sanitary and phytosanitary/technical barriers to trade (SPS/TBT) agreements are not part of the mandated negotiations that began in 2000, they are too important to be left out of any discussion on agricultural trade. As noted in Section II of the agreements, a concerted effort is required to collate the experiences of exporters who know the situation well.

In addition, unfair cases should be identified and remedial action sought in the appropriate WTO fora. Much can be learned from experiences with the export of fruits and vegetables, both fresh and processed, as these products were most affected by the SPS/TBT agreement.

At the same time, as consumer concerns over food quality and safety intensify all over the world, there is no alternative to upgrading the standards of the exported products. Experience in many other countries has shown that product standards on exports can rarely be upgraded without their improvement for the home market first. Much remains to be done in Pakistan in this respect. Finally, Pakistan, like many other developing countries, should continue to voice its dissatisfaction with the slow pace of implementation of the various technical and financial assistance pledges contained in the SPS/TBT agreements.

Food Security Concerns, Including the Marrakesh Decision

Over the past five years, Pakistan has contributed considerably to the debate on food security and the cause of agriculture in developing countries, e.g., in the Committee on Agriculture's Analysis and Information Exchange (AIE) process and through the analysis of SDT provisions. It also played an important role in discussions on the implementation of the Marrakesh Ministerial Decision, which also benefitted other net food importing developing countries. These were important contributions toward ensuring that food security concerns of developing countries are fully taken into account in trade negotiations. With emphasis on the new negotiations to Article 20 of the AOA, which binds the countries to continue the reform process, Pakistan is expected to continue these contributions in the coming years.

TABLE 17A.1. Tariff and nontariff barriers on agricultural commodities (2-digit level)

Heading No.	Description of goods	Tariff rate (%)						Bound rate of	Quantitative restrictions							
									1989		1995		1997		1998	
		1989	1995	1996	1997	1998	1999		Import	Export	Import	Export	Import	Export	Import	Export
01	Live animals	0-40	15-65	15-65	15-65	10-45	10-35	100	SC/NL	BAN	HS/NL	SD	HS/NL	SD	HS/NL	
02	Meat and edible meat offal	80	35-70	35-65	15-65	15-45	10-35	100	SC/NL	BAN	NL	SD	NL	SD	NL	
03	Fish and crustaceans; molluscs and other aquatic invertebrates	0-60	25-70	35-65	15-65	25-45	10-35									
04	Dairy produce; birds' eggs; natural honey; edible products of animal origin; nes products of animal origin, nes	60-100 Rs.10/ kg	25-70	25-65	25-65	25-45	25-35	100	SC/NL	BAN	NL					
05	Products of animal origin, nes	40-80	35-70	35-65	35-65	10-45	10-35	100	SC/NL		NL	NL				BAN
06	Live trees and other plants; bulbs, roots, etc.; cut flowers and ornamental foliage	100	35	35	35	15-35	10-25	100		BAN						
07	Edible vegetables and certain roots and tubers	0-100	10-35	10-35	0-35	0-35	0-25	100								
08	Edible fruit and nuts; peel of citrus fruit of melons	0-100 Rs.12/ kg	30-70	35-65	55-65	15-45	10-35	100								

		Tariff rate (%)							Quantitative restrictions							
									1989		1995		1997		1998	
Heading No.	Description of goods	1989	1995	1996	1997	1998	1999	Bound rate of	Import	Export	Import	Export	Import	Export	Import	Export
09	Coffee, tea, maté, and spices	40-100 Rs.60/kg	15-70	15-65	0-65	15-45	0-35	100		BAN						PR
10	Cereals	0-40	0-65	0-65	0-25	0-25	0-15	100-150	NL	BAN/Quota/RECP		SD/Quota/RECP		SD/Quota/RECP	NL	
11	Products of milling industry; malt, starches; insulin, wheat gluten	40-80 Rs.4/kg	25	25	10-25	10-25	10-25	100				SD		Quota		
12	Oilseeds and oleaginous fruit; miscellaneous or medicinal plants; straw and fodder grains; seeds and fruit; industrial	0-100	10-70	10-65	0-65	0-45	0-35	100	NL		HS/NL		HS/NL	SD	HS/NL	
13	Lac, gums, resins, and other vegetable saps and extracts	0-80 Rs.150/kg	25-70	65	45-65	35-45	25-35	100	NL		NL		NL		NL	
14	Vegetable plaiting materials; vegetable products, nes	20-100 Rs.100/kg	45-70 Rs.110/kg	45-65 Rs.110/kg	45-65 Rs.150/kg	25-45	25-45	100								
15	Animal or vegetable fats and oils and their cleavage fats; animal or vegetable waxes	40-100 Rs.3000-3500/tonne	25-70 Rs.8590-9550/tonne	25-65 Rs.8590-9550/tonne	25-65 Rs.7600-10000/tonne	15-45	10-35	100	SC/NL/T.C.P	BAN	NL		HS/NL	SD	HS/NL	BAN
20	Preparations of vegetable, fruit, nuts or other parts of plants	100	35-70	35-65	45-65	35-45	25-45	100			SC				BAN	

TABLE 17A.1 *(continued)*

Heading No.	Description of goods	Tariff rate (%)						Bound rate of	Quantitative restrictions							
									1989		1995		1997		1998	
		1989	1995	1996	1997	1998	1999		Import	Export	Import	Export	Import	Export	Import	Export
21	Miscellaneous edible preparation	40-125	35-70	35-65	45-65	25-45	25-35	100			NL					
23	Residues and waste from the food industries; prepared animal fodder	20	15-45	15-45	15-45	10-45	10-35	100			NL			SD		SD
24	Tobacco	50	50	70	65	45	35	65			NL					
25	Cotton CEC		0	0	10	10-45	0-35	10				CEC		CEC		CED

Sources: Government of Pakistan 1989-2000; WTO 1995.
Notes: HS denotes restrictions for Health and Safety conditions.
NL denotes Negative List, i.e., items in this list cannot be imported.
TCP denotes Trading Corporation of Pakistan. It is a parastatal body involved in imports of different commodities.
BAN indicates banned items and hence items in this list cannot be imported and/or exported.
RECP denotes Rice Export Corporation of Pakistan.
CECP denotes Cotton Export Corporation of Pakistan.
SD denotes Special Dispensation.
SC denotes Specific Condition which includes taking permission from different ministries as well as meeting specification standards prescribed by the government.
Blank entries mean no restriction.
PR = Partial restriction

TABLE 17A.2. Changes in quantitative restrictions and maximum tariff

Fiscal Year	Items removed from negative list[1]	Items removed from restricted list[2]	Maximum tariff[3] (%)	Import surcharge (%)	Eqra surcharge (%)	License fee (%)
1987-1988	136	10	225	5	5	4
1988-1989	169	51	125	6	5	5
1989-1990	70	20	125	7	5	5
1990-1991	97	43	95	10	5	6
1991-1992	23	17	90	10	5	6
1992-1993	21	11	90	0	5	6
1993-1994	1	1	90	0	5	6
1994-1995	8	—	70	0	0	0
1995-1996	11	—	65	0	0	0
1996-1997	—	17	65	0	0	0
1997-1998	—	—	45 from March 1998	0	0	0
1998-1999	—	2	45 from April 1999	0	0	0
2001-2002	—	—	25 percent	0		

Source: Government of Pakistan 2003.
Notes: [1]Items removed are a mixture of number and categories and, therefore, are only a broad indication.
[2]The concept of a restricted list was abolished in 1994-1995.
[3]Automobiles and alcohol continue to carry tariffs up to 425 percent. Restricted lists contain importable products by specified importers and industrial consumers only.
Sugar seeds, banana suckers, fungicides, herbicides, antisprouting products used for refurbished cylinders, lubricating oils and second-hand/used trawlers are added in the restricted list.

TABLE 17A.3. Nominal protection coefficients by crops

No.		Unit	1990-1991	1991-1992	1992-1993	1993-1994	1994-1995	1997-1998	1998-1999
Wheat									
1	Support price	Rs/tonne	2,800	3,100	3,250	4,000	4,000	6,000	6,000
2	IPP by agricultural prices commission	Rs/tonne	4,818	3,547	5,019	4,385	4,697	9,574	7,055
3	NPC (1/2)		0.58	0.87	0.65	0.91	0.85	0.63	0.85
Rice basmati									
4	Support price	Rs/tonne	7,075	7,700	8,500	9,000	9,720	77.50	
5	EPP by agricultural prices commission	Rs/tonne	11,988	8,796	10,833	8,171	9,743	89.68	
6	NPC (1/2)		0.59	0.88	0.78	1.10	1.00		
Rice coarse									
7	Support price	Rs/tonne	3,750	4,000	4,250	4,528	4,875	3,825	
8	EPP by agricultural price	Rs/tonne	3,983	3,503	4,498	3,709	2,960	3,867	
9	NPC (1/2)		0.94	1.14	0.94	1.22	1.65	0.99	
Seed cotton									
10	Support price	Rs/tonne	6,125	7,000	7,500	7,875	10,000	12,500	
11	EPP by agricultural price	Rs/tonne	10,362	10,996	9,935	9,486	12,408	21,471	
12	NPC (1/2)		0.59	0.64	0.75	0.83	0.81	0.582	
Sugarcane									
13	Support price	Rs/tonne	385	421	440	462	515	875	
14	EPP by agricultural price	Rs/tonne	577	581	481	458	N.A.	713	
15	NPC (1/2)		0.67	0.72	0.91	1.01	Not computed	1.23	

Source: Government of Pakistan 2000.
Note: IPP indicates import parity prices while EPP indicates export parity prices.

NOTES

1. See A. Krueger, M. Schiff, and A. Valdes, "Agricultural incentives in developing countries: Measuring the effect of sectoral and economy-wide policies." *World Bank Economic Review*, 2(3): 255-271, 1988.

2. Base-period AMS, as reported to WTO, was calculated in rupee terms. As from 1996/1997, it has been calculated in dollar terms, since it was claimed that the rupee-denominated AMS gave a distorted picture of domestic support. This change was the subject of considerable discussion in WTO (described as follows).

3. In rupee terms, the AMS fell less markedly, amounting to 25 percent and 54 percent for these two years, respectively.

4. FAO Document Repository 2004.

5. The average values for the two periods were about equal when exceptional years were excluded from the averages, namely 1994 for period 1 and 1996 for period 2.

6. Food excludes fishery products.

7. This sentiment has been expressed by Pakistani officials in many forums, within and outside the WTO. Similar frustration was expressed in Pakistan's statement to the Third WTO Ministerial Conference, at Seattle.

REFERENCES

FAO Document Repository 2004. Agriculture, trade, and food security issues and options in the WTO negotiations from the perspective of developing countries, 2000. Volume II, Country case studies, commodities and trade division, food and agriculture organization of the United Nations. Available online at <http://www.fao.org/documents/show_cdr/asp?url_file=/docrep/003/X8731e/x8731e11.htm.>.

Government of Pakistan 1989-2000. Central Board of Revenue, Ministry of Finance. Pakistan Custom Tariff. Islamabad: GOP.

Government of Pakistan 1989-2000. Gazette of Pakistan. Islamabad: GOP.

Government of Pakistan 1989-2000. Trade Policy. Islamabad: GOP.

Government of Pakistan 2000. Nominal Protection Coefficients by Corps. Agricultural Price Corporation. Islamabad: GOP.

Government of Pakistan 2002. WTO tariff bindings and applied rates for selected major products (percentage, ad valorem). Available online: <http://www.pakistan.gov.pk/fooddivision/informationservices/WTO/wto_tariff_bindings_applied_rates.pdf>.

Government of Pakistan 2003. Ministry of Commerce. Islamabad: GOP.

World Trade Organization (WTO) 1995. Schedule XV-Pakistan. G/SP/10.WTO.

Chapter 18

Food Security in Nepal

Suman Sharma

POVERTY AND ITS DETERMINANTS IN THE CONTEXT OF FOOD INSECURITY

Background

Poverty in Nepal is chronic and widespread, with roughly nine million people currently estimated to be living below the poverty line. However, the poverty incidence of roughly 38 percent, according to the latest government figures, does not convey the depth and severity of poverty in the country (NPC, 2001a). Large disparities in poverty exist across various regions (NESAC, 1998). Furthermore, based on a 1996 survey, poverty is much more severe in rural areas than in urban ones (NPC, 1998). When comparing poverty levels by ecological zone, there is not much variance between the hill and terai zones, but in the mountain zone poverty is much higher. The distribution of poverty across the five development regions indicates that households in the eastern and central development regions are less poor than those in the other regions (NPC, 1998). In summary, in Nepal poverty is much more prevalent and severe in rural areas, particularly in the western, mid-western, and far-western development zones. Because agriculture plays a central role in the lives of the rural population, agricultural production, land ownership, and land quality are the principal determinants of rural poverty (UNDP, 2002).

Low Growth Rate

Two of the characteristics of economic growth in Nepal that have a direct causal effect on the poverty situation are the low growth rate in general and the low agricultural growth rate in particular. During 1976/1977-1995/1996, the overall growth rate was about 4 percent, a rate that only marginally exceeded the population growth rate of 2.37 percent. During the same period, the growth rate in the agricultural sector was even smaller, at less than 2.5 percent, and has shown highly inconsistent behavior.

Due to the pattern of growth in total food grain production, Nepal has slowly but steadily changed from a net exporter to a net importer of grain. This can be seen in the data on the growth rate of the economy and in the nonagricultural and agricultural sectors. The data presented in Table 18.1 indicate that despite the growth of the economy at an annual rate of 4.09 percent between 1990 and 1998, growth varied widely between years. For example, annual growth in real GDP varied from a high of 7.90 percent in 1993/1994 to a low of 2.72 in 1997/1998. Although growth of nonagricultural GDP (NGDP) always remained positive, relatively steady and greater than the population growth rate, growth of agricultural GDP (AGDP) was negative in three out of eight years, had a higher variability, and was less than population growth in five out of the eight years.

Low Agricultural Productivity

In the context of the central role of agriculture in Nepal's economy, a major cause of rural poverty is low agricultural productivity. A few relevant indicators of the centrality of agriculture are: 86 percent of households cultivate some land, 80 percent have some livestock, and 83 percent of the labor force relies mainly on agriculture for employment (CBS, 1996, 1997). Growth of Nepal's economy is determined more by the growth of the agricultural sector since, despite its declining importance,[1] it is still the single largest sector of the economy and source of livelihood for the majority of the population.

TABLE 18.1. Real GDP growth

Year	Agricultural	Nonagricultural	Total
1990/1991	2.15	10.64	6.44
1991/1992	−1.06	9.76	4.62
1992/1993	−0.62	6.47	3.29
1993/1994	7.60	8.12	7.90
1994/1995	−0.33	5.29	2.87
1995/1996	4.42	6.62	5.70
1996/1997	4.13	3.82	3.94
1997/1998	1.04	3.89	2.72
1998/1999	2.37	4.08	3.39
Average	2.17	5.63	4.09

Source: GON 1999.

The data on major food crop production from 1985/1986 to 1998/1999 show that production stagnated, or only marginally increased. The exception was wheat, which had a modest gain (NPC, 2001a). For instance, during 1978/1979 to 1997/1998, the annual growth rates of production in three major food crops, paddy, maize, and wheat, were estimated to be 2.5, 4.0, and 5.9 percent, respectively. During the same period, the area under cultivation for these crops increased by 1.0, 3.8, and 4.0 percent, respectively. The gain in production was therefore due largely to an increase in the area under cultivation rather than an increase in productivity, which for the three crops was only 1.5 , 0.2, and 1.9 percent, respectively (WFP, 2001, p. 12).

The numbers presented in Table 18.2 depict Nepal's lack of progress in its agricultural sector, compared to its neighbors. In the table, three-year average yields of major crops for Nepal and for neighboring South Asian countries in the early 1960s and the late 1990s are compared. The data indicate that amongst the countries in South Asia during the early 1960s, crop yields were highest in Nepal. For instance, crop yields in Nepal were 198 percent higher than in India, 111 percent higher than in Bangladesh, 212 percent higher than in Pakistan, and 108 percent higher than in Sri Lanka. However, the situation in the late 1990s was just the reverse with Nepal having the lowest crop yields. During this period, crop yields in Nepal were

TABLE 18.2. Per hectare yield and growth rates of major crops in Nepal and other South Asian countries (1961-1999)

Yield Countries	1961-1963				1997-1999			
	Paddy	Wheat	Sugar	All	Paddy	Wheat	Sugar	All
Nepal yield (kg/ha)	1,940	1,230	1,979	1,854	2,410.00	1,630.00	3,597.00	2,940.00
Nepal yield as % of								
India	129	146	46	198	83.05	63.17	53.68	46.71
Bangladesh	116	198	53	111	85.81	74.43	84.92	87.05
Pakistan	140	150	61	212	84.38	75.78	74.68	46.32
Sri Lanka	101	—	119	108	74.29	—	66.02	64.91
1961-1963 to 1997-1999								
Growth Rates (%)								
Nepal	—	—	—	—	0.59	0.76	1.63	1.25
India	—	—	—	—	1.79	3.07	1.20	5.28
Bangladesh	—	—	—	—	1.41	3.46	0.34	1.92
Pakistan	—	—	—	—	1.97	2.64	1.07	5.50
Sri Lanka	—	—	—	—	1.43	—	3.26	2.66

Source: NLA, 2002.

only 46.7 percent of yields in India, 87 percent of Bangladesh's, 46.3 percent of Pakistan's, and 64.9 percent of Sri Lanka's. In terms of long-term growth rates in crop yields, Nepal lagged behind all of its neighbors. For the crops compared, yield in Nepal grew by about 1.25 percent per annum while growth rates in India, Bangladesh, Pakistan, and Sri Lanka grew 5.28 percent, 1.92 percent, 5.5 percent, and 2.7 percent, respectively.

Insufficient Access to and Poor Quality of Land

Insufficient access to land and poor land quality are the major factors behind low agricultural productivity. Small landholdings are a reality in Nepal given the high density of population in cultivated areas and the social custom of intergenerational land transfer.[2] Almost 70 percent of landholdings are less than one hectare in size while 40 percent are smaller than 0.5 hectare. Data indicate that the smaller landholdings are usually too small to produce enough grain for a household to survive through the year. Land productivity, the other key determinant, is considerably lower for the poor (UNDP, 2002). Among small farmers, land belonging to the relatively poorer households is about half as productive as the land belonging to the relatively nonpoor. Also, lands owned by the poor receive hardly any irrigation, very little chemical fertilizer, low amounts of improved seeds, low levels of technology, and insufficient levels of extension and support services, such as livestock insurance (NPC, 2001a). The distribution of wealth is also highly unequal. The bottom 20 percent of households receive only 3.7 percent of the total national income, while the top 10 percent receive close to 50 percent (Perry, 2000).

High Population Growth Rate and Adverse Effect on Environment

Another factor that has a direct bearing on the food security and poverty nexus is the high population growth rate and its detrimental effect on the environment. The high population growth rate has put tremendous pressure on the land.[3] This has at times led to measures such as the clearing of forested areas, intensified use of publicly owned land, and farming of marginal areas. Over the years, these activities have had a profoundly adverse effect on the environment which has in turn had a negative impact on the poor, as they usually rely most on "new" and publicly owned land. Cultivated land per capita decreased from 0.173 hectares in 1971 to 0.133 hectares in 1981, and to 0.126 hectares in 1991. This decline has adversely affected food security due to low land productivity.

FOOD SECURITY

Status and Trends

Poverty, low agricultural productivity, the lack of access to quality land, and the heavy dependence on agriculture have all contributed to and been perpetuated by food insecurity. Furthermore, lately the food insecurity problem has escalated in remote areas due to the unrest that the Maoist insurgency has caused.

A number of indicators substantiate Nepal's food insecurity problem in recent years (Perry, 2000). Until the 1970s, agricultural exports generated a significant amount of foreign exchange earnings. However, since the 1990s, the situation has reversed and deficits have been recorded in the agricultural trade balance. Lately, though the situation has improved, data indicate that nutritional deficiency affects almost half of all children. Since 1990, even though nutritional deficiency among children under five years of age appears to be slightly declining, undernutrition remains a severe problem (Mishra, 2001). Also, the scale of deficiency in micronutrients is widespread. Almost 40 percent of the population consumes less than the recommended 2,250 calories daily per person. Finally, based on recent government information, 45 districts out of the total 75 are classified as food deficit. During the past few years, severe food shortages have been reported in districts of the northwestern Karnali region, where food prices are much higher, normal marketing channels are nonexistent, and transportation is often extremely restricted. Finally, according to the 1997 Food Balance Sheet of FAO, which examines the food production estimates of about ten important crops, Nepal was in deficit by about 107,000 metric tonne (MT). However, some caution must be used in analyzing these data (Perry, 2000). FAO data also suggest that during the past forty years, the availability of cereals has been favorable. For instance, on average, the per person availability of cereals increased from 175.2 kilograms in 1961 to 187.8 kilograms in 1997. Furthermore, the Food Balance Sheet data do not consider production of other grains in addition to cereals, other crops besides the ten that are monitored, food produced off the farm, and serious distributional issues of food grain.

Food-Grain Trade, Production, and Availability

As mentioned earlier, Nepal was formerly a food surplus country and net food exporter until the 1970s. However, due to the failure of the food sector to keep pace with population growth, Nepal started experiencing a food deficit in the 1980s and is now a regular importer of foods. The data in Table 18.3 show a trade balance of grain in the 1990s according to which Nepal has been a net importer of cereals (food grains without pulses). Even when pulses are considered, the country has been a net importer of grain except for 1996/1997 and 1997/1998, when exports of pulses were high to offset the negative trade balance of cereals.

In Table 18.4 the food self-sufficiency data for cereals in the 1990s are presented. The food deficit, due to inadequate production, stood at about

TABLE 18.3. Trade balance of grain (1993-1998) (NRs. million)

Commodities	1993/1994	1994/1995	1995/1996	1996/1997	1997/1998	1998/1999
Import						
Rice	617.7	481.6	816.9	379.1	168.3	1,715
Wheat	4.3	9.7	4.6	4.2	2.5	0.2
Pulses	0.0	134.7	119	149.8	181.6	145.8
Total	622	626	940.5	533.1	352.4	1,861
Export						
Rice	0.0	0.0	0.0	0.1	8.0	74.1
Pulses	347.3	456.9	663.4	1,039	1,057.1	1,191.2
Maize	0.0	0.0	0.0	5.8	4.4	0.1
Total	347.3	456.9	663.4	1,044.9	1,069.5	1,265.4
Balance (Export-imports)						
W/O pulses	−622.0	−491.3	−821.5	−377.4	−158.4	−1,641
W pulses	−274.7	−169.1	−277.1	511.8	717.1	−595.6

Source: NLA, 2002.

TABLE 18.4. Food production and deficit data for cereals

	Food production (thousand MT)		Deficit	
Year	Production	Requirement	(thousand MT)	Deficit*
1991	3,373	3,562	189	5.31
1992	3,292	3,634	342	9.41
1993	3,585	3,724	139	3.73
1994	3,398	3,883	485	12.49
1995	3,917	3,948	31	0.79
1996	3,973	4,079	106	2.60
1997	4,027	4,178	151	3.61

Source: NLA, 2002.
*Deficit is defined as percentage of requirement.

189,000 MT in 1991/1992. This increased to about 485,000 MT in 1994/ 1995, with a small decline in 1993/1994. In 1995/1996, the food deficit declined sharply to about 31,000 MT. However, it increased and reached 151,000 MT in 1997/1998. These annual fluctuations are directly determined by variations in the production of paddy. Although the deficit in terms of the percentage of requirement never exceeded 12.5 percent at the macro level, several and rather high variations occurred across regions, including ecological belts. In the 1990s, the food deficit averaged 4.5 percent of consumption requirements. The food deficit situation improved in the late 1990s as the average was around 2 percent of the requirement during 1995/1997 compared with 8.5 percent during 1992/1994.

Irrespective of the changes at the national level, a food deficit analysis of ecological zones shows that the mountains and the hills have always experienced significant food shortages. For instance, during 1990/1991-1997/ 1998, food production in the mountains remained short of requirement to the extent of 37 to 45 percent. In the hills, during the same period, the food deficit remained between 15 and 20 percent of the requirement. The terai, on the other hand, has always had a surplus in food production which amounted to around 19 percent in 1990, 15 percent in 1995, and slightly over 17 percent in 1997 (NLA, 2002).

Government Food Aid, and Subsidy Policies and Programs

Experts have claimed that Nepal does not really have a "food aid policy." In other words, food security has never been a major issue in government policy (Mishra, 2001). Nevertheless, the government has been using the following targeted food assistance measures (Perry, 2000):

- a transport subsidy on food grain costs for people residing in remote areas;
- food-for-work programs to provide rural employment opportunities for the poor;
- midday meal programs in primary schools to increase enrollment, particularly of girls; and
- assistance in natural and other disasters.

Some subsidies are also available for agriculture and food distribution. A review of the programs shows that subsidies are limited to being allocated for: price and transport of chemical fertilizer, transport of food to remote and food-deficit districts, interest on credit, and capital for small and shallow tube-well irrigation, mini-micro hydel schemes, and bio-gas. The Agriculture Perspective Plan (APP) recommendation and the government decision to discontinue the subsidy on fertilizer are widely contested. Also, with the liberalization of the fertilizer business, the quality of chemical fertilizer supplied by the private sector is debated more than its price and availability.

As for the food aid programs, the government has been subsidizing food in targeted districts through the Nepal Food Corporation (NFC). However, given that in 1997 only 28 percent of its total sales were in remote areas, the effectiveness of the targeting program can be questioned. The fact that large numbers of public employees consume the food distributed in remote areas has made the NFC's performance even more suspect. In addition, the NFC had been in trouble for quite some time due to the immense politicization of its day-to-day functioning. In this regard, allowing privatization of the NFC is good for efficiency reasons. Some concerns, however, cannot be ignored, such as whether the private sector would indeed be motivated to provide food grains in remote districts. The NFC's activities are examined in more detail in the following subsection.

Other major food programs that are currently in operation are the Rural Community Infrastructure Works (RCIW), the Primary School Feeding Program, and relief and emergency operations. The RCIW provides seasonal employment to villagers for building sustainable infrastructure with special preference given to community programs that benefit and are managed by women. Every year, about 30,000 unskilled laborers participate in this project and are paid family food rations and cash for the work performed. RCIW has been favorably evaluated mainly because of the benefits derived from improvements in physical infrastructure and the provision of temporary food security to the targeted poor. Furthermore, unlike the earlier food-for-work programs, the RCIW has made attempts to ensure community participation not only in the identification of the type of infrastructure needed, but also in its management (Perry, 2000).

Similarly, the Primary School Feeding Program aims at encouraging enrollment and regular attendance, particularly of girls, by providing regular midday meals at government primary schools. The targeted districts under the program are food deficit, have high educational needs, and participate in the Basic Primary Education Project (BPEP). The program has been favorably evaluated due particularly to the improved attendance and lower repetition rates it has generated. The program has also lowered the opportunity cost of sending children to primary schools. In addition, under relief and emergency operations, the government, with the support of the World Food Program, has been providing basic and supplementary food rations to registered refugees in the country (Perry, 2000).

Activities of the Nepal Food Corporation

The Nepal Food Corporation (NFC) is a public sector institution involved in food-grain procurement and distribution in the country. Created in 1974 as a separate institution, its activities have been centered on the purchase of major food items (rice, wheat, maize, pulses, and edible oil) from surplus regions to fill supply gaps in the deficit regions. At present, NFC is also involved in maintaining a buffer stock, creating go-down (storage) facilities, and handling food subsidy and emergency food supplies. In the late 1990s, the government felt it necessary that NFC be made more efficient in handling grain by limiting its functions to the most essential activities, and concentrating its work in remote and inaccessible areas. This was mandated because the corporation was incurring losses on a continuous basis. Its accumulated loss amounted to NRs. 884 million until 1989/1990, which swelled to NRs. 904.7 million by 1995/1996 (APROSC, 1998). Accordingly, the government, while negotiating the Second Agricultural Program Loan (SAPL) with the Asian Development Bank (ADB), accepted a plan to reform NFC. Based on this, it was agreed that NFC would (1) withdraw from the subsidized distribution of food grains to urban and accessible areas, (2) approve a time-bound action plan for organizational reform to phase out subsidized grain distribution, and (3) limit activities of NFC to the delivery of grain to remote areas. The reform process is still underway.

Table 18.5 presents the grain distributed by the NFC in different areas of the country between 1994/1995 and 1998/1999. The corporation distributed 67,100 MT of grain in the country in 1994/1995. These amounts gradually declined in subsequent years to 25,000 MT in 1998/1999 even though one of the basic purposes for the establishment of the corporation was to provide food to remote and food-deficit districts and thereby help improve the food security situation in these areas. Its operations in different parts of

TABLE 18.5. NFC sales of grain by areas (thousand MT)

Areas	1994/ 1995	1995/ 1996	1996/ 1997	1997/ 1998	1998/ 1999
Most remote	9.7	8.0	8.6	10.0	9.9
Moderately remote	8.3	6.6	6.0	7.8	4.7
Accessible hills	42.9	26.8	14.5	20.1	9.7
Terai	6.2	9.4	1.5	10.0	0.8
Total	67.1	50.8	30.6	47.9	25.0

Source: Official records of NFC 2000.

the country suggest that the corporation has in fact been operating more in accessible areas (64 percent in 1993/1994 and 38.8 percent in 1998/1999). Furthermore, it has been charged that the major part of the food that was distributed to remote areas went to public employees (APROSC, 1998). Another interesting feature of NFC's operation is that the amount of food distributed in most food-deficit and moderately food-deficit areas (38 districts) in 1994 reduced the deficit by less than 4 percent. Thus, it is often argued that the NFC's operation in grain in remote areas served more of a political purpose than the improvement of the food security situation (ANZDEC, 2002).

Since the implementation of the reform plan in 1997, the proportion the NFC distributes to remote areas has increased. The proportion for the most remote areas increased from about 14.5 percent in 1994/1995 to about 39.6 percent in 1998/1999 and to moderately remote areas it increased from about 12.4 percent in 1994/1995 to about 18.8 percent in 1998/1999. Although distribution in the accessible hill areas declined from 63.9 percent in 1994/1995 to about 38.9 percent in 1998/1999, in the terai it declined from about 9 to about 3 percent in this period (NLA, 2002).

LIBERALIZATION POLICIES, THE AGRICULTURE PERSPECTIVE PLAN, AND THEIR IMPACT ON AGRICULTURE

Liberalization Policies

Liberalization policies were initiated in Nepal in the mid-1980s in view of the large and growing balance of payments and budgetary deficits. During the initial years, the major instruments used for the stabilization program were reduction of budgetary deficits through revenue mobilization, restrained domestic borrowings, and exchange rate adjustment. Prior to 1995/1996, the government had not directed liberalization policies to benefit poor and disadvantaged people. It was only in the 1995/1996 budget that the government considered for the first time channeling the gains of economic liberalization to this group. However, the major objective reflected in the budget was to achieve economic reform while maintaining macroeconomic stability (MOF, 1995). It was not until the mid-1990s that the government realized that economic liberalization would achieve the intended goal of maximizing the financial resources only when the majority of rural poor benefited.

The Agriculture Perspective Plan

As discussed earlier, agricultural productivity in Nepal has remained stagnant. Nepal has almost the lowest yield of basic food crops in South Asia compared to the early 1950s and 1960s when it had the highest yield levels. In response to this situation, the Agriculture Perspective Plan (APP), a plan with a 20-year time horizon, was developed in 1995. In fact, the agricultural strategy of the Ninth Plan (1997-2002) is based on the APP. The APP strategy aims to accelerate the agricultural growth rate to obtain strong multiplier effects on growth and employment, both in the agricultural and nonagricultural sectors. Besides the concern for growth, the APP envisages increasing agricultural growth in a regionally balanced manner with the objectives of promoting efficiency through specialization according to regional comparative advantage and ensuring that the poor of all the regions can benefit from the growth process. The strategy of the APP is to utilize technology and ensure an adequate supply of four critical inputs: shallow tube-well irrigation, fertilizer, agricultural roads, and research and extension of services (Chapagain, 2000).

Implicitly, the APP fully recognizes the importance of a liberal economic policy. It has also identified the government's important role in creating a conducive economic environment for farmers and businesspeople through a combination of appropriate policies, direct involvement in the promotion of technology generation and utilization, and creation of the critical physical and institutional infrastructure. For example, as a part of the liberalization process, fertilizer trade was deregulated and privatized through a gradual removal of subsidies. Another result of this policy was the rationalization of government support for farmer-managed irrigation systems and subsidies on shallow tube wells (Chapagain 2000).

The Impact on Agriculture

The major reform package that the government has adopted in the agriculture sector is specified in the Second Agricultural Program Loan (SAPL), which aims to promote agricultural productivity through addressing policy and institutional impediments. The major issues that the SAPL addressed included deregulation of the fertilizer subsector, organizational reform of the Agriculture Inputs Corporation, and the promotion of competitive agriculture produce markets by reform of the NFC. The liberalization policies and the APP had various impacts (ANZDEC, 2002):

- *Growth of agricultural GDP:* During the early 1990s, the per capita growth rate in agriculture was –0.5 percent and increased to an average growth rate of 0.7 percent in the late 1990s. In this decade, of all the main sectors, agriculture was the only one with a growth rate that improved.

- *Improved trade:* During the Ninth Plan period, agricultural trade as a percentage of agricultural GDP increased from an average of 9.1 percent in the early 1990s to 13 percent in the late 1990s. Furthermore, during the Plan period, the agricultural trade position with India improved from a deficit position equivalent to NRs. 1,849 million in 1995/1996 to a surplus position of NRs. 180 million in 1999/2000. It appears that the APP's emphasis on high-value products was responsible for this improvement.

- *Increased productivity:* Although the measurement of productivity presents several difficulties due to conceptual issues and data inavailability, it is heartening to note that agricultural productivity appears to have improved during the late 1990s compared to the early 1990s (ANZEDC, 2002). However, productivity is still low by international standards.

- *Better access to fertilizer market:* For more than two decades, the government has been responsible for importing fertilizer and distributing it at a subsidized rate. However, during this period supply usually fell short of demand. It was also revealed that fertilizer being supplied through the Agriculture Input Corporation (AIC) reached mostly large and better-off farmers. Around 1998, the private sector began to enter the fertilizer market and since then it appears that the supply of urea is readily available to farmers. One study has reported that as a result of liberalization in the fertilizer industry, an increasing rate of fertilizer use has been observed, and reductions have not occurred as a result of price increases (Upadhya, 2000). Moreover, in the fertilizer subsector, the market reform measures have improved the availability of fertilizer for all farmers, especially the poorest ones.

- *Organizational reform of NFC:* As part of the NFC Reform Action Plan some important reform measures have been undertaken. For example, a few offices have been closed and 305 temporary employees terminated. The NFC continues to supply subsidized grain to remote areas up to the amount of subsidy permissible (NRs. 225 million per year or equivalent to $3 million US). In order to avoid a high impact on local market prices, the NFC has adopted a policy of setting the final price of grain delivered under this program at 20 percent below the prevailing market price in these areas. Moreover, the NFC has been directed to maintain a food reserve of 40,000 MT. As for the maintenance cost of the reserve, the government subsidizes half of the cost while the NFC is required to bear the remainder.

MAJOR ISSUES IN THE CURRENT POLICIES AND PROGRAMS

In light of the previous discussion, the following major issues can be identified regarding the government's current food policies and programs:

Faulty Targeting

The NFC is naturally constrained in reaching the genuinely food-insecure households because of the enormous cost of transportation involved in its operations. As pointed out earlier, grain distributed by the NFC reportedly goes to public-sector employees. This is because the NFC's assistance is based on spatial targeting according to which everyone in a district classified as remote can access food. When food arrives at the district center, government employees generally receive the information before the needy people. Finally, the government's annual transport subsidy of $3 million US is insufficient to supply food quotas to all the remote districts. Thus improved targeting is needed.

In other food programs, such as the Rural Community Infrastructure Works (RCIW), a self-targeting mechanism targets districts based on their food-deficit status. Lately, the RCIW is focusing more on mid- and far-western districts and phasing out its operation in the terai districts. However, critics argue that much misappropriation of funds occurs in food accounting under the project (Perry, 2000). The annual report of the project once stated that even though the RCIW has a built-in self-targeting mechanism, some districts were included mainly due to political influence.

Cost Ineffectiveness

The entire food distribution program of the NFC is regarded as highly cost ineffective. As stated earlier, the annual government subsidy to the NFC is too inadequate to transport food to all remote districts. Even when compared to South Asian Association for Regional Cooperation (SAARC) countries, food programs are far more cost ineffective due mainly to expensive internal transport, and storage and handling costs. A recent study has estimated that the district-by-district internal transport, storage, and handling costs have varied from about US$7 per ton to US$1,176 (ANZDEC, 2002).

Other Key Issues Facing NFC

In nonremote areas, the NFC's distribution of grain represents a very small proportion of total grain sales and therefore has no significant impact on the market price for grain in these areas. For example, in the fiscal year 1999/2000, the NFC's food-grain procurement and sales in nonremote areas were 29,000 MT and 9,400 MT, respectively. Therefore, it is not clear why the NFC is engaged in grain distribution of such small quantities in those areas including 20 terai districts. But the grain sales in 12 districts classified as remote were sufficient to influence the local market's food-grain prices. This is beneficial for solving the food deficit in those areas, however, it is also likely to depress production in these remote areas. Furthermore, there appears to be no clear guidelines for the release of grains from the National Strategic Food Reserve (NSFR) and the monitoring of grain distribution (ANZDEC, 2002).

Fertilizer Use

During the first three years of the Ninth Plan (1997-2002), the fertilizer subsidy was eliminated in phases. The monopoly of the AIC was broken and the private sector became involved in both imports and the distribution of fertilizer in Nepal. Based on official figures on fertilizer imports, the private sector's share of total imports increased from 20 percent in 1997/1998 to 60 percent in 2000/2001. However, official figures show that the average distribution of fertilizer in the late 1990s was lower than that of the early 1990s, having declined from an average of 165,000 MT to 133,000 MT. This casts doubt on the effectiveness of the reform in the fertilizer sector. On the other hand, during the late 1990s, productivity of cereal crops increased. If fertilizer use has declined as implied by official figures, one would have expected a decline or at most a marginal yield improvement. But this did not happen. One important reason could be the heavy dependence of agriculture production on the monsoons.

The key problem appears to be the absence of official figures on fertilizer use in Nepal. The official figures show only imports and distribution from AIC and the registered private sector. The unofficial imports and distribution from the nonregistered private sector and unregulated inflow of fertilizers from India are not taken into account. For instance, it is estimated that official figures of fertilizer imports are lower than actual rates by about 20 percent due to unofficial fertilizer imports from India. Also, household fertilizer use is not currently monitored and made part of official statistics.

Due to fertilizer price differences between Nepal and India, substandard fertilizer is being brought into Nepal. Moreover, imports of low-quality fertilizer have increased in the post-liberalization period. Imports of substandard fertilizer have hampered the credibility of the private sector, just at the time when it is needed the most. Also, the complete removal of the urea subsidy in Nepal would make urea prices in Nepal 100 percent higher than those in Indian markets, thereby seriously hampering the intentions of those who expect to sell Indian fertilizer at cost. Without any incentive for the private sector to supply fertilizer to remote hilly districts serious doubt would be cast on its future role. Although fertilizer supply is good in the terai, leaving it entirely to the market has already encouraged imports of low-quality fertilizer. Thus, the government must take necessary actions for quality control. Moreover, data have confirmed that the fertilizer use rates in Nepal are the lowest in South Asia (ANZDEC, 2002). This could be an additional area that requires investigation.

The Impact on Household Income and Food Security

The food security programs of the government are designed around "pocket packages" rather than addressing the overall structural issues associated with food insecurity. For example, food insecurity, in addition to other factors,

> is also a result of unequal distribution of assets and wealth between communities and families, ecological belts and administrative regions. The government's food security efforts have not been effective in tackling these underlying issues and root causes . . . [the government's] targeted food subsidies have not made an impact on food security. (Perry, 2000, p. 40)

However, recent approaches of the WFP, similar to RCIW projects, have attempted to address issues such as community participation as food production and food security issues at the household and community levels are given greater consideration.

Impact of Cheap Rice from India

Because rice prices in the Indian bordering towns are cheaper than those in Nepal, lately an increasing amount of cheap rice is being imported. This problem has increased in magnitude in recent years and is often perceived as a great threat to farmers in Nepal. Based on Customs Department figures, in 1999/2000, Nepal imported 235,740 MT of rice at an average price of NRs. 11.09 per kilogram. Production of rice in the same time was 2.430

million MT. Thus the official figures indicate that 9.6 percent of total available production of rice was imported from India. If there was a bumper crop both in Nepal and India, such an unprecedented import of rice would depress prices in Nepal below the level they would have been without imports. Such a situation is not appealing to Nepali farmers. A recent study made comparisons of prices in the terai and in the bordering Indian markets during 1999/2000. The comparison shows that prices in the border towns in India were 12 percent lower than those in the terai, thereby suggesting strong incentives for exports from India to the terai (ANZDEC, 2002).

NOTES

1. The share of agriculture in GDP declined from about 47 percent in 1990/1991 to about 39.5 percent in 2000.

2. By tradition, land is passed on from parent (father) to son(s) in equal proportions if there is more than one son.

3. The latest estimate of population growth rate in Nepal is 2.27 percent per annum.

REFERENCES

ANZDEC (2002). Nepal Agriculture Sector Performance Review (ADB TA No. 3536-NEP), Draft Final Report, Volume 1: Main Text, ANZDEC Limited, New Zealand (prepared for Ministry of Agriculture and Cooperatives, Nepal and Asian Development Bank). Manila: ADB.

APROSC (1998). A Comprehensive Study on the Future Role of the Nepal Food Corporation. Kathmandu: Agricultural Projects Services Centre.

Asian Development Bank. 2002. Asian Development Outlook 2002. New York: Oxford University Press Inc.

CBS (1996 and 1997). Nepal Living Standards Survey Report—1996 and 1997 (Main Findings: Volume One and Two). Nepal: Central Bureau of Statistics, HMG/N.

Chapagain, D. (2000). An Overview of Liberalization and Its Effects on Agriculture and Poverty in Nepal. In *Do We Need Economic Reforms Phase II?* Kathmandu: Institute for Integrated Development Studies.

Government of Nepal, Ministry of Finance. 1999. *Economic Survey 1998/1999.* Kathmandu: GON.

Mishra (2001). Country Report Nepal (IDT/MDG Progress on International Millennium Declaration Development Goals). Submitted to the United Nations Development Programme, Nepal.

MOF (1995). Budget Speech of Fiscal Year 1995/96, Ministry of Finance. Kathmandu: HMG/N.

Nepal Food Corporation (NFC) (2000). Official Records of NFC 2000. Kathmandu: NFC.

NESAC (1998). *Nepal: Human Development Report—1998.* Kathmandu: Nepal South Asia Centre.

NLA (2002). Policy and Strategy for Poverty Alleviation and Sustainable Household Food Security for Nepal, draft version, National Labour Academy, Kathmandu, submitted to the Food and Agriculture Organization, Regional Office, Bangkok.

NPC (1998). *Ninth Plan (1997-2002).* National Planning Commission, His Majesty's Government of Nepal.

NPC (2001a). Interim Poverty Reduction Strategy Paper (I-PRSP), draft, National Planning Commission, His Majesty's Government of Nepal.

NPC (2001b). *Mid-Term Evaluation of the Ninth Plan (1997-2002)* (in Nepali). National Planning Commission, His Majesty's Government of Nepal.

Perry, S. (2000). Enabling Development: Food Assistance in Nepal, Final Draft, submitted to the World Food Program, Nepal.

UNDP (2002). Nepal Human Development Report 2001. Kathmandu: United Nations Development Programme.

Upadhya, S. (2000). Farmers in Flux: A Participatory Study of Fertilizer Use in the Context of Economic Liberalization, draft, study carried out as part of the ActionAid research on Globalization and Poverty.

WFP (2001). Nepal Food Security and Vulnerability Profile 2000. World Food Programme Nepal.

Chapter 19

Household Food
and Nutrition Security in India

Abusaleh Shariff

INTRODUCTION

Only four decades ago India was struggling to meet the food requirements of its 440 million people. Presently, with a population of one billion it
has surplus food stocks. Yet many millions in India still go hungry, and over
half the women and 60 percent of children in India are anemic and undernourished. This incongruence between adequate food supply at the macro
level and food insecurity and undernutrition at the micro level among a
large part of the population can be attributed to various supply-and-demand
constraints, including lack of purchasing power among poor people, and inefficiency in the delivery of public food programs. Few people outside the
government are aware that India has the world's largest public food and nutrition security program for all age groups. This program encompasses a
public distribution system, with an elaborate structure of production incentives, marketing support, buffer stocking and distribution, and nutrition
schemes, which the Integrated Child Development Services (ICDS) scheme
coordinates. Although focusing on young children, the ICDS also targets
expectant and lactating mothers and adolescent girls. The midday meal
(MDM) programs for schoolchildren, food grants to the destitute, food-for-
work (FFW) programs for the adult working population, and food grants to
those on old age pension are other important food subsidization programs.

About 70 percent of the population both in rural and urban areas is food
deficient because they have inadequate access to calories and micronutrients. More than half of the population seems to face chronic food insecurity

Acknowledgements: An earlier version of this chapter was presented in a workshop
on "Welfare, Demography and Development" organized by Cambridge University,
Cambridge, September 11-12, 2001. I thank P. K. Ghosh and S. K. Mondal for research
support.

in India. The deficits in calories, protein, fat, and iron are dangerously high within this group (Shariff and Mallick, 1999). The causes for micro- or household-level food insecurity have been due to either the lack of purchasing power, high food prices, food shortage, poorly developed markets, or some combination of these factors. Natural calamities, such as floods, cyclones, and drought, have also brought food insecurity. The first set of causes require a long-term strategy of employment generation for the poor, and making essential foods available through reasonable rates and credit provision. Natural disasters need to be handled through emergency aid, imports, and food-for-work schemes.

In spite of a decline in macro-level food insecurity, manifested in an increase in per-capita per-day availability of food grain, household food insecurity of the poor has remained essentially the same. Poverty still persists both in its absolute and relative dimensions, in spite of a steady decline in poverty ratios especially during the past two decades. Its alleviation continues to be a major challenge even fifty years after independence.

THE STRUCTURE OF FOOD AND NUTRITION INSECURITY

Food security encompasses more than mere availability of food. Other issues can be categorized into: (1) chronic food insecurity, (2) nutritional insecurity, (3) insecurity caused by lack of food absorption, and (4) transitory food insecurity.

Chronic Food Insecurity

Chronic food insecurity results when the energy derived from food is always less than the minimum prescribed amount. For India, the minimum daily calorie requirement is 2,400 kcal for rural areas and 2,100 kcal for urban ones. Both supply- and demand-side factors are involved. On the supply side the factors are food production, imports, distribution, and population growth. The demand-side factors are purchasing power, product and/or price subsidies, and public support in the form of programs, such as the ICDS, food for work, and others.

India's economy no longer experiences shortages. This is amply clear from the nearly 60 million tonnes (about 30 percent of annual production) of food-grain stock that the Food Corporation of India (FCI) has accumulated. However, substantial quantities of food grain are wasted due to poor conditions of storage, preservation, processing, and distribution. Furthermore, in the poorer and less favored areas, such as the eastern and central parts of India, production is often low. One way of ensuring adequate food security then is to augment production in these areas. This is one way to insulate the poor from the shocks of food shortages as prices rise and to provide relief beyond that which the targeted public distribution system offers. Policies and programs that deliver knowledge and inputs to farmers so that they may be motivated to cultivate pulses, vegetable oils, and other crops, which are rich in nutrient content, are imperative.

Demand for more nutritional, value-added, and processed foods is likely to increase, given the declining rate of population growth and increasing levels of disposable income. The demand for milk and meat will in turn require disproportionate amounts of cereal production. However, even if there is sufficient supply and availability, hunger will not be eradicated until people have enough purchasing power to buy food.

Nutritional Insecurity

Apart from calories, certain essential nutrients are required, and the need is different depending on age. Inadequate intake of nutrients, especially at the early stages of life, might lead to disease and disability for life and over several generations. Several factors influence nutritional intake: culture, diet preferences, nutritional knowledge and practices, and intrahousehold distribution. Among the chronically food insecure, some groups are especially at risk. The most vulnerable are pregnant and nursing mothers, unborn babies, and children under five. The level of malnutrition among young children in India is among the highest in the world. Low birth weight is the most important indicator of undernutrition both at the household and national levels. More than 60 percent of newborns in India are underweight. To combat this situation, the government of India has developed the ICDS program consisting of three components and covering some 45 million children and mothers. In this program, about 98 million schoolchildren get a daily snack or meal in school and roughly 200 million people are benefiting to some degree from subsidized food distribution.

Lack of Food Absorption

Food absorption is a major problem in developing countries. Dreze and Sen (1990) point out that many other factors affect nourishment such as health care availability, education, sanitary conditions, clean water, and eradication of infectious disease.

The problem can be understood as food absorption insecurity, a condition in which the body is not in a position to absorb the nutrients from the food taken due to the factors mentioned. Eliminating the problem of hunger, nutrition, and food insecurity thus depends on the whole development process. In remote rural areas, communities have little access to safe water, sanitation, and adequate health care. Worm infestations and malaria are common in India. In sickness, nutrient reserves in the body and the capacity to absorb new nutrients decrease rapidly. Deficiency in food can combine with inadequate hygiene or health care in a "malnutrition-infection nexus" and result in disability or death. In addition, the increasing use of chemicals in agriculture and food processing has added new concerns for health.

Transitory Food Insecurity

This refers to a temporary shortfall in food consumption for households that normally are able to meet their minimum dietary needs but are unable to cope with shocks such as natural disasters, food price hikes, and civil conflict. Severe and frequently recurring shocks can turn transitory food insecurity into chronic situations of poverty and undernutrition.

THE VULNERABLE GROUPS

Those who lack land and other assets, live in drought-prone areas, belong to tribal groups, and live in forested areas, belong to the scheduled castes, or live at high altitudes have higher degrees of food insecurity. Female-headed households are also highly vulnerable and food insecure. Among the poor, infants, children, the old, and women have higher levels of undernourishment, and the infirm, chronically sick, and disabled are others who have high degrees of food insecurity. Malnutrition persists among these groups, and the differences in severity across the states are only slight.

The vulnerable have been significantly affected by the rise in food prices. There was a more than normal increase in food prices especially in rice, wheat, and pulses during 1999 in India (FAOSTAT, 2004). Second, while the share of coarse cereals has declined substantially the price differentials between coarse cereals and wheat and rice have also declined. Assurances must be made that a certain proportion (about one-third) of the total food production is of indigenous varieties of rice and other traditional cereals; the prices of these cereals must be maintained at about half the cost of wheat.

In urban areas, slum dwellers, who constitute about 30 percent of the urban population, are highly vulnerable, and many workers in the *bidi* (Indian cigarette), incense stick, brick, construction, and garment industries, as well as rickshaw pullers, head loaders, and petty traders become food insecure due to fluctuations in demand for their labor (see Table 19.1). An increase in

TABLE 19.1. Crude estimates of the food insecure in India

	Households in millions
Total number of households	181.0
Rural Households	126.7
Food-insecure wage workers	44.3
Number of farming households	81.9
Marginal farmers	29.0
Small farmers	4.6
Artisans and small business	6.4
Rural food-insecure households	84.3
Urban Households	54.3
Food-insecure slum households	8.1
Food-insecure other urban areas	6.8
Urban food-insecure households	14.9
Total food-insecure households	99.2
Total food-insecure population	546 million people
Ultrapoor and vulnerable	59.5
Ultrapoor and vulnerable population	327 million people

rural to urban migration from the states of Bihar, eastern parts of Uttar Pradesh, Orissa, and Gujarat has also affected food insecurity. Often these migrants, who are in family units, are the worst affected. Finally, artisans who are in traditional occupations and who generate small incomes face food insecurity. Over 99 million households are food insecure in India. The ultrapoor and vulnerable are estimated to be about 60 million households or 327 million people.

Household and Gender Dimensions

Gender differentials with respect to a number of household-level parameters such as food and nutrition intake, and access to health care and primary education, are high in most South Asian economies. Women and girls in India, regardless of class, caste, religion, age, and region, have relatively lower access than men to food within the household. Even if they are the producers and managers of food products at home, women restrain themselves from consuming adequate food so as to enable the men and male children to consume more. Often their consumption is grossly inadequate. The situation is worse during pregnancy, childbirth, and lactation (Shariff, 1993). A number of cultural practices prevent pregnant and lactating women from eating decent and locally available cheap food. In fact, pregnant women are made to eat less than normal food intake during pregnancy for the fear that it may lead to abnormal fetal growth and cause a difficult delivery or death of the mother. Similarly, lactating mothers avoid a large number of food items, such as vegetables and fruits, because it is believed that they are harmful to infants.

Scrimshaw (1997) describes the consequences of malnutrition including inadequate iodine intake during pregnancy, iron deficiency anemia in infancy, and early protein-energy malnutrition in infancy and early childhood, which not only contribute to increased morbidity and mortality in childhood, but also result in stunting, and thus lower physical work capacity of adults.

The intensity of food insecurity at the household level and the incidence of stress caused by food shortages can be gauged by the prevalence of anemia among women and low birth weights (see Tables 19.2, 19.3, 19.4, and 19.5).

According to Dreze and Sen (1990), the incidence of anemia measured as below 11 grams per deciliter was about 50 percent in Andhra Pradesh, Tamil Nadu, and Haryana. As a result of the gender differentials, the life expectancies of women have been relatively low, until recently. These differentials are based on the disparity between women and men in terms of the command within the household over resources and commodities.

Dietary Practices

Food habits are regulated by a number of cultural practices. These practices not only determine whether a household is vegetarian or nonvegetarian, but also regulate diet according to class, caste, and religious identity. Often, locally available foodstuffs, such as coarse cereals and local varieties of vegetables and fruits, are the least preferred although they often have relatively higher nutritional value and micronutrients than nonlocal foods.

Patterns in Nutritional Intake

Calories

The nutritional pattern based on the national sample survey (NSS) consumer expenditure survey data for 1987/1988 and 1993/1994 suggests that calorie consumption was lower in both rural and urban areas compared to the recommended levels (Tables 19.6 and 19.7). The calorie deficits are

TABLE 19.2. Prevalence of anemia among ever married women, 1998-1999

State	% of ever married women with anemia	% of ever married women with		
		Mild anemia	Moderate anemia	Severe anemia
India	51.8	35	14.8	1.9
North				
Haryana	47.0	30.9	14.5	1.6
Himachal Pradesh	40.5	31.4	8.4	0.7
Punjab	41.4	28.4	12.3	0.7
Rajasthan	48.5	32.3	14.1	2.1
Central				
Madhya Pradesh	54.3	37.6	15.6	1
Uttar Pradesh	48.7	33.5	13.7	1.5
East				
Bihar	63.4	42.9	19	1.5
Orissa	63	45.1	16.4	1.6
West Bengal	62.7	45.3	15.9	1.5
Assam	69.7	43.2	25.6	0.9
West				
Gujarat	46.3	29.5	14.4	2.5
Maharashtra	48.5	31.5	14.1	2.9
South				
Andhra Pradesh	49.8	32.5	14.9	2.4
Karnataka	42.4	26.7	13.4	2.3
Kerala	22.7	19.5	2.7	0.5
Tamil Nadu	56.5	36.7	15.9	3.9

Source: National Family Health Survey (NFHS-2), 2000.
Note: The hemoglobin levels are adjusted for altitude of enumeration area and for smoking when calculating the degree of anemia hemoglobin.

higher among the bottom 20 percent and the next lowest quintile both in rural and urban areas. By 1993/1994, however, in rural areas, the calorie deficits among the bottom 20 percent increased from 34 to 44 percent below the recommended levels and for the next lowest quintile from 20 to 31 percent. In urban areas the respective deficits increased only marginally. Calorie intake deficiencies and excesses for rural and urban areas in the major states for the years 1971/1972 and 1993/1994 are given in Tables 19.8 and 19.9.

TABLE 19.3. Prevalence of anemia among children age 6 to 35 months, 1998-1999

State	% of children with anemia	% of children with		
		Mild anemia	Moderate anemia	Severe anemia
India	74.3	22.9	45.9	5.4
North				
Haryana	83.9	18.0	58.8	7.1
Himachal Pradesh	69.9	28.7	39.0	2.2
Punjab	80.0	17.4	56.7	5.9
Rajasthan	82.3	20.1	52.7	9.5
Central				
Madhya Pradesh	75.0	22.0	48.1	4.9
Uttar Pradesh	73.9	19.4	47.8	6.7
East				
Bihar	81.3	26.9	50.3	4.1
Orissa	72.3	26.2	43.2	2.9
West Bengal	78.3	26.9	46.3	5.2
Assam	63.2	31.0	32.2	0.0
West				
Gujarat	74.5	24.2	43.7	6.7
Maharashtra	76.0	24.1	47.4	4.4
South				
Andhra Pradesh	72.3	23.0	44.9	4.4
Karnataka	70.6	19.6	43.3	7.6
Kerala	43.9	24.4	18.9	0.5
Tamil Nadu	69.0	21.9	40.2	6.9

Source: National Family Health Survey (NFHS-2), 2000.
Note: The hemoglobin levels are adjusted for altitude of enumeration area and for smoking when calculating the degree of anemia hemoglobin.

Protein

The protein deficit among the bottom 20 percent of the rural population increased from 25 to 37 percent below the recommended level between 1971/1972 and 1993/1994. However, protein supply increased from 42 to 62 percent among the top 20 percent in rural areas. Deficiencies and excesses of proteins for major states in rural India for the years 1971/1972 and 1993/1994 are presented in Tables 19.10 and 19.11.

TABLE 19.4. Prevalence of malnutrition among children ages below three years by states, 1998-1999

State	Weight for age (% below)		Height for age (% below)		Weight for height (% below)	
	−3 SD	−2 SD	−3 SD	−2 SD*	−3 SD	−2 SD*
India	18.0	47.0	23.0	45.5	2.8	15.5
North						
Haryana	10.1	34.6	24.3	50.0	0.8	5.3
Himachal Pradesh	12.1	43.6	18.1	41.3	3.3	16.9
Jammu and Kashmir	8.3	34.5	17.3	38.8	1.2	11.8
Punjab	8.8	28.7	17.2	39.2	0.8	7.1
Rajasthan	20.8	50.6	29.0	52.0	1.9	11.7
Central						
Madhya Pradesh	24.3	55.1	28.3	51.0	4.3	19.8
Uttar Pradesh	21.9	51.7	31.0	55.5	2.1	11.1
East						
Bihar	25.5	54.4	33.6	53.7	5.5	21.0
Orissa	20.7	54.4	17.6	44.0	3.9	24.3
West Bengal	16.3	48.7	19.2	41.5	1.6	13.6
Assam	13.3	36.0	33.7	50.2	3.3	13.3
West						
Gujarat	16.2	45.1	23.3	43.6	2.4	16.2
Maharashtra	17.6	49.6	14.1	39.9	2.5	21.2
South						
Andhra Pradesh	10.3	37.7	14.2	38.6	1.6	9.1
Karnatka	16.5	43.9	15.9	36.6	3.9	20.0
Kerala	4.7	26.9	7.3	21.9	0.7	11.1
Tamil Nadu	10.6	36.7	12.0	29.4	3.8	19.9

Source: National Family Health Survey (NFHS-2), 2000.
Note: Each index is expressed in standard deviation units (SD) from the international reference population median.
*Includes children who are below −3 SD from the international reference population median.

TABLE 19.5. Prevalence of malnutrition among children age 1 to 4 years according to place of residence in India and selected states, 1992-1993

States	Underweight (weight-for-age below) 2SD of median			Stunted (weight-for-age below) 2SD of median			Wasted (weight-for-age below) 2SD of median		
	Rural	Urban	Total	Rural	Urban	Total	Rural	Urban	Total
India	59.9	45.2	53.1	54.1	14.8	52.0	18.0	15.8	17.5
Andhra Pradesh	52.1	40.0	49.1	—	—	—	—	—	—
Assam	51.8	37.3	50.4	53.5	39.6	52.5	11.4	5.6	10.8
Bihar	64.1	53.8	62.6	61.8	55.2	60.9	22.7	16.3	21.8
Gujarat	45.8	40.5	44.1	44.6	41.6	43.6	20.3	16.1	18.9
Haryana	39.4	33.0	37.9	48.0	42.4	46.7	5.7	6.4	5.9
Himachal Prade	48.3	30.2	47.0	—	—	—	—	—	—
Jammu & Kashmir	—	—	44.5	—	—	40.8	—	—	14.8
Karnataka	—	—	54.3	—	—	47.6	—	—	17.4
Kerala	30.6	22.9	28.5	29.6	21.5	27.4	11.5	12.0	11.6
Madhya Pradesh	59.4	50.1	57.4	—	—	—	—	—	—
Maharashtra	57.5	45.5	52.6	50.8	39.1	46.0	21.5	18.3	20.2
Orissa	—	—	53.3	—	—	48.2	—	—	21.3
Punjab	47.4	40.0	45.9	40.4	38.4	40.0	21.4	14.3	19.9
Rajasthan	41.1	43.9	41.6	43.0	43.5	43.1	17.7	29.1	19.5
Tamil Nadu	52.1	37.3	46.6	—	—	—	—	—	—
Uttar Pradesh	—	—	49.8	—	—	49.2	—	—	16.2
West Bengal	—	—	56.8	—	—	43.2	—	—	11.9

Source: Measham and Chatterjee, 1999.

Fat

Overall deficits in fat consumption continue to be very high at 16 percent in rural areas. In urban areas, the deficit is marginal at 3 percent below the recommended level. The bottom 20 percent in rural areas experienced a deficit of 26 percent below the prescribed level and in urban areas, a deficit of 22 percent. The top 20 percent however, enjoyed an excess of 20 percent in rural areas and 22 percent in urban India.

Iron

Iron deficiencies were 3.6 and 14.3 percent below the prescribed level for the total population in rural and urban areas, respectively. The bottom

TABLE 19.6. Per capita per day consumption of calories, protein, fat, carbohydrates, iron, and calcium in India

Population	Calories (2,400 k cal)		Protein (60 grams)		Fat (40 grams)	Carbohydrate (399 grams)	Iron (28 mg)	Calcium (400 mg)
	1987-1988	1993-1994	1987-1988	1993-1994	1993-1994	1993-1994	1993-1994	1993-1994
Rural India								
Below 20	1,577	1,351	45	38	14	269	20.2	326.8
Next 20	1,930	1,661	54	46	19	327	23.9	463.5
Next 40	2,263	2,291	64	64	32	437	28.0	689.6
Top 20	2,957	3.325	85	97	61	598	35.5	1,119.1
All	2,146	2,100	61	59	24	399	27.0	674.0
Urban India								
Below 20	1,498	1,483	42	41	18	287	18.5	340.4
Next 20	1,751	1,736	49	48	27	325	21.6	503.0
Next 40	2,047	1,992	58	56	40	351	24.5	762.1
Top 20	2,510	2,392	73	69	62	386	29.5	1,292.0
All	1,889	1,924	54	54	37	341	24.0	735.0

Source: Computed from Shariff and Mallick, 1999.

TABLE 19.7. Percentage deviation from the recommended level in the per capita per day consumption of calories and nutrients in India

Population	Calories (2,400 k cal)		Protein (60 grams)		Fat (40 grams)	Carbohydrate (399 grams)	Iron (28 mg)	Calcium (400 mg)
	1987-1988	1993-1994	1987-1988	1993-1994	1993-1994	1993-1994	1993-1994	1993-1994
Rural India								
Below 20	−34.29	−43.71	−25.00	−36.67	−26.0	−32.58	−27.86	−18.30
Next 20	−19.58	−30.79	−10.00	−23.33	−21.0	−18.05	−14.64	15.75
Next 40	−5.70	−4.54	6.67	6.67	−8.0	9.52	0.00	72.25
Top 20	23.21	38.54	41.67	61.67	20.0	49.87	26.78	194.92
All	−10.58	−12.50	1.67	1.67	−16.0	0.00	−3.57	68.50
Urban India								
Below 20	−28.67	−29.38	−30.00	−31.67	−22.0	−15.84	−34.11	−14.90
Next 20	−16.62	−17.33	−18.33	−20.00	−13.0	−4.69	−23.03	25.74
Next 40	−2.52	−5.14	−3.33	−6.67	0.0	2.93	−12.50	90.52
Top 20	19.52	13.90	21.67	13.04	22.0	11.66	5.25	223.00
All	−10.05	−8.38	−10.00	−10.00	−3.0	0.00	14.28	83.75

Source: Computed from Shariff and Mallick, 1999.

TABLE 19.8. Per consumer unit deficiencies and excesses of calories in rural India, 1971-1972 and 1993-1994

States	Below 20	20-40	40-60	60-80	Top 20	Below 20 as % to top 20	Deviation from national mean
Andhra Pradesh							
1971-1972	−41	−18	2	30	82	33	−58
1993-1994	−30	15	−8	3	21	58	−124
Assam							
1971-1972	−57	−29	-8	19	47	29	−59
1993-1994	−37	−22	−11	−2	15	55	−277
Bihar							
1971-1972	−41	−15	12	49	169	22	8
1993-1994	−24	−2	7	18	41	54	−46
Gujarat							
1971-1972	−47	−16	−1	31	78	30	98
1993-1994	−39	−24	−17	0	19	52	−213
Haryana							
1971-1972	−25	−6	14	50	115	35	928
1993-1994	−32	−16	−8	9	41	48	426
Himachal Pradesh							
1971-1972	—	−21	9	36	91	—	466
1993-1994	−24	−14	−3	8	30	58	233
Jammu and Kashmir							
1971-1972	−21	3	22	44	99	40	766
1993-1994	−9	1	8	10	34	68	471
Karnataka							
1971-1972	−41	−16	6	32	102	29	115
1993-1994	−30	−15	−5	9	25	56	−108
Kerala							
1971-1972	−61	−42	−23	2	45	27	−701
1993-1994	−50	−35	−25	−14	10	45	−232
Madhya Pradesh							
1971-1972	−18	9	31	75	183	29	−463
1993-1994	−22	−7	3	14	35	57	14
Maharashtra							
1971-1972	−44	−22	−5	19	80	31	−157
1993-1994	−30	−17	−11	−3	20	58	−256
Orissa							
1971-1972	−39	−15	5	37	161	24	−191
1993-1994	−17	1	11	18	37	60	57
Punjab							
1971-1972	−29	5	1	33	107	34	987
1993-1994	−37	−22	−15	1	26	50	324
Rajasthan							
1971-1972	−17	−2	−37	60	128	36	489
1993-1994	−21	−7	4	19	42	56	407

States	Below 20	20-40	40-60	60-80	Top 20	Below 20 as % to top 20	Deviation from national mean
Tamil Nadu							
1971-1972	−48	−27	−7	43	64	31	−330
1993-1994	−39	−24	−15	−4	16	53	−336
Uttar Pradesh							
1971-1972	−28	1	22	52	77	41	474
1993-1994	−18	−4	6	18	41	58	216
West Bengal							
1971-1972	−49	−26	−8	16	37	37	−413
1993-1994	−26	−11	0	10	32	56	50
All India							
1971-1972	−37	−13	8	38	97	32	0
1993-1994	−25	−11	−2	8	29	58	0

Source: Computed from data in National Sample Survey Organisation, Department of Statistics. Government of India 1996.

TABLE 19.9. Per consumer unit deficiencies and excesses of calories in urban India, 1971-1972 and1993-1994

States	Below 20	20-40	40-60	60-80	Top 20	Below 20 as % to top 20	Deviation from national mean
Andhra Pradesh							
1971-1972	−47	−29	−9	8	45	37	−82
1993-1994	−21	−9	1	13	29	61	−87
Assam							
1971-1972	−56	−37	−17	−1	19	37	−83
1993-1994	−26	−12	−3	3	28	57	1
Bihar							
1971-1972	−48	−27	−10	12	29	41	64
1993-1994	−18	−5	5	16	36	61	125
Gujarat							
1971-1972	−42	−33	−16	9	49	39	−74
1993-1994	−31	−16	−3	8	29	54	−51
Haryana							
1971-1972	−48	−31	−14	8	38	38	90
1993-1994	−29	−13	−4	6	18	60	74
Himachal Pradesh							
1971-1972	—	—	−9	−2	34	—	−376
1993-1994	−22	−13	−9	0	10	71	372

TABLE 19.9 (*continued*)

States	Below 20	20-40	40-60	60-80	Top 20	Below 20 as % to top 20	Deviation from national mean
Jammu and Kashmir							
1971-1972	−35	−19	−5	12	32	50	73
1993-1994	−43	−16	−9	3	16	49	408
Karnataka							
1971-1972	−32	−25	−7	12	36	50	−327
1993-1994	−25	−12	1	12	29	58	−57
Kerala							
1971-1972	−52	−32	−10	19	66	29	−596
1993-1994	−35	−17	−6	8	35	48	−97
Madhya Pradesh							
1971-1972	−48	−27	−12	5	46	36	151
1993-1994	−20	−8	3	13	25	64	14
Maharashtra							
1971-1972	−39	−25	−14	1	28	48	−197
1993-1994	−21	−12	−6	1	21	65	−110
Orissa							
1971-1972	−48	−27	−12	9	26	41	38
1993-1994	−19	−6	1	10	22	66	212
Punjab							
1971-1972	−53	−38	−24	−2	31	36	107
1993-1994	−31	−19	−9	3	23	56	27
Rajasthan							
1971-1972	−39	−16	−14	10	28	47	307
1993-1994	−24	−9	2	8	21	63	162
Tamil Nadu							
1971-1972	−46	−25	−10	15	61	33	−460
1993-1994	−30	−12	1	12	37	51	−176
Uttar Pradesh							
1971-1972	−36	−17	−3	10	36	47	−83
1993-1994	−18	−6	0	10	34	61	73
West Bengal							
1971-1972	−52	−28	−12	4	27	38	−268
1993-1994	−23	−11	−2	6	23	63	45
All India							
1971-1972	−48	−30	−17	2	29	40	0
1993-1994	−26	−11	−2	7	26	59	0

Source: Computed from data in National Sample Survey Organisation, Department of Statistics, Government of India 1996.

TABLE 19.10. Per consumer unit deficiencies and excesses of protein in rural India, 1971-1972 and 1993-1994

States	Below 20	20-40	40-60	60-80	Top 20	Below 20 as % to top 20	Deviation from national mean
Andhra Pradesh							
1971-1972	−43	−18	0	27	73	33	−3
1993-1994	−38	−26	−19	−9	9	57	−12
Assam							
1971-1972	−63	−35	−15	11	43	26	−7
1993-1994	−43	−31	−21	−12	6	54	−15
Bihar							
1971-1972	−37	−9	22	64	170	23	7
1993-1994	−22	0	9	20	44	54	0
Gujarat							
1971-1972	−42	−9	4	32	82	32	6
1993-1994	−38	−22	−16	−1	16	54	−6
Haryana							
1971-1972	−15	7	26	70	141	35	39
1993-1994	−23	−5	3	24	59	48	23
Himachal Pradesh							
1971-1972	—	−11	22	49	95	—	22
1993-1994	−16	−6	5	17	41	60	14
Jammu and Kashmir							
1971-1972	−11	17	27	41	104	44	24
1993-1994	4	11	18	18	42	73	20
Karnataka							
1971-1972	−41	−16	3	44	83	32	1
1993-1994	−32	−19	−10	2	18	58	−7
Kerala							
1971-1972	−65	−49	−31	−11	39	25	−26
1993-1994	−57	−43	−34	−23	5	41	−12
Madhya Pradesh							
1971-1972	6	36	68	138	302	26	53
1993-1994	−20	−6	7	20	45	55	4
Maharashtra							
1971-1972	−26	−15	5	28	86	40	2
1993-1994	−27	−16	−10	−3	19	62	−6
Orissa							
1971-1972	−43	−23	0	26	146	23	−10
1993-1994	−30	−15	−5	2	22	57	−9
Punjab							
1971-1972	−12	14	10	42	115	41	35
1993-1994	−31	−15	−8	11	40	49	18
Rajasthan							
1971-1972	−3	16	42	84	128	42	26
1993-1994	−10	7	20	38	62	56	24

TABLE 19.10 *(continued)*

States	Below 20	20-40	40-60	60-80	Top 20	Below 20 as % to top 20	Deviation from national mean
Tamil Nadu							
1971-1972	−47	−31	−13	59	46	36	−10
1993-1994	−47	−34	−25	−13	6	50	−17
Uttar Pradesh							
1971-1972	−16	18	41	68	99	42	26
1993-1994	−9	5	15	29	53	59	14
West Bengal							
1971-1972	−47	−28	−11	11	31	40	−13
1993-1994	−36	−23	−12	−2	21	53	−7
All India							
1971-1972	−30	−7	15	49	111	33	0
1993-1994	−25	−12	−3	8	31	57	0

Source: Computed from data in National Sample Survey Organisation, Department of Statistics, Government of India 1996.

TABLE 19.11. Per consumer unit deficiencies and excesses of protein in urban India, 1971-1972 and1993-1994

States	Below 20	20-40	40-60	60-80	Top 20	Below 20 as % to top 20	Deviation from national mean
Andhra Pradesh							
1971-1972	−56	−41	−25	−9	25	36	−13
1993-1994	−34	−22	−12	−1	19	56	−9
Assam							
1971-1972	−65	−49	−32	−13	8	32	−10
1993-1994	−35	−22	−12	−5	24	52	−5
Bihar							
1971-1972	−47	−23	−7	15	30	41	2
1993-1994	−11	2	12	23	44	62	5
Gujarat							
1971-1972	−47	−36	−19	1	27	42	−5
1993-1994	−31	−16	−5	2	19	58	−3
Haryana							
1971-1972	−62	−23	−4	18	47	26	8
1993-1994	−19	−2	6	17	29	63	8
Himachal Pradesh							
1971-1972	−100	−100	−22	−19	12	—	1
1993-1994	1	11	11	22	28	79	14

States	Below 20	20-40	40-60	60-80	Top 20	Below 20 as % to top 20	Deviation from national mean
Jammu and Kashmir							
1971-1972	−36	−22	−12	1	23	52	−5
1993-1994	−28	4	9	26	41	51	15
Karnataka							
1971-1972	−41	−40	−27	−10	36	44	−14
1993-1994	−29	−18	−6	3	20	59	−5
Kerala							
1971-1972	−69	−56	−40	−16	22	26	−25
1993-1994	−43	−25	−14	2	30	44	−5
Madhya Pradesh							
1971-1972	−40	−18	−4	10	34	45	3
1993-1994	−13	−1	8	15	28	68	4
Maharashtra							
1971-1972	−40	−27	−17	−9	22	50	−4
1993-1994	−21	−14	−9	−2	15	69	−2
Orissa							
1971-1972	−54	−37	−22	−2	17	39	−8
1993-1994	−24	−9	1	13	28	60	0
Punjab							
1971-1972	−46	−30	−17	4	33	41	4
1993-1994	−24	−11	−1	12	33	57	6
Rajasthan							
1971-1972	−25	−4	2	26	35	56	11
1993-1994	−5	10	21	26	34	70	12
Tamil Nadu							
1971-1972	−63	−48	−37	−18	22	30	−23
1993-1994	−42	−26	−14	−3	21	48	−10
Uttar Pradesh							
1971-1972	−30	−10	4	13	31	54	3
1993-1994	−4	8	11	20	44	66	8
West Bengal							
1971-1972	−61	−36	−25	−9	10	35	−11
1993-1994	−28	−14	−4	5	24	58	−1
All India							
1971-1972	−49	−31	−19	−2	24	41	0
1993-1994	−24	−10	−2	7	26	60	0

Source: Computed from data in National Sample Survey Organisation, Department of Statistics, Government of India 1996.

20 percent of the population has recorded a very high level of iron deficiency, at 27.9 percent below the recommended level in rural areas and 34 percent below in urban areas. Only the top 20 percent in rural and urban parts enjoyed an excess of 27 percent and 5 percent, respectively.

Coping Strategies

The coping strategies to mitigate food shortages or the risk of them are extensive and complex, and differ substantially between rural and urban areas and within rural areas among agroclimatic regions and between different population groups. The strategy may be as simple as borrowing grain from kin or neighbors, or it may involve the more serious step of withdrawing children from school and making them work long hours to increase household income. When a household experiences a food shortage, it probably means that there are low or no cash reserves. This in turn implies that the household, especially its women and children, are denied health care. Families who own properties may resort to liquidating utility assets and then productive assets as well. Often, households sell their highly valued livestock to ward off temporary shocks and food shortages. Large-scale indebtedness and sale or mortgaging of agricultural lands and property are the extreme response to such a crisis next only to permanent migration to other parts of the country in search of employment.

Food and Nutrition Programs

Figure 19.1 illustrates the policy efforts undertaken in India since independence for achieving food security. To reduce malnutrition among the rural poor the government launched the *Antyodaya Anna Yojana* program in December 2000. This program consisted of identifying 10 million poor families and providing them with 25 kg of food grain per month at a price below market value of Rs. 2 per kg for wheat and Rs. 3 per kg for rice. The Department of Rural Development launched another program, *Annapoorna Anna Yojana,* in 2000-2001. In this program, 10 kg of food grain per month are given free to those who, although eligible for old-age pensions, remain uncovered under the National Old Age Pension Schemes (NOAPS). However, only 70 percent of the grain allocated for this purpose (1.1 million tonnes out of 1.6 million tonnes) has been procured by the states, according to the Union Minister of Consumer Affairs and Public Distribution System (PDS). In the union budget 2001-2002, the program was modified to extend the coverage to those persons who are covered under NAOPS (approximately 6.9 million) beyond the 1.4 million initially targeted under the *Annapoorna* scheme.

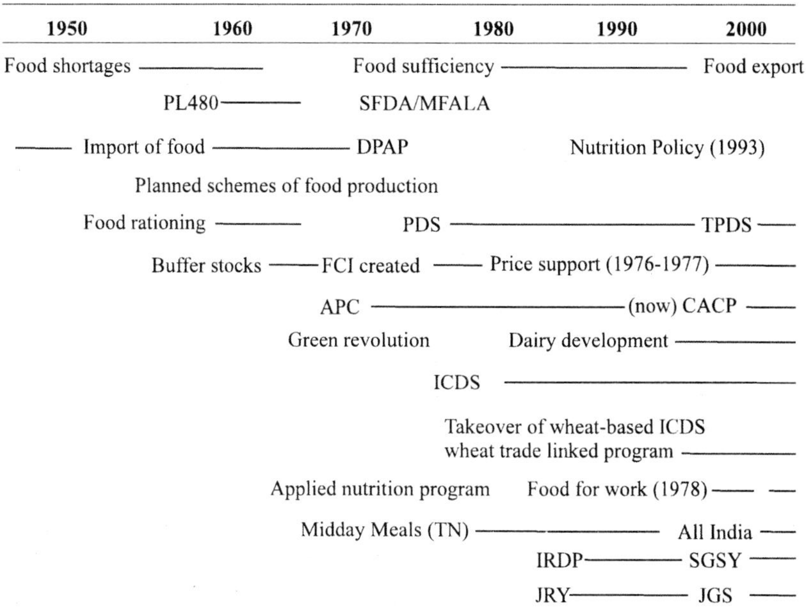

1950	1960	1970	1980	1990	2000

Food shortages ———————— Food sufficiency ————————— Food export

PL480——— SFDA/MFALA

——— Import of food ————————— DPAP Nutrition Policy (1993)

Planned schemes of food production

Food rationing ———— PDS ————————————— TPDS ——

Buffer stocks ——— FCI created ——— Price support (1976-1977) ————

APC ————————————— (now) CACP ——

Green revolution Dairy development —————

ICDS —————————————

Takeover of wheat-based ICDS
wheat trade linked program —————

Applied nutrition program Food for work (1978) ——— —

Midday Meals (TN) ————————— All India ——

IRDP————— SGSY ——

JRY————— JGS ——

PL480	Public Law (of the United States) number 480
SFDA	Small Farmers Development Agency
MFALA	Marginal Farmers ALA
DPAP	Drought-Prone Area Program
PDS	Public Distribution (of food) System
TPDS	Targeted Public Distribution (of food) System
FCI	Food Corporation of India
APC	Agricultural Prices Commission
CACP	Commission for Agricultural Costs and Prices

FIGURE 19.1. A time line of major policy initiatives for food security in India.

CONCLUSION

- Nutrition supplementation, midday meals, ICDS, dietary supply in hospitals, vitamin A supplementation, salt iodination, and iron and folic acid prophylaxis are only some of the important programs that have been put into place for a food security strategy. Market intervention strategies to contain prices and arrest the reduction in food supplies, which result from a lack of demand, are also needed. Understanding the local grain market and distribution networks, the local futures market, agricultural credit, and other facets is necessary. Although strategies to augment income levels are not adequate to eliminate undernutrition, employment generation as a tool needs to be strengthened. Finally, evaluating income transfer, subsidized food, and nutrition programs is important to improve their efficiency and targeting.

- The sustainability of agricultural practices is a crucial issue. Increased food production in the future must come not from area expansion, but from enhanced productivity through improved and sustainable agricultural practices. Innovative sets of practices are needed that include traditional skills, novel cropping systems, precision farming methods, and advances in science and technology. The ongoing National Watershed Development Program in Rainfed Areas, has an integrated strategy for the optimum utilization of water resources for crop production in poorly endowed areas. The program is based on community participation and "ownership." In view of the fact that extensive areas of the country are becoming prone to soil degradation, community-based programs for conservation and enhancement of land-water resources need to be developed through microstudies and microplanning.

- Recognizing the importance of regionally differentiated strategies, the recently introduced Macro Management Scheme is designed to assist states in identifying and implementing area-specific interventions. The government of India is planning to launch a program for harnessing the ample groundwater resources of the eastern region. This program will assist in utilizing the enormous untapped potential of this region, which is expected to result in additional food grain production of about 4 million tonnes per year. Among the various strategies, diversification toward horticulture ought to be promoted. Not only does horticulture improve the productivity of land in terms of biomass production and returns per unit area, but it also provides increased employment opportunities in the production of eco-foods, biofertilizers, biopesticides, bioprocessing, health foods, and herbal medicines.

- The livestock and fisheries sectors provide other avenues for supplementing farmers' incomes and generating gainful employment in the rural sector particularly among landless, small and marginal farmers, and women. India has already become the world's largest milk-producing country and reached 81 million tonnes in production in 2002. India is now the sixth largest producer of fish in the world and second largest producer of freshwater fish after China. Animal husbandry and fisheries are likely to assume increasing importance in the future. Already this sector accounts for 25 percent of agricultural GDP and its growth far outstrips that of agriculture. Furthermore, the returns from this sector are extremely rewarding for small and marginal households. A major inroad can be made on rural poverty through strengthening this sector. However, the present availability of animal protein in the Indian diet needs to be increased, for those who consume animal meat, at least twofold from the present level of 10 grams, with special emphasis on maintaining proper levels for growing children and nursing mothers.

- Unfortunately, about 10 percent of food grain, 30 to 40 percent of horticultural produce, and around 10 to 12 percent of livestock and fisheries stock gets spoiled or damaged due to inadequate postharvest and processing facilities. India can ill afford to incur such huge losses. The processing of agricultural produce in the country in comparison with other developing countries needs to be developed.

- Agriculture is capital starved. The share of agriculture in gross domestic capital formation (GDCF) has been steadily declining. Agriculture accounts for only 25 percent currently, compared to 35 percent in 1980-1981. Considering that about two-thirds of the population in India depends on agriculture for its livelihood, investment in agriculture needs to be stepped up on a massive scale. The areas of investment that have immediate impact on agricultural production and development are irrigation, technology, infrastructure, and education. The private sector too should be allowed to invest in agriculture.

- The liberalization of world trade in agriculture has opened up new vistas of growth. India has a competitive advantage in agricultural exports because of its near self-sufficiency in terms of inputs, relatively low labor costs, and diverse agroclimatic conditions. The government has to strike a balance between the welfare of farmers and that of consumers in pursuing the opportunities that global trade offers.

- The high cost of food stock management is a major problem. An urgent need exists to ensure that the central government's annual food stock subsidy, which has gone up five times in a decade, is better targeted. One way to reduce costs is to decentralize the buying and distribution of food. Instead of having one agency, the FCI, buy food at one price and then distribute it to the states, giving individual states the ability to purchase and distribute food would be preferable. The Government of India has a proposal for such a scheme in the future. This measure will encourage state governments to take creative steps to buy the cheapest food they can get from the nearest sources and distribute it in the most efficient way possible. This will ensure that the central government subsidy can benefit more poor people than is the case currently.

- The PDS has not lived up to the people's expectations especially in the poorer north and northeastern states. The limited offtake in these states, where the majority of poor live, points to serious deficiencies in the administrative capability of the system. Even though the penetration of PDS is very high in rural India (92 percent), the actual use of PDS seems to be very low, only 35 percent (NCAER, 1999). Indications show that use of PDS varies by religious affiliation and education level. The relationships must be better understood to ensure that subsidized grain is reaching the most needy.

- The implementation of free food provided to schoolchildren under the midday meal scheme leaves much to be desired. The panchayati raj (village government) institutions and other citizen's organizations have to play an active role in making these schemes successful. The use of surplus food stocks in innovative ways for promoting female literacy programs and school attendance should be explored.

- The World Food Programme is piloting a program that focuses on women as the key agents of change. Women's educational levels, nutritional status, and control over food resources and assets determine household food security. Consequently, the problem of food insecurity can be addressed from various angles. Three different food-based interventions—supplementary feeding, food-for-work, and incentive schemes for girls' school attendance—are all being promoted in carefully identified and targeted communities. Supplementary feeding ensures that small children receive nutritious food and alleviates the burden on mothers to obtain food. Food-for-school attendance acts as an incentive to send girls to school. Food-for-work offers employment to women as well as to men within the community and helps to check migration to already overcrowded urban areas. Moreover, assets created through community work such as irrigation channels, wells, or protected watersheds improve the livelihood of not only the family, but of the entire community as well.

REFERENCES

Dreze Jean and Amartya Sen. 1990. *The Political Economy of Hunger, Selected Essays.* Oxford: Clarendon Press.

Food and Agricultural Organization. (FAO). 2004. FAOSTAT Online; FAO statistical database. Available online at <http://apps.fao.org>.

Institute for Population Sciences. 2000. National Family Health Survey (NFHS-2). India 1998-1999. Mubai: Author.

Measham, Anthony R. and Meera Chatterjee. 1999. *Wasting Away: The Crisis of Malnutrition in India.* Washington, DC: The World Bank.

National Council of Applied Economic Research (NCAER). 1999. *Human Development Indicators, 1999.* New Delhi: Oxford University Press.

National Family Health Survey (NFHS). 2000. *India.* Mumbai, India: International Institute for Population Sciences.

National Sample Survey Organisation. 1996. Report No. 238 (27th Round; 1971-1972) and No. 405 (50th Round; 1993-1994). New Delhi: Department of Statistics, Government of India.

Payne, Philip and Michael Lipton. 1994. *How Third World Rural Households Adapt to Dietary Energy Stress: The Evidence and the Issues,* Washington, DC: International Food Policy Research Institute.

Planning Commission (GOI). 2001. Toward Hunger Free India. New Delhi: M.S. Swaminathan Research Foundation and UN World Food Programme.

Scrimshaw, Navin S. 1997. The Lasting Damage of Early Malnutrition. In *Ending the Inheritance of Hunger, Lectures given by Robert William Fogel, Navin S. Scrimshaw and Amartya Sen.* WFP/UNU Seminar, Rome, May 31.

Shariff, Abusaleh. 1993. *Factors Affecting Child Health: Search for Maternal Education Effects in Rural Gujarat.* Working Paper No. 47, Gujarat Institute of Development Research, Gota, Ahmedabad.

Shariff, Abusaleh and Ananatha C. Mallick. 1999. Dynamics of Food Intake and Nutrition by Expenditure Class in India. *Economic and Political Weekly,* 34(27): 1790-1800.

PART VII:
EMERGING ISSUES

Chapter 20

Emerging Issues in Trade and Technology: Implications for South Asia

Per Pinstrup-Andersen

South Asians account for 22 percent of the world population, 44 percent of the world's poor people (those earning less than the equivalent of a U.S. dollar per day), 40 percent of all food-insecure people in the developing world, and over half of the malnourished preschool children in the world. South Asians earn just 4 percent of global income. Finding the means by which the people of this region of the world can free themselves from poverty and hunger without harming the environment is therefore not only a pressing regional issue, but it has a major bearing upon how to achieve a more prosperous and peaceful world.

Agriculture remains the key to sustainable development in South Asia. About 70 percent of the region's population lives in rural areas, and rural poverty rates are substantially higher than those of the cities. Agriculture provides the livelihood of about two-thirds of South Asia's workforce, and accounts for 25 percent of the regional gross domestic product. Also, this is a region of many small farms. The average size of the 125 million South Asian farms is just 1.6 hectares, and 80 percent of the holdings have an average size of 0.6 hectares. In Bangladesh, fully 96 percent of the farms have an average size of 0.3 hectares, and might well be considered gardens in other parts of the world. Those farms provide food, feed, and fiber for the region's 1.3 billion people (Gulati, 2001).

What happens to agriculture and to small farms will make a big difference for the people of the region. National and regional public policies related to trade and technology should seek to shape these forces in ways that will promote broad-based agricultural growth and sustainable natural resource management, while minimizing negative impacts. Bilateral and multilateral development assistance agencies need to support the appropriate policies.

Agricultural growth catalyzes equitable development in poor countries where much of the populace is rural. Agriculture's linkages to the nonfarm economy generate employment, income, and growth economywide (Delgado et al., 1998). Gains in agricultural productivity can lead to lower food prices which will benefit poor consumers in the cities and countryside alike. Increasing productivity in agriculture can also slow the pace of rural-to-urban migration and help meet growing food demand due to population growth and urbanization. Moreover, a healthy agricultural economy offers farmers incentives to conserve the natural resource base.

MAKING GLOBALIZATION WORK FOR POOR PEOPLE

Many sustainable development advocates are concerned that trade liberalization in general, and agricultural trade liberalization in particular, will hurt poor people, as larger-scale producers come to dominate export crop opportunities and cheap imported produce from developed countries (often produced and exported with heavy subsidies) wipes out small landholders. Even analysts with a more optimistic view agree that globalization is presently occurring in a highly inequitable context and under rules that have biases against poor people and countries. However, these optimists would argue, in contrast to some critics of globalization, if governments in developed and developing countries undertake appropriate policy changes, globalization *can* be made to work for poor and hungry people (Díaz-Bonilla and Robinson, 2001).

With respect to agricultural trade, many commodities produced domestically in South Asian countries enjoy comparative advantages in home markets, and many are competitive, or could become so, in global markets. With appropriate policies, these advantages can be enhanced to the benefit of poor rural people. However, much depends on the willingness of developed countries to open their markets to developing countries' products, and their ability to reduce, or better still, remove tariff escalation against higher value commodities and processed goods, and reduce trade-distorting subsidies on their domestic agriculture and exports. India is playing a very prominent role as a leader of the developing countries in the current round of global agricultural trade negotiations through the World Trade Organization (WTO). By putting "food and livelihood security" on the table as a "nontrade concern" to be addressed in the negotiations, India has helped to steer the discussions toward a clearer focus on the potential impact upon poor people. Pakistan and Sri Lanka have also played an active role in the negotiations thus far, pressing for expanded access for developing country agricultural exports in developed country markets.

Varying patterns of domestic and international competitiveness exist among the region's major agricultural products. Despite controls and regulations on food grains, Indian rice and wheat are very competitive with imports, particularly since the productivity gains of the green revolution resulted in large real price declines. Both India and Pakistan are significant rice exporters. South Asian rice is likely to remain competitive, especially in Asian markets, given lower freight costs in comparison to the United States. India and Pakistan also export cotton. The region has long enjoyed a strong position as an exporter of tea, coffee, spices, and jute, but low prices for some of these commodities make expansion of production uneconomic. The region could increase exports of both fresh and processed fruits and vegetables, but expansion will require improvements in infrastructure, storage, transport, processing, and the ability to meet sanitary and technical requirements in developed-country markets. The same applies to fish exports, where there is also potential competitive advantage (Gulati and Hoda, 2005).

Until recently, the larger countries of the region—India, Pakistan, and Bangladesh—as well as Sri Lanka, pursued protectionist trade policies, aimed at promoting domestic, import-substituting industries. Since the late 1980s, most countries in the region have liberalized foreign trade, including agricultural trade. Sri Lanka has probably gone the farthest with trade liberalization among South Asian countries. Both Sri Lanka and Pakistan are now major wheat importers. In India, liberalization of the industrial sector after 1991 improved the terms of trade between agriculture and manufacturing, and spurred private investment in nongrain agriculture, particularly in horticulture products, livestock products, and fisheries, all of which are internationally competitive or have the potential to become so. Beginning in 1994, India extended trade liberalization to agricultural products, and there are now few restrictions on exports. However, restrictions remain on imports of grain, oilseeds, and edible oils (Gulati and Hoda, 2005).

Because Bangladesh liberalized grain imports in the early 1990s, it was able to maintain stable rice prices after the 1998 floods devastated local crops with imports from India, thereby reducing the need to seek external food aid. In recent years, food assistance supplies have waxed and waned according to domestic supply conditions in the United States, the largest donor country, and are often scarce at times of increased need (Del Ninno et al., 2001).

If India remains hesitant about completing the liberalization of agricultural trade, it may lose out on opportunities that globalization offers while seeking to avoid the very real risks. Now that China has joined the WTO and is deepening its engagement with globalization, the biggest risk for India may be getting left behind.

Policies that enhance the productivity of the region's small farms will go a long way toward determining whether small landholders will gain from enhanced agricultural export opportunities. These policies include ensuring that poor farmers and less-favored areas have access to infrastructure (especially all-weather roads and storage facilities), inputs, credit, and markets.

Unless developed countries are willing to open their markets to temperate-zone agricultural exports from developing countries and end tariff escalations against processed and higher-value products, however, the benefits that developing countries and the poor people who live in them will derive from globalization will be limited. In addition, South Asian countries face high tariff barriers in some other developing countries' markets for such key exports as mangoes, tea, and cashews.

If developed countries want developing countries to continue to open their markets for agricultural products and other goods and services, they must in turn reduce the substantial subsidies provided to developed country farmers. Global agricultural subsidies total $360 billion annually, or a billion dollars a day, and developed-country governments pay out 80 percent of the total. Their subsidies amount to almost six times the annual level of official development assistance provided to the developing world. A large number of developing countries are among the 140 members of the WTO, so agricultural trade negotiations can no longer be limited to discussions among the United States, the European Union, and Japan. The Cairns Group of nonsubsidizing agricultural exporting countries, which includes both developed and developing nations, played an important role in the Uruguay Round agricultural negotiations in the 1980s and 1990s, and may have an even more pivotal role in the current talks. A coalition including the Cairns countries and developing countries with large agricultural sectors, such as Pakistan, Bangladesh, India, and China, would be particularly effective in challenging the current distorted patterns of global agricultural trade. The developed countries cannot expect the developing world to endorse one-sided agricultural trade liberalization ad infinitum.

In 2001, India proposed that the concept of "special and differential treatment of developing countries" in the WTO agreements should recognize the need for developing countries "to tackle their special concerns such as food and livelihood security while reforming agricultural trade." The African group has similarly proposed that trade liberalization take into account "such non-trade concerns as food security, sustainable development, and poverty alleviation." Pakistan and Sri Lanka have also supported efforts to expand the application of the "special and differential" principle to agriculture (Hoda and Gulati, 2005, p. 130). Given the importance of agriculture in the economies of the poorest developing countries, especially in Africa, and its potential to influence poverty reduction, it is clear that some balance between complete trade liberalization and an ability to promote and protect domestic agriculture without violating WTO rules will be needed for some time. Some of the smaller and poorer countries of South Asia, such as Nepal and Bhutan, might also require such special treatment.

In India's case, however, the country can take the measures needed to achieve food security under the existing WTO rules affecting developing countries. Such measures include input subsidies and product-specific price supports up to the level of 10 percent of the value of production. Reform of input subsidies would advance food security. Programs designed to invest in human resources, improve safety nets, and invest in public goods, such as agricultural research, which would increase poor farmers' productivity and reduce their risks, is likewise permissible under the existing WTO agriculture rules. India's remaining barriers to agricultural imports, particularly high tariff barriers to imported cereals and edible oils, do not advance food security much. Indian cereal production is sufficiently competitive domestically not to require high protection. Although further investment and policy reform is needed to expand domestic edible oil production capacity, Indian consumers, and especially poor consumers, would benefit from liberalization of the import market for edible oils.

Most of the countries in the region have significant foreign exchange from exports of textiles and garments. These industries create opportunities for higher-value-added exports and employment for poor people. Although working conditions vary from country to country, as does the ability of workers to form organizations to press for improvements, earnings are generally higher than in traditional occupations, especially for women, who account for a high proportion of the workforce.

In this area, as in agriculture, the ability of poor South Asians to gain from globalization is severely constrained by ongoing protectionism on the part of developed countries. The relevant WTO agreement requires the wealthy countries to phase out their quota systems for textile and clothing imports by 2005, but protected domestic textile industries and the associated trade unions remain politically powerful, and it is uncertain whether developing countries will be able to succeed in getting trade barriers dismantled.

The South Asian Association for Regional Cooperation (SAARC) has long pushed for freer trade within the region. However, political disagreements and limited trade complementarities have hampered progress. Studies have found that creation of a regional free trade area would significantly expand intra-regional trade (Rosegrant and Hazell, 2000).

At present, the region's two largest countries, India and Pakistan, engage in little trade with each other. Expanding economic ties might serve to reduce tensions between these neighbors.

THE ROLE OF SCIENCE AND TECHNOLOGY

Technology has certainly played an important role in development in South Asia. This region was once written off by the international community as a collection of "hopeless basket cases," doomed to the Malthusian nightmare of too many people and not enough food. Some pundits of the 1960s and 1970s even argued that the world should adopt "triage" or "lifeboat ethics," and leave South Asia and Africa to whatever fate had in store. Happily, wiser heads prevailed, both in the international community and in the region. The countries of the region adopted appropriate policies and modern agricultural technology, and the results have been impressive. In India, the population rose by 67 percent between 1970 and 1995, but cereal production grew by 88 percent. In Bangladesh, Bhutan, Nepal, Pakistan, and Sri Lanka, the population rose 88 percent; cereal production kept up, growing 89 percent. Cereal yields rose 88 percent in India and 54 percent in the other five countries. Per capita income grew 82 percent in India and 60 percent in the other countries, while per capita calorie consumption rose 14 percent in India and 4 percent in the other countries (Asian Development Bank, 2000). Indeed, there is currently enough food available in the region to provide every South Asian with more than enough calories (over 2,400 per day) to meet minimum requirements, if the food was distributed according to need (Rosegrant et al., 2001).

Equity improvements have occurred as well as growth: the regional poverty rate fell from 59 to 43 percent during 1970-1995 (Asian Development Bank, 2000). The proportion of South Asians who are chronically undernourished fell from 38 percent in 1980 to 24 percent recently (FAO, 2000, 2001). In the case of India, the country no longer faces food shortages and dependence on external food aid, but instead must tackle the problem of what to do with a 60-million-metric-ton stock of surplus of wheat and rice (Ministry of Finance, 2002).

Science and technology, if applied within a framework of appropriate policies, can do a great deal to advance food security, agricultural growth, equity, and sound natural resource management. Technological change can be risky. Thirty-five years ago, South Asian countries took bold steps to make sure that they could feed themselves, and launched the green revolution, making investments in irrigation and providing farmers in irrigated areas with access to high-yielding rice and wheat seeds, fertilizer, and pesticides. Initially there was skepticism about whether the technology would mostly benefit richer farmers who enjoyed good access to inputs. In fact, the benefits were widely shared (Asian Development Bank, 2000). Small-scale farmers as well as larger producers gained from increased yields and lower unit costs of production due to adoption of new agricultural technology.

Landless rural people found new employment opportunities on and off the farm, and consumers benefited from substantially lower food prices. Agricultural growth stimulated growth throughout the economy, as rural demand for goods and services grew with rural incomes (Hazell and Haddad, 2001).

In India's Tamil Nadu state, for example, adoption of high-yielding green-revolution-grain varieties has meant not only increased yields and cheaper, more abundant food, but also income gains for small and large farmers alike, as well as nonfarm poor rural households. Higher incomes have improved nutrition. These gains resulted not just from new agricultural technology, but also from: state antipoverty policies that included nutrition and other social welfare programs that were well targeted to reach poor people; investment in human capital, agriculture, and rural development; and measures to ensure equitable access to resources.

In some other parts of South Asia, increased *inequality* followed adoption of green-revolution-grain varieties. However, this resulted not from factors inherent in the technology, but from policies that did not focus on equity and human capital development. Even in these areas, rural landless laborers usually found new job opportunities due to increased agricultural productivity, particularly if physical infrastructure and markets were well developed.

On the other hand, inequalities between well-endowed and resource-poor areas of the region increased as a result of factors that were indeed inherent in new technology. Successful adoption of the earliest green revolution crop varieties depended on access to water, fertilizer, and pesticides. More recent agricultural research efforts have focused on developing high-yielding varieties bred for pest and stress resistance, and therefore not dependent on pesticides or irrigation water (Hazell and Ramasamy, 1991).

Productivity increases resulting from the adoption of green revolution technologies saved forests and marginal lands from cultivation because of higher yields on existing lands. However, excessive or improper use of inputs sometimes harmed the environment (Lipton with Longhurst, 1989).

Agricultural research alone will not foster sustainable development, but it can play an important role in a larger strategy for eradicating poverty and hunger while protecting the environment. Poor rural people in South Asia and other developing regions face many problems, and research is essential to generate solutions. Research and development (R&D) that creates technology to increase productivity, such as environmentally friendly, yield-increasing crop varieties or improved livestock, is especially important.

Low agricultural productivity results in high unit costs of food, higher poverty, food insecurity, poor nutrition, low farmer and farmworker incomes, little demand for goods and services produced by poor nonagricultural rural households, and urban unemployment and underemployment. Poor farmers in most developing countries cannot expand cultivated area without further damaging the environment. Research and development must be supported in order to sustain and increase yield on existing land. Pre- and postharvest crop losses from pests and weather result in low and fluctuating yields, incomes, and food availability, and reduce potential output value by up to 50 percent (Oerke et al., 1994). For example, poor South Asian cotton farmers have faced devastating infestations of bollworms that have developed resistance to frequent pesticide sprayings. Low soil fertility and lack of access to plant nutrients, along with acid, salinated, and waterlogged soils also contribute to low yields, production risks, and natural resource degradation. These problems affect many poor farmers in India, Pakistan, and Nepal, where excessive water use and poor management of irrigation systems have contributed to soil degradation. Inadequate infrastructure, poorly functioning markets, and lack of access to credit, extension services, and technical assistance add impediments. Secure access to land, as well as to water, is critical.

R&D and extension services need to focus on women farmers as well as men. Women play important roles as producers of food, managers of natural resources, income earners, and caretakers of household food and nutrition security. Research has shown that giving women the same access to physical and human resources as men increases agricultural productivity dramatically. In Bangladesh, International Food Policy Research Institute (IFPRI) research has found that when women control assets, more household income is spent on children's clothing and education, and the rate of illness among girls is reduced (IFPRI, 2000).

Although private sector research may produce some of the knowledge and technology small farmers need, poor farmers usually offer business and industry too little profit potential to attract private investment in R&D. Moreover, much of the knowledge and technology needed by poor farmers is of a "public goods" nature. Public goods have two characteristics: the consumption of the goods by an individual does not detract from that of another individual; and, second, it is impossible or at least very difficult to exclude anybody from consuming the goods. Most knowledge derived from publicly funded research fulfills the first condition: if I know something it does not prevent others from knowing the same thing. Also, once the knowledge is obtained, whether it was purchased or obtained for free, it can be used over and over again without paying the original owner. Public goods production is usually unattractive to profit-making entities unless there is public funding available.

The gains to society and to poor people from pro-poor agricultural research are high. Social rates of return on agricultural research investment is more than 20 percent per year, compared to long-run real interest rates of 3 to 5 percent for government borrowing (Alston et al., 2000). Nevertheless, developing countries are seriously underinvesting in public agricultural research. The average annual growth rate of public agricultural research expenditures in the developing world has declined since the 1970s. In India, public agricultural R&D spending only accounted for 0.5 percent of agricultural GDP in the 1990s, compared to 0.6 percent for all developing countries, 2.6 percent for all developed countries, and 3 percent in the United States. Public agricultural research needs to focus more on addressing the problems of poor farmers and regions, as larger farms and better-off regions are likely to attract private research investment (Hazell, undated; Gulati and Hoda, 2005).

There are several ways to expand small farmer-oriented research. Governments can increase public-sector research and convert some social benefits to private gains, e.g., by purchasing exclusive rights to new technology and providing it for free or at a nominal charge to small farmers (Sachs on Development, 1999). Other arrangements include public-private joint ventures, research foundations, competitive funds, and production levies.

If agricultural research is to be relevant to poor farmers, it must be carried out in a manner that encourages farmer participation in identifying problems and farm-level constraints. Such participatory research, in which farmers and scientists work as partners, rather than in a patron-client relationship, not only focuses on the needs of poor farmers, but it can enhance farmers' own capacity to innovate and solve problems. Moreover, the partnership is a two-way street: farmers' knowledge about local crop landraces can benefit crop breeders enormously.

Farmers must be able to choose agricultural practices and technologies including agroecological methods, conventional research methods, and molecular biology research methods, such as genetic engineering. Moreover, researchers must link these approaches more closely with indigenous knowledge (Pinstrup-Andersen, 2001). Farmers in India's semiarid tropical zones, for example, have on their own devised approaches to soil conservation, and this knowledge could be integrated into research programs aimed at sustainable intensification of agriculture in these resource-poor areas (Kerr, 2000).

In 52 developing countries organic and agroecological approaches have been used successfully to boost food crop production (Pretty, 2001). These approaches aim to reduce the amount of external inputs that farmers have to use, relying instead on available farm labor and organic material, as well as on improved knowledge and farm management. In a large number of projects, the agroecological approach has increased productivity and contributed to more efficient water use, improved soil quality, and effective pest and weed control with few or no chemicals. In rice farming areas, elimination of synthetic pesticides permit fish, shrimp, and crab farming in paddies, thereby enhancing protein and micronutrient supplies. One of the great strengths of the agroecological paradigm is that it promotes sustainable management of natural resources and active participation by farmers in identifying problems as well as designing and implementing appropriate solutions at the farm and community levels. Its greatest weaknesses are relatively low labor productivity and yields. In areas with limited rainfall, which includes many poor South Asian farmers, scarcity of biomass and high demands for alternative uses of biomass (for fodder and fuel, for example) limit the potential of many organic approaches to land management. In such circumstances, technologies and policies for conserving water and profitably increasing the production of useful biomass, such as promotion of woodlots, should have high priority.

Conventional approaches to agricultural R&D are increasingly integrating aspects of agroecological approaches. For example, in Tamil Nadu one village that had adopted green-revolution-rice varieties in the 1960s, continued to plant newer, high-yielding varieties as they became available. In recent years, however, farmers in the village have moved away from heavy use of synthetic pesticides and mineral fertilizers. They now rely primarily on biological pest management methods and integrated approaches to plant nutrition that judiciously combine organic and inorganic fertilizers (Rajarathinavelu, 2001).

Conventional R&D may also draw on molecular biology techniques to enhance the efficiency of crop breeding (Pingali, 2001). Researchers have not yet fully exploited the productivity gains achievable from conventional plant breeding. New high-yielding varieties of rice, wheat, and a number of other crops are in the research pipeline at national and international agricultural research institutions.

Public agricultural R&D should also focus on development and dissemination of technologies and natural resource management practices that are environmentally sound. Some of these technologies already exist and include precision farming, crop diversification, integrated pest management, pest-resistant crop varieties, and improved soil and water management practices. Some of these, if managed appropriately, can increase yields while reducing natural resource degradation. Conventional and molecular biology-based approaches to research should be used to develop pest- and disease-resistant and drought- and salt-tolerant crop varieties that do not depend on application of chemicals. In South Asia's dryland areas, diseases and pests often wipe out the groundnut, pigeon pea, and cotton crops upon which small landholders depend for their livelihoods (Hazell, 2005; Paarlberg, 2001).

South Asia again must choose whether to adopt new agricultural technology that many critics consider risky. Contentious public debate over the environmental and socioeconomic risks of modern agricultural biotechnology have meant long delays in approval of the commercial release of genetically modified crops in India. In March 2002, the government's Genetic Engineering Approval Committee granted permission to Indian farmers to grow genetically modified cotton commercially, four years after the first field trials (ISAAA, 2002). Extensive research on additional applications of modern agricultural biotechnology is underway both through the public Indian Council of Agricultural Research (ICAR) and the domestic and international private sector. Opposition has centered primarily on the involvement of foreign companies, such as Monsanto, in efforts to commercialize genetically modified crops (Paarlberg, 2001).

Despite the antagonism toward genetically modified cotton in India, use of similar cotton seeds (containing genes from the soil bacterium *Bacillus thuringiensis,* or Bt, which produce a toxin that kills the cotton bollworm) in China has had significant benefits. Use of the seeds has reduced synthetic pesticide use on cotton farms, lowered labor costs, increased profits, reduced collateral damage to nontargeted species and water resources, and, most important, dramatically reduced pesticide poisoning cases among farmers and agricultural laborers. It has boosted the incomes of small producers significantly (Paarlberg, 2001). In South Africa, Bt cotton has led to substantial yield increases where it is commercially available (Njobe-Mbuli, 2000). Indian cotton farmers had long lobbied for government approval to grow Bt cotton. It might well benefit other South Asian cotton farmers as well.

ICAR research is funded by Indian taxpayers, not foreign companies, and focuses on the needs of poor farmers and consumers (e.g., providing high-protein and insect-resistant rice varieties, as well as insect-resistant cotton, oilseed crops, and potatoes). ICAR often collaborates with public international agricultural research centers such as the International Rice Research Institute and the International Crop and Research Institute for the Semi-Arid Tropics. It is hoped that some of the results of this research will find their way into farmers' fields sooner rather than later. As required before commercial introduction of genetically modified crops, India has already developed an impressive system to assess and manage risks to human health and the environment. This includes officials charged with responsibility for environmental protection, agriculture, health, and science and technology, along with university scientists. This system is adequate to address legitimate biosafety concerns (Paarlberg, 2001).

Slow progress in enacting intellectual property rights legislation relating to plant varieties has impeded both private sector research and commercialization of domestic and imported seeds derived from biotechnology. The recent passage of the 2001 Protection of Plant Varieties and Farmers' Rights Legislation by the Indian parliament created an intellectual property regime that may serve as a model for other developing countries, in South Asia and elsewhere, to meet their obligations under the WTO and the new International Treaty on Plant Genetic Resources for Food and Agriculture. It seeks to balance the need to offer incentives for innovation to private plant breeders with provisions on benefit sharing for individual and community holders of traditional knowledge and on the rights of farmers to save and exchange seeds (Ministry of Finance, 2002; Ramanna, 2002). Such a system should be effective in promoting both agricultural innovation and the equitable and sustainable management and conservation of India's agricultural biodiversity.

In addition to pro-poor agricultural research, R&D should focus on improving poor rural people's access to information and communications technology, which is developing rapidly in South Asia. The development of wireless broadband can greatly expand access to telephones and the Internet. Using a hub-and-spoke approach, one village with full Internet access can disseminate information and receive requests from other villages that have wireless access to the hub, thereby saving on costs. Digitized radio broadcasts via satellite can now cheaply reach large numbers of people. These technologies offer poor communities vastly improved access to timely market information, awareness of relevant government policies, and access to new agricultural know-how. This can contribute to higher incomes for poor farmers and the nonfarm rural poor engaged in handicraft production, with additional positive impacts on household food security and nutritional status. For poor Indian fisherfolk, cell phones already make it possible to compare prices in different markets for their catch. Technologies aimed at collecting geographically referenced data can be enormously useful in food security research and policymaking, e.g., improved targeting of safety-net programs and early famine warning (Chowdhury, 2001).

Many poor rural communities in South Asia and elsewhere consume little electricity. Extending electricity grids to meet their energy needs may prove more costly and take longer than harnessing new and renewable sources of energy already available in these communities—wind, solar, and biomass—through renewable energy technologies. These include biogas plants, solar lanterns, solar home lighting systems, improved cook stoves, improved kerosene lanterns, solar water pumping systems, solar water heating systems, and water mills. India's Tata Energy and Resources Institute (TERI) has installed or disseminated renewable and energy-efficient technologies in about 100 villages in different parts of India by setting up demonstration projects. A sustainable approach to poverty alleviation that uses energy technology interventions would need to adapt to the specific needs and capabilities of communities. This would include maintaining technologies for continued use, ensuring that financing is available to improve access to such solutions, ensuring that market mechanisms are in place to provide the technologies, and formulating appropriate public policy to facilitate communities' adoption of the technologies (Pachauri and Mehrotra, 2001).

CONCLUSION

Globalization can be shaped to benefit poor people and support sustainable development. Domestic policies can help ensure that small farmers and other rural poor have a stake in export opportunities via access to infrastructure, inputs, credit, markets, and organizations such as cooperatives. However, unless developing countries and their allies can also convince the developed countries to reduce their barriers to exports and their agricultural and export subsidies none of this will be realized.

Appropriate policies and institutions are also needed to exploit opportunities for both public- and private-sector R&D investments with high social returns in South Asia. Recently developed new technology in such areas as molecular biology, information and communications, and energy may be important elements of a strategy to reduce poverty, food insecurity, and unsustainable use of natural resources. However, these technologies must be adapted to the conditions within which small farmers and poor consumers operate and poor people themselves must be active participants in the adaptation process.

REFERENCES

Alston, J.M., C. Chan-Kang, M. Mara, P.G. Pardey, and T.J. Wyatt. 2000. *A meta-analysis of rates of return to agricultural R&D: Ex pede herculem?* Research Report 113. Washington, DC: IFPRI.

Asian Development Bank. 2000. *Rural Asia: Beyond the green revolution.* Manila: Asian Development Bank.

Chowdhury, N. 2001. Information and communications technologies. In Pinstrup, P. (Ed.), *Appropriate Technology for Sustainable Food Security*, Policy Brief 7 of 9, 2020 Vision Focus 7. Washington, DC: IFPRI.

Delgado, C.L., J. Hopkins, and V.A. Kelly. 1998. *Agricultural growth linkages in Sub-Saharan Africa.* Research Report 107. Washington, DC: IFPRI.

Del Ninno, C., P.A. Dorosh, L.C. Smith, and D.K. Roy. 2001. *The 1998 floods in Bangladesh.* Research Report 122. Washington, DC: IFPRI.

Díaz-Bonilla, E. and S. Robinson (Eds.) 2001. *Shaping globalization for poverty alleviation and food security*. 2020 Vision Focus 8. Washington, DC: IFPRI.

FAO (Food and Agriculture Organization of the United Nations). 2000. *Agriculture toward 2015/2030: Technical interim report*. Rome: FAO.

———. 2001. *The State of food insecurity in the world, 2001*. Rome: FAO.

Gulati, A. 2001. The future of agriculture in south Asia: W(h)ither the small farm? Presentation at the IFPRI 2002 Vision Conference on Sustainable Food Security for All by 2020, Bonn, Germany, September 4-6.

Gulati, A. and A. Hoda. 2005. *Negotiating beyond Doha*. Delhi: Oxford University Press.

Hazell, P. 2005. Future challenges for the rural South Asian economy. In Babu, S.C. and A. Gulati, *Food Security and Economic Reforms: The Impact of Trade and Technology in South Asia* (pp. 175-185). Binghamton, NY: The Haworth Press, Inc.

Hazell, P. and L. Haddad. 2001. *Agricultural research and poverty reduction. 2020 Vision Food, Agriculture, and the Environment*. Discussion Paper 34. Washington, DC: IFPRI.

Hazell, P.B.R. and C. Ramasamy (Eds.) 1991. *The green revolution reconsidered*. Baltimore and London: The Johns Hopkins University Press for IFPRI.

Hoda, A. and A. Gulati. 2005. Indian agriculture, food security, and the WTO-AOA. In Babu, S.C. and A. Gulati, *Food Security and Economic Reforms: The Impact of Trade and Technology in South Asia* (pp. 115-136). Binghamton, NY: The Haworth Press, Inc.

IFPRI. 2000. *Women: The key to food security*. Issue Brief 3. Washington, DC: IFPRI.

International Service for the Acquisition of Agri-biotech Applications (ISAAA). (2002). CropBioTech Update. Available online at <www.isaaa.org>.

Kerr, J. 2000. Development strategies for semiarid South Asia. In Pender, J. and P. Hazell (Eds.), *Promoting Sustainable Development in Less Favored Areas*. 2020 Vision Focus 9. Washington, DC: IFPRI.

Lipton, M. with R. Longhurst. 1989. *New seeds and poor people*. London: Unwin Hyman.

Ministry of Finance, Government of India. 2002. Economic survey, 2001-2002. Available online at <http://www.indiabudget.nic.in/es2001-02/general.htm>.

Njobe-Mbuli, B. 2000. Biotechnology for innovation and development. In Persley G.J. and M.M. Lantin (Eds.), *Agricultural biotechnology and the poor* (pp. 115-117). Washington, DC: Consultative Group on International Agricultural Research.

Oerke, E.-C., H.-W. Dehne, F. Schonbeck, and A. Weber. 1994. *Crop production and crop protection: Estimated losses in major food and cash crops*. Amsterdam: Elsevier.

Paarlberg, R.L. 2001. *The politics of precaution: Genetically modified crops in developing countries*. Baltimore and London: The Johns Hopkins University Press for IFPRI.

Pachauri, R.K. and P. Mehrotra. 2001. Alternative energy resources. In Pinstrup, P. (Ed.), *Appropriate technology for sustainable food security*, Policy Brief 8 of 9. 2020 Vision Focus 7. Washington, DC: IFPRI.

Pingali, P. 2001. Conventional approaches. In *Sustainable Food Security for All by 2020: Proceedings of an International Conference*. September 4-6, 2001. Bonn, Germany. Washington, DC: IFPRI.

Pinstrup-Andersen, P. 2001. *Appropriate technology for sustainable development*. 2020 Vision Focus 7. Washington, DC: IFPRI.

Pretty, J. 2001. Agroecological approaches. In *Sustainable Food Security for All by 2020: Proceedings of an International Conference* (pp. 125-127). September 4-6, 2001. Bonn, Germany. Washington, DC: IFPRI.

Rajarathinavelu, K. 2001. Conventional approaches. In *Sustainable Food Security for All by 2020: Proceedings of an International Conference*. September 4-6, 2001. Bonn, Germany. Washington, DC: IFPRI.

Ramanna, A. 2002. A Tragedy of the Anticommons? India's IPR Policies in Agriculture. Environment and Production Technology Division Seminar at IFPRI, March 27.

Rosegrant, M.W. and P.B.R. Hazell. 2000. *Transforming the rural Asian economy: The unfinished revolution*. Oxford: Oxford University Press.

Rosegrant, M.W., M.S. Paisner, S. Meijer, and J. Witcover. 2001. *Global food projections to 2020: Emerging trends and alternative futures*. Washington, DC: IFPRI.

Sachs on Development: Helping the World's Poorest. 1999. *The Economist*. August 14, pp. 17-20.

WTO Watch. 2002. Available online at <http://www.tradeobservatory.org>.

World Trade Organization. 2002. Available online at <http://www.wto.org/english/tratop/_e/agric_e/negs_bkgrnd05_intro_e.htm#currentnegs.>.

Chapter 21

Converting Policy Research into Policy Decisions: The Role of Communication and the Media

Klaus von Grebmer

A major challenge in achieving food security in South Asia is addressing the dichotomy between academic research and policy decision making. The work that researchers conduct often does not reach policymakers because of inadequate and inappropriate communication. The link between research and decision-making systems remains weak. This chapter explores the strategies for effective policy communication and provides some suggestions on how policy researchers can reach policymakers and the media to enhance the impact of their findings.

COMPETENCE IN RESEARCH AND COMMUNICATIONS LEADS TO IMPACT

Policy research is not an end in itself. It is financed and undertaken to contribute to the progress of humanity and this honorable goal can only be achieved if the results of the research are communicated to policymakers and to those who influence them, i.e., the media. For research to have an impact outside academic circles, researchers have to leave their "ivory towers." State-of-the-art research needs state-of-the-art communications to achieve impact.

Research on the production, distribution, and consumption of food can:

- confirm the appropriateness of policy actions taken,
- indicate that policy actions are needed to reduce risks/costs or increase benefits,
- show in advance the probable outcomes of alternative policies,

- synthesize information on how other policymakers have coped with an issue, and
- alert policymakers to major threats.

THE RIGHT INFORMATION IN THE RIGHT FORM AT THE RIGHT TIME

In general, the needs of policymakers are simple: they want the right information, in the right form, at the right time. But what sounds like a truism is difficult and sometimes cumbersome to put into practice. Consumers of research results are not alike; their communication needs can differ tremendously. The right form in which to convey information depends on a policymaker's background, perspective, and political context. However, policymakers do have a common preference: they are more likely to read research results and policy implications that are timely and clearly and succinctly presented.

The right time depends very often on what stage the policymaking process is in. In general terms, it can be divided into eight stages:

1. developing the policy agenda,
2. identifying the specific objectives and policy options,
3. evaluating the options,
4. advancing recommendations,
5. building a consensus,
6. legislation,
7. implementation, and
8. policy evaluation and impact assessment.

Research results that feed into the process during stages 1-4 are likely to have the best chance of finding their way into the consensus-building, legislation, and implementation stages.

Very often, when research results are published, policymakers find them indigestible. This is unfortunate but understandable because the results have been written for a different target group, namely other researchers. All researchers have to document in a detailed way the scientific methods with which they obtain their results. It is their fundamental ethical obligation to rigorously examine and publish the results and methodology of reported research. Researchers' commitment to objectivity and disciplinary and scientific practice also obliges them to use and describe the latest scientific methods. This is in fact how science corrects mistakes and ever more closely approximates truth and understanding. Relentless double-checking and independent third-party evaluations are the cornerstones of the scientific process. Most policymakers, however, will not read lengthy research reports, especially when these are written in a language with a different target group in mind.

Academic institutions that are not concerned about policy impact can be satisfied with the publication of a research report, but for the International Food Policy Research Institute, with its vision to develop "sustainable options for ending hunger and poverty" (IFPRI, Mission Statement. Available online at: <http://www.ifpri.org>), the challenge does not end with published research. Research reports have to be simplified and condensed in close cooperation with the researchers and presented in a way that is appealing to "insiders." This group includes, e.g., policy advisers who give their recommendations to policymakers, experts in the donor community, and any other group that has a professional interest in a research issue. The simplification and condensation process is not an easy task and many researchers believe that it is a threat to the scientific appeal of their published work.

The "translation process" has to go even further than simplification when we want to reach policymakers directly. Publications have to get and hold their attention otherwise they will not bother to read them. Policymakers have to deal with a variety of issues, and several hundred documents pass over their desks daily. The increasingly scarce resource of time makes competition for their attention even stiffer. In order to attract and hold policymakers' attention, issues and findings have to be well-presented. Research results that are easy to understand, take account of the political arena, and offer immediate help in pending policy decisions will interest policymakers the most.

Research can influence the policy process if the information presented to the policymaker:

- gives him or her a good understanding of the magnitude and dynamic of the problem at stake;
- explains the causes of the problem (e.g., poverty is a function of . . . ; malnutrition is a function of . . .); and
- recognizes the political context, outlines the basic actions that can be taken, and indicates the outcome.

These are the important elements of "the right information in the right form at the right time." Research results that could have improved political decisions and could have made a tremendous impact in reducing malnutrition and poverty are often of no value once the political process is underway. As the Prussian military thinker Carl von Clausewitz once stated: "You can conquer back lost land, but not lost time."

When research findings attract the attention of policymakers, they are more likely to be integrated into policy decisions, especially if they support political agendas in a timely fashion. Once findings lead to policy action, research has achieved its goal. It has had impact for the benefit of the poor.

Contrary to this approach, which is cognizant of policymakers' needs, some people still promote the "container theory" of communication. This theory assumes an ideal situation: a "sender" packs the information he or she wants to convey into a container and passes it on to a "receiver" who unpacks it and immediately understands the full content. Companies whose presidents communicated with their staff in this way have gone bankrupt and political leaders who followed these principles either succeeded as dictators with a lack of a sense of reality or worked themselves out of a job. Research results and the basic concepts behind them must be explained clearly, and reinforced on a person-to-person basis with ample room for dialogue and discussion.

REACHING OUT TO THE MEDIA

Another important approach to conveying research results to policymakers is via the media. Once the mass media take up an issue, the likelihood that policymakers will become interested in it increases dramatically. If a policy issue attains a high public profile, dealing with it generally boosts the personal or party profile of the policymakers involved. Furthermore, since policymakers read and listen to influential news outlets, IFPRI research that gets prominent news coverage will reach policymakers.

In order for the mass media to take up a food policy issue, it has to be inherently attractive or it has to be presented in a way that makes it attractive. To increase the likelihood that issues and research findings make it into the media we have to meet and satisfy the needs of this particular market. Inherently attractive topics for the mass media include politics, health, sex, and sports. The likelihood of news coverage increases when information is sensationalistic or deviates from the norm. It further increases when information is fresh and media outlets are first in line to report it. Competition among the media for audiences compels them to search for material that is interesting, surprising, or controversial, in short—"newsworthy."

No one will read a newspaper or watch television news that reports, e.g., "All flights to and from India have arrived on time today, our government is doing a great job and all our political parties are just fine." A truism in journalism school is that a "dog bites man" story will rarely get attention, whereas a "man bites dog" story would easily get it.

If research results on food policy issues are not new or surprising, the likelihood that they are going to make headlines in the media is low. Even if the research results are new and surprising, thorough marketing is necessary to get them into the media. Therefore, research results have to be further condensed and simplified and have to be put into a media-friendly context: research on water issues is more likely to make headlines when there is a water crisis and heads of state are meeting to discuss the issue. Research on the importance of cocoa production in certain countries is not an interesting topic per se; however, if it can be placed in the context of high chocolate consumption on Valentine's Day, the media are far more likely to pick it up. The more the media report about an issue, the higher it will climb on the political agenda, and the higher it is on the political agenda the greater will be the impact achieved by research.

In times of tight budgets, several European and Asian countries have cut their funding for international agricultural research or consider doing so. Some of these cuts have had an easy ride in the political arena, because international food security has not been high on the national agenda. If the general public does not know about the tremendous benefits of agricultural research, and if many members of parliament do not know about them, budget reductions will not meet heavy political resistance and can be engineered quite easily by a small but dedicated interest group. In the end, it is the poor in developing countries who will suffer from this failure to communicate research results.

To avoid this, communication to policymakers and the media must be effective. Armed with research results they find useful and understandable, policymakers are more likely to make decisions that contribute to the increase of national and international wealth and development. Effective communication of state-of-the-art research to policymakers and the media will help to ensure that decisions on food policy issues do not become prey to populism or irrationality.

Policymaking in South Asia, as in many developing countries, remains ad hoc and does not fully use research-based information. This is partly due to the poor linkage between the policy research community and the decision makers. Development of appropriate communication skills among researchers and analysts will bridge this gap. The important role policy communication has in enhancing the poverty-reducing impact of social sciences research can hardly be overemphasized.

Index

Page numbers followed by the letter (t) indicate tables; those followed by the letter (b) indicate boxed text; and those followed by the letter (f) indicate figures.

Order a copy of this book with this form or online at:
http://www.haworthpress.com/store/product.asp?sku=5302

ECONOMIC REFORMS AND FOOD SECURITY
The Impact of Trade and Technology in South Asia

_____ in hardbound at $79.75 (ISBN: 1-56022-265-5)

_____ in softbound at $59.95 (ISBN: 1-56022-257-3)

Or order onlIne and use special offer code HEC25 in the shopping cart.

COST OF BOOKS_____

POSTAGE & HANDLING_____
(US: $4.00 for first book & $1.50
for each additional book)
(Outside US: $5.00 for first book
& $2.00 for each additional book)

SUBTOTAL_____

IN CANADA: ADD 7% GST_____

STATE TAX_____
(NJ, NY, OH, MN, CA, IL, IN, & SD residents,
add appropriate local sales tax)

FINAL TOTAL_____
(If paying in Canadian funds,
convert using the current
exchange rate. UNESCO
coupons welcome)

☐ **BILL ME LATER:** (Bill-me option is good on
US/Canada/Mexico orders only; not good to
jobbers, wholesalers, or subscription agencies.)

☐ Check here if billing address is different from
shipping address and attach purchase order and
billing address information.

Signature_____

☐ **PAYMENT ENCLOSED: $**_____

☐ **PLEASE CHARGE TO MY CREDIT CARD.**

☐ Visa ☐ MasterCard ☐ AmEx ☐ Discover
☐ Diner's Club ☐ Eurocard ☐ JCB

Account #_____

Exp. Date_____

Signature_____

Prices in US dollars and subject to change without notice.

NAME_____

INSTITUTION_____

ADDRESS_____

CITY_____

STATE/ZIP_____

COUNTRY_____ COUNTY (NY residents only)_____

TEL_____ FAX_____

E-MAIL_____

May we use your e-mail address for confirmations and other types of information? ☐ Yes ☐ No
We appreciate receiving your e-mail address and fax number. Haworth would like to e-mail or fax special
discount offers to you, as a preferred customer. **We will never share, rent, or exchange your e-mail address
or fax number.** We regard such actions as an invasion of your privacy.

Order From Your Local Bookstore or Directly From
The Haworth Press, Inc.
10 Alice Street, Binghamton, New York 13904-1580 • USA
TELEPHONE: 1-800-HAWORTH (1-800-429-6784) / Outside US/Canada: (607) 722-5857
FAX: 1-800-895-0582 / Outside US/Canada: (607) 771-0012
E-mailto: orders@haworthpress.com

For orders outside US and Canada, you may wish to order through your local
sales representative, distributor, or bookseller.
For information, see http://haworthpress.com/distributors

(Discounts are available for individual orders in US and Canada only, not booksellers/distributors.)

PLEASE PHOTOCOPY THIS FORM FOR YOUR PERSONAL USE.

http://www.HaworthPress.com   BOF04